S0-DUE-659

PUBLIC CONSTRUCTION CONTRACTS AND THE LAW

PUBLIC CONSTRUCTION CONTRACTS AND THE LAW

Henry A. Cohen

NEW YORK
F. W. DODGE CORPORATION
A McGRAW-HILL COMPANY

Printed and bound in U.S.A.

Library of Congress Catalog Card No. 61-14015

THIS BOOK IS DEDICATED TO MY WIFE

Esther

FOR HER THOUGHTFUL AND PATIENT ENCOURAGEMENT

Foreword

This book is intended as a guide to the basic rules, reliable procedures, and statutory requirements governing public construction contracts. Because the rules are often elaborately drawn up to minimize the opportunities for fraud, their application is quite difficult, and even careful formulation of the contract documents cannot always avoid controversy. The contents of this work have therefore been arranged to chart a clear path for those most concerned—contractors, engineers, architects, suppliers of equipment or materials, the surety, public officials and agencies, and their lawyers.

In recent years, a thorough understanding of the principles discussed in this book has become essential to the prudent expenditure of public funds. For the sake of simplicity and conciseness, no attempt has been made to discuss legal principles except as they affect public agencies. Many highly regarded volumes discuss the more fundamental concepts of contract law, and to recapitulate them would be superfluous. What the author has attempted to do here is to present the general principles of public construction contracts together with illustrations to show their application. To this end, relevant court decisions have been cited. Most of these are drawn from the jurisdictions of the various states; a special chapter considers the procedures and principles of federal construction contracts.

This book can hardly claim to provide the complete answer to all the administrative and legal problems that may be encountered in awarding and executing public construction contracts. In particular, it cannot be a substitute for legal counsel in situations requiring professional skill and knowledge. It is hoped, however, that it will offer a key to sound practices, and that it will encourage the further development of standard procedures, to the benefit of the public.

Contents

Foreword vii

1 The ABC of competitive bidding 1

1.1 Principles, standards, and purposes 2
1.2 Exceptions to the rule 4
1.3 Bidder to know law 7
1.4 Government agency to provide specifications 10
1.5 Patented material 13
1.6 Bidding restrictions 15
1.7 Advertised notices for bids and their contents 20
1.8 Advertising for bids 22

2 Guidebook for contract awards 27

2.1 Correct computation of bids 27
2.2 Bidder's right to an award 28
2.3 Single bids 29
2.4 Tie bids 30
2.5 Rejection and postponement of bids 32
2.6 Bid deposits 33
2.7 Proposals—signatures and forms 37
2.8 Firm bids 44
2.9 Informal bids and irregular bids 45
2.10 Variance by bidder and stipulations 49
2.11 Modification of bid by telegram 50
2.12 Words, figures, and "no charge" 52
2.13 Unbalanced bids 53
2.14 Alternate bids 57
2.15 Failure to bid on alternates 60
2.16 Joint adventures and joint ventures 61

3 Mistakes in bids and specifications 64

3.1 Conditions for release from bid 65
3.2 Stating claim of mistake 66
3.3 Honest mistake in a bid 68
3.4 Right of rescission 69
3.5 Obvious error in bid 70
3.6 Principles of bidder's claim of mistake 71
3.7 Misinterpretation of specifications 76
3.8 Mistakes by the government agency 77

4 Lowest responsible bidder 80

4.1 Definition of lowest responsible bidder 80
4.2 Determination of irresponsibility 84
4.3 Judgment and discretion of awarding officer 85
4.4 Administrative decision and court judgment 87
4.5 Influence of surety bond 88
4.6 Prequalification 89
4.7 Contractor's statement 92
4.8 Licenses, taxes, and fees 93
4.9 Summary of rules and requirements in fifty states 106

5 Warranty of plans and extra and additional work 108
5.1 Contractor's obligations and warranty of plans 108
5.2 Contractor's procedure 113
5.3 Responsibility for defects in specifications 115
5.4 Meaning of approximate quantities 118
5.5 Additions to scope of contract 121

6 How to read a public construction contract 136
6.1 Validity of a public contract 136
6.2 Government and power to make contracts 138
6.3 Government's power to discontinue a construction contract 141
6.4 Definition of public works 146
6.5 Words and their meaning in public construction contracts 147
6.6 The intention or "meeting of the minds" in the written contract 148
6.7 Choice of words and usage and custom 153
6.8 The meaning of the term "working days" 157
6.9 Validity of decisions by the engineer in charge 159
6.10 Third party beneficiaries 161
6.11 Negligence and liability for damages 169
6.12 Municipal zoning regulations and state construction projects 176

7 Obligations of the contractor and the surety 181
7.1 General principles of performance 181
7.2 Substantial performance 182
7.3 Anticipated default in completion 189
7.4 Use of surety bonds 190
7.5 Meaning of statutory bond 195
7.6 Liability of public official for failure to require surety bond 196
7.7 Subrogation and surety rights 197
7.8 Surety bonds, labor, and materialmen 203
7.9 Sub-subcontractors and payment bonds 208
7.10 Release of surety from liability 213

8 Liquidated damages and delay 218
8.1 Liquidated damages vs. penalties in contracts 218
8.2 Liquidated damages 220
8.3 Delay 229
8.4 No damages for delay clause and its application 235
8.5 Contractor's rights upon delay in progress payments 239

9 Federal construction contracts 243

9.1 General principles of federal contracts and jurisdiction of Court of Claims 244
9.2 Local regulations and federal construction contracts 249
9.3 Standard federal construction contract documents 253
9.4 Prescribed use of Standard Forms 257
9.5 Standard Form 23A, "General Provisions (Construction Contracts)" 258
9.6 The Davis-Bacon Act 286
9.7 Eight Hour Laws, Anti-Kickback Act, and Walsh-Healy Act 290
9.8 Renegotiation of contracts 293
9.9 Renegotiation Act of 1951 295

A Summary of state requirements 299

B Standard federal forms 340

Abbreviations 367

Table of cases 370

Index 395

chapter **1**

The ABC of competitive bidding

Competitive bidding for public construction contracts has been an express requirement in public administration since about 1845, and practically all levels of government follow this method. The double objective of competitive bidding is to stimulate competition and to prevent favoritism and fraud when contracts are awarded. Competitive bidding is a functional procedure in government that is concerned with a price for producing a certain project or work of construction. The supply of public work and the demand for it by contractors are factors in establishing the price, just as supply and demand of a commodity govern its price. Therefore, responsible public officials must achieve a realistic equilibrium between supply and demand in order to serve the interest of the taxpayers. Such a balance will provide ideal competition for quality performance.

A private individual has the right to seek the lowest price for the construction work that he wants to undertake, even to the extent of specifying inferior quality. The public official, however, is required to look for a fair and reasonable price that will produce a project of high quality for the benefit of the health and safety of the people of his community. The preparation of suitable specifications by the responsible government agency will assure comfortable housing for the citizens, adequate schools, modern public roads, and serviceable sanitation, as well as a pure water supply.

1.1 Principles, standards, and purposes

Competitive bidding after public advertising for proposals by contractors is a fundamental time-honored procedure that assures the prudent expenditure of public money.

The bidding procedure for public work is governed by the express legislative policy of the government body or unit involved—federal, state, or municipal. That procedure is administered within the scope of the applicable public policy pursuant to standards established by the responsible agency or official. Those standards must be reasonable in their requirements and their application at each step in the procedure must be uniform and consistent.

The integrity of the administration of competitive bidding and the opening of sealed proposals in public depends principally upon the responsible public officer adhering to the established and promulgated rules that govern the procedure. If the notice to bidders provides that proposals must be delivered at a stated place, up to a certain hour, then the delivery of the proposal elsewhere or at any later time requires that the proposal must not be considered in the competition for the particular letting. When a bidder participates in a public letting, he is entitled to an assurance that all other bidders for the same contract will be bound by the same rules that apply to him.

Because the contractor who seeks to be the successful bidder engages in a business commitment at the public letting, the transaction at that point assumes serious proportions. Profit-making is the motive of the contractor who strives to stay in business, and the jobs of his workmen are affected by his success or failure in public competition.

The government official who conducts the letting must always control and supervise the procedure from the time of the advertisement or invitation for bids through the receipt of the sealed proposals, the opening in public, and the award. This control and supervision insures that all procedural steps, both major and minor, will be performed uniformly and fairly.

The provisions for competitive bidding found in statutes relating to the expenditure of public money for construction work are written to prevent extravagance or improvidence because of favoritism or fraud. Effectively observed competition for public construction contracts is the respected procedure that promotes a real and an honest cost basis for the accomplishment of the work for which the contractors are asked to submit proposals.

The object and purpose of a statutory requirement for competitive bidding is:

"to protect the public against collusive contracts; to secure fair competition upon equal terms to all bidders; to remove not only collusion, but temptation for collusion and opportunity for gain at public expense; to close all avenues to favoritism and fraud in its various forms; to secure the best values for the (public) at the lowest possible expense; and to afford an equal advantage to all desiring to do business with the (public) by affording an opportunity for an exact comparison of bids.

"Laws of this kind requiring contracts to be let to the lowest bidder are based upon public economy, are of great importance to the taxpayers, and ought not to frittered away by exceptions.

"They originated, perhaps, in distrust of public officers whose duty it is to make public contracts, but they also serve the purpose of affording to the business men and taxpayers . . . affected by them a fair opportunity to participate in the benefits flowing from such contracts, which are nowadays amongst the most important items of the present day business world.

"In so far as they thus serve the object of protecting the public against collusive contracts and prevent favoritism toward contractors by public officials and tend to secure fair competition upon equal terms to all bidders, they remove temptation on the part of public officers to seek private gain at the taxpayers' expense, are of highly remedial character, and should receive a construction always which will fully effectuate and advance their true intent and purpose and which will avoid the likelihood of same being circumvented, evaded or defeated." (*Webster* v. *Belote,* 138 So. 721, Fla., 1931, which cites *Hunnan* v. *Bd. of Education,* 25 Okla. 372; *Fones Hardware Co.* v. *Erb,* 54 Ark. 645; *Anderson* v. *Fuller,* 41 So. 684; *Mazet* v. *City of Pittsburgh,* 137 Pa. 548; and *Wells* v. *Burnham,* 20 Wis. 112.)

The purpose of competitive bidding was discussed in a leading case on the subject in *Harlem Gaslight Co.* v. *Mayor of New York,* 33 N.Y. 309, 329, 1865, and the comment is pertinent to the statements quoted in the preceding paragraph. In that case, the city officials decided to provide street lighting. They granted a public utility corporation the privilege of occupying the beds of the streets with the corporation's mains, but upon the express condition that it should supply the gas for street lighting purposes at a stated charge. When the validity of the arrangement for buying the gas for the municipal lighting was questioned, the court commented that if proposals had been sought by advertising there could have been but a single offer at best, because the only facilities laid down in the streets belonged to the corporation and there was no one else who had the ability or the means to supply the gas for the street lighting purpose. The decision also pointed out that there are certain exceptions when due consideration is given to the purpose of competitive bidding when the related statutes require that procedure. As to such requirements of public bidding, the court added the following:

"They are not to have such a construction as to defeat this purpose, to impede the usual and regular progress of the public business, or to deprive the inhabitants, even temporarily, of those things necessary and indispensable to their subsistence, their health, or the security and protection of their persons or property. Contingencies may arise when services, materials, and property above the prescribed value, may be immediately needed, and when competitive offers and written contracts would be unserviceable or impossible. In such a case the statutes would not apply, because such application could not have been intended. Whenever the nature of the service, or of the property needed for the public use, or the time within which it must be had to prevent irreparable mischief under competitive offers is impossible, then the provisions of the acts referred to (i.e., requiring competitive bids) cannot apply, because such could not have been the intention of the lawmakers, and such emergencies were not among the mischiefs which the provisions referred to were designed to correct." The court decided that the arrangement for buying the gas was valid, particularly because the conditions for providing street lighting constituted an exception to the requirement for competitive bidding.

1.2 Exceptions to the rule

There are two recognized exceptions to the rule of competitive bidding when there is a general statutory requirement for that procedure in entering into public contracts: (1) where it is not desirable or not possible to advertise for particular work; and (2) when emergency conditions must be remedied. Examples of the first exception to the rule are when: (1) specialized skill and professional services are required; (2) a governmental unit such as the police department needs land in a particular locality; (3) a certain type of building is needed; or (4) a particular article is necessary to accomplish a certain purpose.

The exception to the constitutional or statutory rule that requires advertising and awards to the lowest responsible bidder is found in contracts for professional services of engineers, lawyers, architects, physicians, accountants, and surveyors. Significant comment on this exception is found in *Hunter* v. *Whitaker,* 230 S.W. 1096, Tex., 1921. This case involved a statute that required open competitive bids for contracts by a municipality of more than a certain sum of money. The court said: "To hold that the act would require that the services of a man belonging to a profession such as that of the law, of medicine, of teaching, civil engineering, or architecture, should be obtained by a county only through competitive bidding would give a ridiculous meaning to the act, and requires an absurdity. . . . Such a construction would require the selection of attorneys, physicians, school-teachers and civil engineers by competitive bids, the only test being the lowest bid for the services of such men. Such a test would probably be the best that could be conceived for obtaining the services of the least competent man, and would be most disastrous to the material interests of a county." The services of professional skill have been excepted from the rule of competitive bidding that governs most public contracts in almost all states, as disclosed by the following selected citations: *Spangler* v. *Leitheiser,* 37 At. 832, Pa., 1897; *State ex rel. McMahon* v. *McKenzie,* 29 Ohio C.C. 115, 1906; *Horgan & Slattery* v. *New York,* 114 App. Div. 555, N.Y., 1906; *Stratton* v. *Allegheny County,* 91 At. 894, Pa., 1914; *Braaten* v. *Olson,* 148 N.W. 829, No. Dak., 1914; *Re Cook County,* 177 N.W. 103, Minn., 1920; *Franklin v. Horton,* 116 At. 176, N.J., 1922; and *Rollins* v. *Salem,* 146 N. E. 795, Mass., 1925.

In the case of *Mallon et al.* v. *Bd. of Water Commissioners, et al.,* 128 S. W. 764, Mo., 1910, the local purchasing agent was required to buy materials and supplies by competitive bidding, except in emergency and not to exceed $500. The water commissioners were empowered to select water meters and to install them in residences at the expense of the consumer. Not more than $500 worth of the particular meters were bought at any one time, and the purchases were made without competitive bids. The lawsuit was to restrain the water commissioners from selecting one meter to the exclusion of any other kind preferred by the consumer. The court said: "It seems to us that a number of kinds, some good, some excellent, and others bad, would lead to confusion and uncertainty in results to be had from them. It appears clear that uniformity is a reasonable requirement, and that the board has a right to prescribe that a certain kind of meter be installed at the expense of the consumer." However, the court decided that upon the express requirements of the local statute, there was no emergency as to the use or need of water meters, but, rather there was a constant need for them "and from time to time continuously to make orders for meters by private purchase is clearly a violation of the plain terms of the law requiring public competition." The water commissioners should have obtained competitive bids on meters that produce the information required and so specified; then an appropriate selection could have been made after all manufacturers had had an opportunity to compete for the business in the manner prescribed by the local statute.

As to the second exception to the rule of competitive bidding in cases of emergent conditions, a determination of an emergency is not an easy task. Special circumstances govern each case, and, whereas the definition of "emergency" implies a sudden or unexpected necessity requiring speedy action, it is not practical to state a general rule for a determination of "emergency."

An example of an emergency that warranted the application of the exception to the rule of competitive bidding was disclosed where a city charter empowered the city council to authorize a contract without advertising for bids for actual emergency work in connection with a contract that had been made to dredge a harbor channel. The dredging had polluted the water at a bathing beach, and the operation of the pipe line and the leakage therefrom obstructed traffic. In a court action to test the validity of the contract to correct the situation, it was decided that the facts disclosed a condition that was within the meaning of "emergency" work as authorized by the city charter, and the omission of competitive bidding was approved in this case. (*Los Angeles Dredging Co.* v. *Long Beach,* 291 P. 839, Cal., 1930.)

Another lawsuit disclosed a situation that did warrant competitive bidding because, in the opinion of the court, no real emergency existed. A contract for the construction of a bridge was made without first getting open competitive bids; the local law provided for competition for public construction projects, except in cases of emergency. The local council insisted that time was of the essence in providing the structure for public travel, although all of the means of crossing the river that had previously served the people were still intact and were adequate. The principal reason for declaring an emergency was the increase in the local population to the point where the number of inhabitants to be served was generally outgrowing the existing facilities. The court ruled against the action of the local council, and said: "Under these circumstances it seems worse than idle to argue that the rush was so great as not to brook of a delay of three weeks, the time required to advertise for competitive bids." (*Green* v. *Okanogan County,* 111 P. 226, Wash., 1910.) In this case, too, it was disclosed that the local charter also provided that a contract for any emergent public work had to be based upon a declaration of such emergency by the council. This language did not vest the public body with discretion to award a contract without competitive bidding. In such a case the courts have the power to review the facts of the so-called "emergency" and to approve or disapprove the action of the board.

The board of supervisors in the case of *San Christina Invest. Co.* v. *San*

Francisco, 141 P. 384, Cal., 1914, was authorized to pass an ordinance for emergency work without competitive bids when the character of the emergent condition is recited in the ordinance. The emergency was created, according to the recital in the ordinance, after a great fire, which caused considerable damage to streets and sewers, requiring extensive and costly repairs of those facilities. The tax rate had to be increased to provide the necessary funds. The court decided that the language of the related statute did not give the supervisors the discretion and authority to declare conclusively as to the existence of a great emergency or necessity. The intent of this statute is to permit a court to consider a protest against an award of a contract without public competition and to determine if the declaration in the ordinance as to the particular condition presents an emergency.

The adequate type of statutory discretion, however, was vested in the board of public works in the case of *Stern* v. *Spokane et al.,* 111 P. 231, Wash., 1910. In this case, the law required competitive bids except: "If the mayor and city council shall by resolution declare an emergency to exist." The language relating to an emergency would have the effect of nullifying the general requirement in the statute for advertisement and competitive bidding. In such a case the courts cannot interfere because the public officers acted within the limits of their discretion and authority. The declaration of emergency complied with the statutory requirements that justify an award without competitive bidding.

The possibility of a damaging frost was approved by the court as an emergency whereby material for use on a greenhouse in a public park could be purchased without competitive bidding. (*Sheehan* v. *New York,* 37 Misc. 432, N.Y., 1902.)

Public officials are expected to observe the implication to advertise for proposals when the statute on the subject requires an award to the lowest bidder, even though the advertisement is not expressly called for by the same statute. (*Galbreath* v. *Newton,* 45 Mo. App. 312, 1870.)

1.3 Bidder to know law

The bidder who offers to enter into a contract with a state or municipality must see that the requirements of competitive bidding have been observed by the public officials. The bidder is supposed to know the law, and the source of that information is in the related statutes. The laws on the subject of public construction contracts are passed by the respective legislatures to protect the interest of the public. In most jurisdictions the omission of the proper procedural steps leaves the contractor without any remedy to obtain payment for the work he has performed pursuant to the writing of what he assumed to be a valid contract. Whereas in certain jurisdictions there are variations in the application of the rules of law on this subject, those variations are usually recognized and enforced in instances of implied contracts, provided the particular situation meets certain technical conditions. General reliance upon an implied contract is hazardous. The fact that the state or municipality has accepted the work and retains the benefit thereof does not generally clothe the contractor with a remedy in all jurisdictions whereby he can collect for having done his work. The advertisement for proposals and the competitive bidding for the work are of prime importance in valid procedure when the governing laws require such procedure. (*Miller* v. *McKinnon,* 124 P. 2d 34, Cal., 1942; *Johnson County Savings Bank et al.* v.*City of Creston et al.,* 231 N.W.705, Iowa, 1930; *Floyd County* v. *Owego Bridge Co.,* 137 S.W. 237, Ky. 1911; *Brady* v. *New York,* 20 N.Y. 312, 1859.)

When a local charter or other related statutory provision does not call for competitive bids, the courts have upheld the action of the public official in omitting competitive bids. (*Kingsley* v. *City of Brooklyn,* 78 N.Y. 200, 1879.)

Where the local statute governing the improvement of streets provided that the responsible public official "may" advertise for proposals in a newspaper published in the county, he decided that it would not be necessary to insert the notice to bidders in such local newspaper. A taxpayer of the city asked the court to stop the council from making an award because of the omission to advertise for proposals in a local county newspaper. The court upheld the award and stated that the action by the council was valid without the local advertisement. The decision pointed out that the power to make the contract was not dependent upon advertising for proposals in the county newspaper, and that by the permissive word "may" in the statute, the responsible official had the discretion to decide if he would comply with the statutory provision for advertising locally. (*Dillingham* v. *Mayor, etc. of City of Spartanburg,* 56 S.E. 381, So. Car., 1907.)

If a government body enters into a contract without complying with the preliminary procedure when required by the statute that governs the award of public construction contracts, the contract is considered beyond the power of the public agency or department and is void. For example, advertising for bids and competitive bidding are mandatory in those contracts coming within the terms of the applicable statute. The steps in the bidding procedure that are mandatory are also jurisdictional, and a failure on the part of the responsible public officers to follow the procedure creates a void and unenforceable commitment of the governmental body and cannot be ratified. (*Miller* v. *McKinnon,* 124 P. 2d 34, Cal., 1942.)

In the case of *Miller* v. *McKinnon,* a taxpayer sued the contractor to recover money that a government agency paid under a contract entered into without advertising for bids. The court decided that the public body exceeded its power when it made the contract, and under such circumstances there cannot be an implied contract, either. As to the hardship that is caused the contractor in a case of this kind, the court said: "It may sometimes seem a hardship upon a contractor that all compensation for work done, etc., should be denied him, but it should be remembered that he, no less than the officers of the corporation, when he deals in a matter expressly provided for in the charter, is bound to see to it that the charter is complied with. If he neglect this, or choose to take the hazard, he is a mere volunteer and suffers only what he ought to have anticipated. If the statute forbids the contract which he has made, he knows it, or ought to know it, before he places his money or services at hazard. Persons dealing with the public agency are presumed to know the law with respect to the requirement of competitive bidding and act at their peril."

With respect to money paid to a contractor under a void public contract, the general principles governing the right to recover such payments as having been made illegally, are not applied in the same manner in all jurisdictions. In the *Miller* v. *McKinnon* case, recovery was allowed against the contractor. The same result was achieved in *Bangor* v. *Ridley,* 104 At. 230, Me., 1918 and in *Milquet* v. *Van Straten,* 202 N.W. 670, Wis., 1925. On the other hand, courts in other jurisdictions show a tendency to deny the recovery of payments to a contractor when the public body retains the benefits or where the circumstances are such that it is impossible to restore the benefits to the contractor. This was the result in *Vincennes Bridge Co.* v. *Atoka County,* 248 F. 93, Okla., 1917; *Sacramento County* v. *Southern P. Co.,* 59 P. 568, Cal., 1899; *McCarthy* v. *Bloomington,* 127 Ill. App. 215, 1906; *Laird Norton Yards* v. *Rochester,* 134 N.W. 644, Minn., 1912; *Shoemaker* v. *Buffalo Steam Roller Co.,* 165 App. Div. 836, N.Y., 1915; and *State ex rel. Hunt* v. *Fronizer,* 82 N.E. 518, Ohio, 1907.

In the case of *Dovel* v. *Village of Lynbrook,* 213 App. Div. 571, N.Y., 1925 it was decided that the General Municipal Law, or the Village Law under which the village was incorporated, did not require the contracts to be awarded after advertisement and to the lowest responsible bidder. After citing authorities in previous cases the court said "it would seem that the Board of trustees were authorized to act as they did and award the contract to one known to them to be competent and capable to do the work, even though he was not the lowest bidder. In other words, in the absence of some statutory requirement or evidence of bad faith, it was entitled to use its judgment in awarding this contract to the same extent that an individual would be."

1.4 Government agency to provide specifications

Competitive bidding procedure generally provides that prospective bidders must be furnished with the specifications by the public agency seeking proposals for the particular project. If the advertisement for bids for the common public construction project requires the contractors to write the specifications, the basis for determining the lowest responsible bidder will not be developed because the prices submitted in the respective proposals will not be for the production of the same result. An exception to this principle may be found when bids are sought for a highly specialized product or equipment that is unusual and may involve a particular technique to achieve a specified result.

When a certain municipality advertised for proposals to construct a garbage crematory, the bidder was required to specify the system, the size, dimensions, and the manner of construction, along with the number of tons of fuel required, and the number of men necessary for its operation. General plans and specifications were prepared by the public board, but it was left to the bidders to submit the detailed plans and specifications of the building. The local statute provided for competitive bidding and required that a plan or profile should accompany the specifications, along with a description of the kind of material furnished, all to be filed in the office of the public works board. In discussing the general specifications, the court said in *Ricketson* v. *Milwaukee,* 81 N.W. 864, Wis., 1900:

"The general specifications adopted required each bidder to submit with their bids 'complete plans and specifications fully showing and describing the buildings, machinery, and furnaces, and other necessary appurtenances of the entire cremation plant in detail, with all dimensions given.' This was a plunge in the dark. . . . The indefinite character of the specifications, and the absence of plans, had the effect of stifling all competition. Each bidder was called upon to make a proposal, resting largely upon his own judgment, with absolutely no guide as to details. No one could tell which was the lowest bid, because no two would be on the same basis. That fact alone condemns the action taken. If the city has not a sufficiently definite idea of what it wants, to cause proper plans and specifications to be made, then it must wait until further information can be secured, or the plan has become so far developed as to be more than a long-felt want."

An illustration of the exception to the rule requiring specifications to be prepared by the public agency is a project for furnishing and installing equipment required to operate and control a toll system on a highway. In order to attract the most modern procedures and methods of accounting and the latest mechanical techniques, the invitation for proposals included information for bidders which specified the following: (1) general requirements for toll station equipment and office tabulating machines; (2) the number and types of stations where the toll system would be operated; (3) the estimated volume of traffic; (4) the basic charge for vehicles of various classifications; (5) the control of issuance of transit tickets and receipts; and (6) a description of anticipated fraudulent practices. The requirements of suitable accounting systems were outlined for the bidders, with the installation of adequate reporting procedures for individual toll stations and also for consolidated system reports. An alternative plan was also described in the information for bidders, which included automatic devices that could be considered and adopted after satisfactory trial performance. These devices were to count vehicles, detect and record vehicles using the toll highway with a flat-rate permit, compute and record toll fare, and classify vehicles.

Following the voluminous and definite statements of the objectives of the installation and use of the equipment and systems, the bidders were also informed that their proposals must describe the complete system including all phases of the operating procedures, protective features, and the methods of reconciling collectors' accounts, with alternates that would take care of any equipment breakdown. Complete specifications and drawings of all items of the equipment were to be provided by the bidders with full data on the status of development and the results of tests or demonstrations, as well as estimated cost of operation. The information for bidders declared that the award would be based on quality rather than price and that consideration would be given to efficiency of the systems proposed, the degree of revenue protection afforded, ability to handle the traffic, flexibility of arrangements, reliability of equipment and costs. With this mass of detailed information as a guide, competition was encouraged, and whereas the plans and specifications were not prepared by the public agency, the description of the special conditions to be considered and the exactness of the result sought warranted the procedure that was instituted as an exception to the general rule (excerpt from "Information for Manufacturers desiring to submit Proposals for New York State Thruway Toll System").

In considering the preparation of the plans by the public body which is charged with the responsibility of that function, it is the usual but not universal procedure for the public officials to exercise discretionary power in determining whether a public project should be constructed under a single contract for the entire work or by two or more separate parts of the project. In some jurisdictions that discretion is governed or modified by a related statute or a charter provision. For example, section 153.03 of the Ohio Revised Code (Anderson) provides that all contracts for "the erection, alteration or repair of any building in the state, by the state . . . which provide for the installation of gas fitting, steam and hot water heating, ventilating apparatus, steam power plant, or the electrical equipment, and all work kindred thereto . . . when the entire cost of the erection, alteration or repair of the plumbing and gas fitting, steam and hot water heating, ventilating apparatus, steam power plant, or the electrical equipment, and all work kindred thereto, is to exceed one thousand dollars," shall be awarded pursuant to "separate plans, specifications and blue prints for each of the following classes of work:

(A) Plumbing and gas fitting;
(B) Steam and hot water heating, ventilating apparatus, and steam power plant;
(C) Electrical equipment"

so as to permit "separate and independent proposals and bids upon each of the classes of work set forth in divisions (A), (B) and (C)" of section 153.02 of the Ohio Revised Code.

In the case of *Columbus Bldg. & Constr. Trades, etc. Council* v. *Moyer, etc.*, 126 N.E. 2d 429, Ohio, 1955, a notice to bidders for certain work on a state university building stated that "bidders are invited to bid on . . . any combination of divisions" of the work, and "a single bid for any combination of divisions may, however, be lower than the total of several bids for such combination of divisions" and including the three classifications identified in section 153.02 of the Ohio Revised Code. The court ruled that the proposal as contained in the form that was governed by the provisions in the notice for bids was invalid; therefore, the bids were to be rejected because the public officers did not have the discretion to vary the mandatory provisions of the statute.

Another illustration of the statutory requirement is found in section 135 of the State Finance Law of the State of New York. The law requires that the construction of any public building that is estimated to cost more than $50,000 must be undertaken pursuant to separate specifications for classifications of the work to be done, as follows: (1) plumbing and gas fitting; (2) steam heating, hot water heating, ventilating and air conditioning apparatus; and (3) electric wiring and standard illuminating fixtures. Separate and independent bidding is required pursuant to specifications that are prepared on that basis, and all contracts must be awarded for the three subdivisions separately and to the lowest responsible bidders engaged in the stated classes of work. The inclusion of the provision in the statute was intended to discourage and to eliminate a monopoly, whereby a general contractor could, it is alleged, subcontract the separate classes of work among a favored few at a price that could be arbitrarily fixed by the general contractor. Representatives of labor unions were reputedly the sponsors of the legislative provision to protect their members from unemployment if the number of available jobs could be controlled by a general contractor who decided to engage only a few selected subcontractors. From time to time, since the enactment of the statute more than 30 years ago, public officials responsible for the construction of State-owned buildings have attempted either to revise the financial limitation or eliminate the restrictive provision altogether. These officials have shared the opinion that a general contractor could produce the building construction at the same or a lower price than required under the conditions created by the existing statutory provision. They also feel that work would be facilitated because fewer contractors would be engaged; therefore there would be less supervision of separate groups of workers by different employers, which results in inefficiency because of overcrowded working conditions and delays in availability of an area that cannot be utilized by two groups or more at the same time.

A comparable statutory requirement is found in section 66.29 of the Wisconsin Statutes (West), which prescribes that: "On those public contracts calling for the construction, repair, remodeling, or improvement of any public building or structure, other than highway structures and facilities, the municipality shall separately let (a) plumbing, (b) heating and ventilating, and (c) electrical contracts where such labor and materials are called for. . . ."

1.5 Patented material

A conflict of authority exists as to a requirement in a specification to use a patented or monopolized material. For example, the so-called "Wisconsin rule" prohibits any specification of a patented material because it stifles competitive bidding, while the so-called "Michigan rule" permits a specification for patented material. (*Sanborn* v. *City of Boulder*, 221 P. 1077, Colo., 1923.) This distinction is explained in the decision in *Dillingham* v. *Mayor etc. of City of Spartanburg*, 56 S.E. 381, C., 1907, as follows:

"The keystone of the argument in support of the Wisconsin line of cases is that where the statute requires competitive bidding after advertising, as a condition precedent to the power of the municipality to contract for street improvement, the statute is violated when the city ordinance or the contract specifications require the use of a patented or monopolized article, because there can be no real competition when the bidding is practically restricted to the individual or corporation controlling the patent; on the other hand, the fundamental reason supporting the Michigan line of cases is that, even where the statute requires competitive bidding, it is not violated, or does not apply when all the competition is allowed which the situation permits; that a municipality should not be denied the right, for the benefit of its citizens, to avail itself of useful inventions and discoveries, even though protected by patents; and that when a city exercising its power to make the public improvements in good faith decides to contract for the use of patented articles, there is created no monopoly and no abatement in competition beyond what necessarily results from the rights and privileges given the patentee by the Federal Government."

The Department of Public Works in New York State may name a patented material or article in special notes or specifications, but the general specifications expressly provide that the use of a tradename is intended to indicate the degree of excellence or performance that is required, and any equivalent that will produce the same result or objective will be considered when submitted by the successful bidder as a substantial conformance with the specifications. In the absence of fraud on the part of the public official, and also in the absence of a statute to the contrary, a specification calling for such patented material is permissible.

A leading case on the subject of specifying patented materials in public contracts is *Warren Brothers Co.* v. *City of New York,* 190 N.Y. 297, 1907, in which it was held that a reasonable opportunity for competitive bidding is found under circumstances that permit owners of patented and unpatented pavements to join in bidding for a contract. "The dictates of public policy," the court said, "under the circumstances here presented, would seem to require that the owners of patented and unpatented pavements should be on equal terms. It may well be in this age of invention and progress that the wit of man may devise, if it has not already, a smooth and noiseless pavement that is cheaper and more enduring than any now in use. If this proves to be the fact, there is no reason why the inventor and the city should not profit by this situation."

When a common council adopted an ordinance identifying different kinds of pavement to be constructed in the city, "bitulithic pavement" was described as consisting "of a surface of at least two inches of what is known as bitulithic top surface, laid either on concrete or broken stone foundation, as may be specified in the bidding sheet." The city charter empowered the property owners who are liable for the improvement assessment to designate the kind or make of pavement material. The designation was for bitulithic pavement, but the lowest bidder did not confine itself to the specifications, offered an asphalt stone mixture pavement, and attempted to justify the omission to bid on the bitulithic pavement because it was a patented article. The court in *Whitmore et al.* v. *Edgerton etc.*, 87 Misc. 216, N.Y., 1914, said: "The adoption of specifications for a patented pavement does not prevent competition or competitive bidding, and that proposition has been decided many times and in different states. *Holmes* v. *Council Detroit*, 120 Mich. 226; *Baird* v. *Mayor*, 96 N.Y. 567; *Silsby Mfg. Co.* v. *City of Allentown*, 153 Pa. 319." The decision declared the plaintiff's bid to be informal (that is, not adhering to the proper form), and the opinion pointed out that the patentee of the material had opened the way for competitive bidding when it offered the use of the product to any bidder on the whole contract at a fixed price. By such an offer the competitive bidding can be accomplished, but it is not fair competition when the specification requires a material upon which a manufacturer has a monopoly by reason of patents and does not make the product available to all interested parties upon the same terms and in advance of the date of the public letting by the governmental agency.

1.6 Bidding restrictions

Local materials: Preference is sometimes expressed in a specification for public construction for materials produced within the particular state, and in some instances there are even restrictions against materials produced outside of that state. This situation may be considered as restraint of trade and restrictive of competition, but in *Allen* v. *Labsap*, 188 Mo. 692, 698, 1915, the court held that the requirement of stone that is cut and dressed within the state is not a restraint. On the other hand, *People ex rel. Treat* v. *Coler*, 166 N.Y. 144, 1901, ruled that a statutory restriction as to the use of stone that is dressed or carved outside of the State of New York is void, because it is in conflict with the commerce clause of the Federal Constitution. The Court of Appeals wrote of the statutory restriction that: "It is a regulation of commerce between the states which the legislature had no power to make. The citizens of other states have the right to resort to the markets of this state for the sale of their products, whether it be cut stone or any other article which is the subject of commerce. . . . Under the Constitution of the United States, business or commercial transactions cannot be hampered or circumscribed by state boundary lines, and that is the effect of the statute in question. . . . (*People* v. *Hawkins*, 157 N.Y. 1; *People* v. *Buffalo Fish Co.*, 164 N.Y. 93)."

Certain taxpayers: The validity of an award to the low bidder, a foreign corporation, was questioned in court by the resident contractor (*Schrey et al.* v. *Allison Steel Mfg. Co.,* 255 P. 2d 604, Ariz., 1953) because the related statute governing awards granted a 5 per cent preference to satisfactory contractors who had paid county and state taxes for two successive years immediately prior to submission of the proposals. The court said: "All discrimination or inequality (in legislation) is not forbidden. Certain privileges may be granted some and denied others under some circumstances, if they be granted or denied upon the same terms and if there exists a reasonable basis therefor. When presented with a law showing partiality, we are always inevitably led into the troublesome problem of classification. The principle involved is not that legislation may not impose special burdens or grant special privileges not imposed on or granted to others; it is that no law may do so without good reason. A principle which none can dispute is that a statute may be allowed to operate unequally between classes if it operates uniformly upon all members of a class, provided the classification is founded upon reason and is not whimsical, capricious or arbitrary." The court pointed out that there is only a qualified right to contract, "not a right to contract as one wishes, and that such right may be restricted if there is a reasonable basis for the restriction. (*State* v. *Senatobia Blank Book & Stationery Co.,* 76 So. 258, Miss., 1917; *Heim* v. *McCall,* 239 U.S. 175, 1915.) The legislature may regulate the letting of public construction contracts, and the State may make a contract with any one it pleases, unless it violates the federal or the state constitution." The decision stated that the preference law involved in this case was constitutional and the contract had to be awarded to "the lowest responsible bidder, giving effect to the 5 per cent differential in determining who is such bidder."

Employers of unionized workmen: In recent years, as the organization of labor unions has expanded, some municipalities include with their specifications and contracts for public work a restrictive provision that is commonly referred to as a "men and means clause". This clause is usually stated as follows: "The contractor and his subcontractors shall not employ on the work any men or means whose employment may cause strikes, stoppages or other similar troubles by workmen employed either by the contractor, his subcontractors, by other contractors or their subcontractors, or by other workmen whose services affect the progress of the work hereunder." (From New York City Department of Public Works form, article 34, Chapter VIII.) Obviously, the language of the provision does not expressly restrict the performance of the work to members of labor unions, but the contractor agrees that he will engage to complete the job promptly by avoiding controversy between employees in the various classifications of workmen who are hired for the work.

The question concerning the power of public officials to discriminate between firms employing union and those employing nonunion workmen has been considered, from time to time, by the courts with particular reference to a determination of the lowest responsible bidder. In a lawsuit as to the validity of the requirement of a union label upon city printing, the judicial comment was "The citizen may be rich or poor in purse; union or nonunion upon the labor question; Catholic, Protestant, Jew, or infidel in matters of religion; Republican, Democrat, or Prohibitionist in political affiliation; but, by the stand of constitutional and statutory right, he is neither more nor less than a citizen of the State, entitled to an equal opportunity thereon according to the capacity and ability with which nature may have endowed him. In denying him that opportunity, a double wrong is perpetrated, first, upon the individual who is entitled to be considered upon his personal merits uninfluenced by these extrinsic considerations; and secondly, upon the State at large, whose expenses are multiplied and whose integrity is jeopardized by a system of favoritism, the demoralizing effect of which is patent to every thoughtful student of public affairs." The decision was that the ordinance requiring the union label on city printing was beyond the administrative powers of the city council. (*Miller* v. *Des Moines,* 122 N.W. 226, Iowa, 1909.) Other and similar decisions on the same requirement by local laws that contracts are not to be awarded for public work to anyone but those employing union labor, because such laws are discriminatory, contrary to public policy, and tended to create monopolies, are found in *Holden* v. *Alton* 53 N.E., Ill., 1899 and *Marshall & Bruce Co.* v. *Nashville,* 71 S.W. 815, Tenn., 1903.

A board of education specification in connection with a public construction contract to be awarded to the lowest responsible bidder, that only union labor would be employed by the contractor, was declared by the court to be unlawful. The decision pointed out that the restriction constituted a discrimination between various classes of citizens, and the specification would restrict competition, thereby making the contract more costly. The court also denied the right of the board of education to agree with labor organizations for the insertion of the restrictive stipulation. (*Adams* v. *Brenan,* 52 N.E. 314, Ill., 1898.) The same judicial conclusions in other States are found in the following citations: *Atlanta* v. *Stein,* 36 S.E. 932, Ga., 1900; *Goddard* v. *Lowell,* 61 N.E. 53, Mass., 1901; *Wright* v. *Hoctor,* 145 N.W. 704, Neb., 1914;

Lewis v. *Board of Education,* 102 N.W. 756, Mich., 1905; *State of Ohio ex rel. United District Heating Inc.* v. *State Office Bldg. Commission,* 179 N.E. 138, Ohio, 1931.

The prevailing view is that a contract based on competitive bids and then awarded only to those bidders who employ union labor is void, but there is some authority to the contrary. For example, a statute provided that: "All orders awarded or contracts made by the State purchasing agent must be awarded to the *lowest responsible bidder, taking into consideration the location of the institution or agency.*" Advertisements for competitive bids for a group of construction contracts for plumbing and heating work on four state buildings and a connecting tunnel resulted in bids on a portion of the work of $17,700 from contractor A and $17,400 from contractor B. Although B's bid was lower, an award was made to A, and a taxpayer asked the court to stop the award because it was not made to the lowest responsible bidder. The contract had been substantially completed when the lawsuit was started. The court pointed out that the statute required the purchasing agent to consider certain specified matters. Such consideration included the fact that the site of the work was a place where labor is highly organized into unions, and the principal contract of the group of the buildings was being performed by a contractor employing union labor, while contractor B, in the case before the court, maintained an "open shop" and employed nonunion labor. Time was most important in completing the work, and after considering the bids, the purchasing agent, under the statutory direction for such consideration, determined that if the award was made to the lowest bidder, contractor B, difficulties would arise between the laborers engaged on the different projects with resultant strikes or walkouts and great delay in completion of the projects. The court decided that the provisions of the related statute did not require an award to the lowest bidder but placed responsibility on the purchasing agent to determine the ability of contractor B to perform and complete the work promptly and satisfactorily. The small "discrepancy here appearing, $300 on a $17,000 contract, is not an unreasonable consideration as relates to this location, and the rule announced would be equally applicable if the principal contractors were 'open shop' and the losing bidder on the subcontract was 'union'." (*Pallas* v. *Johnson,* 68 P. 2d 559, Colo., 1937.)

In *Taylor* v. *Board of Education,* 253 App. Div. 653, N.Y., 1938, affirmed 278 N.Y. 641, the courts upheld the awarding officers in a taxpayer's action to declare an award illegal because the Board had deliberately adopted the policy of awarding contracts for electrical work in improvements for school buildings to those contractors who employ exclusively members of a certain local labor union. The local statute required awards to be made to the lowest responsible bidder after advertising and competitive bidding. The court said that for the taxpayer's action to succeed "there must be shown not only illegal action, but one in some way injurious to municipal and public interests which if permitted to continue will in some manner result in increased burdens upon and dangers and disadvantages to the municipality and to those who are taxpayers." It was held that the rejection of the bids of nonunion employers was done in good faith by responsible public officials who had the right to exercise "proper discretion which would avoid increased burdens and threatened dangers to the municipality."

In *Burland Printing Co., Inc.* v. *LaGuardia,* 9 N.Y.S. 2d 616, 1938, the court declared as reasonable a contract provision that required the use of the union label on city work. "This clause," the opinion stated, "in the light of modern conditions is entirely reasonable."

The power of a public body to refuse an award solely because nonunion labor may possibly be employed and the equities of the employer and of organized labor are discussed in *State of Ohio ex rel. United District Heating Inc.* v. *State Office Bldg. Commission,* 179 N.E. 138, Ohio, 1931. After a bidder had been notified that a resolution had been adopted by the commission to accept his proposal as being the lowest bid, he was asked whether he employed union labor. Upon his answer that he conducted an "open shop" without discrimination as to an employee's affiliation or nonaffiliation with a labor union, the commission notified the contractor that the resolution for an award to him had been rescinded "because the contractor did not employ exclusively union men." The contractor sued to compel an award to him because of his rights to contract under the Fourteenth Amendment of the Constitution of the United States. In denying the right of the commission to discriminate against the bidder for the reason given, the court commented:

"The claim is made that costly delays and added expenses may occur because of possible trouble if this contract be not awarded to the bidder employing union labor. This claim assumes that a great state cannot control its laws requiring public bidding; cannot protect its citizens from unconstitutional discrimination. If such discrimination be permitted, all the laws controlling public bidding and requiring awards to be made to the lowest bidder have no potency. The state would be helpless.

"But let us assume that the shoe had been placed on the other foot, assume that public officers, anticipating labor troubles, would refuse to award a bid to a contractor employing union labor. What would be the answer of the respondents to that proposition and what would be the answer of the dissenting member of this court? In such event organized labor would protest, and rightly so; and this court would scrupulously protect it from such unconstitutional discrimination. In the case of *LaFrance Electrical Construction & Supply Co.* v. *International Brotherhood of Electrical Workers,* 108 Ohio St. 61, 140 N.E. 899, this court was called upon to protect and did protect the lawful rights of union labor. In the course of that opinion it was said at page 95 of 108 Ohio St. 61, and at page 908 of 140 N.E. 899: 'Equality of justice demands that in any controversy the rights of all parties be scrupulously maintained. The right of workmen to be employed, irrespective of union membership, must be maintained; the right of the employer to conduct his business without illegal interference must be upheld; and legal means employed by strikers must not be curtailed.' "

The court consisted of seven judges, six of whom concurred in the decision that ruled that the commission was not empowered to restrict the bidding to employers of union labor only. The dissenting opinion pointed out that the commission desired to save the "state the expense and avoid the delays incident to the lawful differences between workingmen," and the commission adopted what they considered a wiser course when the restriction was made to permit bids from employers who operate a "union shop" exclusively. He commented on the various classifications of labor who were already engaged in the other parts of the construction of a state office building, all of whom are members of labor unions. He emphasized the duty of the commission to award to the "best" bidder not merely the "lowest" bidder, and in carrying out this responsibility "this court may not lightly declare that the commissioners have abused a discretion."

1.7 Advertised notices for bids and their contents

Notice to bidders for the purposes of the requirements of the competitive bidding procedure is a solicitation of proposals, and whereas the notice does not carry with it any promise nor commitment of a contract, it should be informative and definite enough to identify the general nature of the work and the place where the plans and specifications are available for distribution and for the use of prospective bidders. In many jurisdictions the contents of the advertisement are specified by a statute or a local charter. But, in any event, the advertisement should state sufficient information to aid a contractor in deciding whether or not the project is attractive enough to induce him to prepare and to submit a bid.

A suggested checklist of the basic information that should be included in the published notice should contain the following: (1) a brief description of the work of construction or improvement; (2) a reference to the drawings and specifications therefor and where they may be seen and obtained; (3) the time when and the place where the proposals invited by the advertisement will be received; (4) the requirement of a bid deposit; (5) the requirements as to surety bonds for completion and for labor and material together with a statement of the amount of each of such bonds; and (6) any other pertinent information to assist the prospective bidder.

The following form was copied from the papers in the project that is identified as "Specification No. 15415-C" of the New York State Architect's Office, and is recommended as an adequate and definite notice to guide a prospective bidder in his determination as to the preparation and the filing of a proposal for the work:

Rehabilitate Roofing, Etc.
State Armory
643 Park Avenue, New York City

NOTICE TO BIDDERS

Sealed proposals covering *Construction Work for Rehabilitation of Roofing and Appurtenant Work, State Armory, 643 Park Ave., New York City, in accordance with Specification No. 15415-C and accompanying drawings,* will be received by Henry A. Cohen, Director, Bureau of Contracts, Department of Public Works, 12th Floor, The Governor Alfred E. Smith State Office Building, Albany, N.Y., *on behalf of the Executive Department, Division of Military and Naval Affairs, until* 2:00 P.M., *Advanced Standard Time, which is* 1:00 P.M., *Eastern Standard Time,* on Wednesday, *August 3, 1960,* when they will be publicly opened and read.

Each proposal must be made upon the form and submitted in the envelope provided therefor and shall be accompanied by a certified check made payable to the State of New York, Commissioner of Taxation and Finance, in the amount stipulated in the proposal as a guaranty that the bidder will enter into the contract if it be awarded to him. The specification number must be written on the front of the envelope. The blank spaces in the proposal must be filled in, and no change shall be made in the phraseology of the proposal. Proposals that carry any omissions, erasures, alterations or additions may be rejected as informal. The State reserves the right to reject any or all bids. Successful bidder will be required to give a bond conditioned for the faithful performance of the contract and a separate bond for the payment of laborers and materialmen, each bond in the sum of 100 per cent of the amount of the contract. Drawings and specification may be examined free of charge at the following offices:
State Architect, 270 Broadway, New York City.

State Architect, 4th Floor, Arcade Bldg. 486-488 Broadway, Albany 7, N.Y.
District Supervisor of Bldg. Constr., State Office Building, Albany 7, N.Y.
District Supervisor of Bldg. Constr., Genesee Valley Regional Market, 900 Jefferson Road, Rochester 23, N.Y.
District Engineer, 65 Court St., Buffalo, N.Y.
State Armory, 643 Park Ave., New York City

Drawings and specifications may be obtained by calling at the Bureau of Contracts, (Branch Office), 4th Floor, Arcade Bldg., 486-488 Broadway, Albany 7, N.Y., or at the State Architect's Office, 18th Floor, 270 Broadway, New York City, and by making deposit for each set of $15.00 or by mailing such deposit to the Albany address. Checks should be made payable to the State Department of Public Works. Proposal blanks and envelopes will be furnished without charge. The State Architect's Standard Specifications of Jan. 2, 1960 will be required for this project and may be purchased from the Bureau of Finance, Department of Public Works, 14th Floor, The Governor Alfred E. Smith State Office Building, Albany, N.Y., for the sum of $5.00 each.
DATED: 7-5-60

1.8 Advertising for bids

Duration requirements: The provisions of a statute or local charter as to advertising a notice to contractors for bids are mandatory; their observance is not discretionary, but is a limit to the right of the awarding officer to make a contract under such provisions. The purpose of the notice is to safeguard the public against official indifference, and such restrictive statutes, being mandatory, cannot be circumvented by public officials. The failure to observe the requirements as to advertising a notice for proposals is the same as if no notice whatever had been given. "The principle of competition—all important in the eye of the law—could thus be stifled, if not altogether eliminated" (*Comstock* v. *Eagle Grove City,* 111 N.W. 51, Iowa, 1907).

The statutory provisions as to the number of times or the duration of time of an advertisement for bids must be substantially observed, although the provisions are, as a rule, construed liberally. The purpose of a requirement for advertisement is to afford prospective bidders a reasonable time in which to inspect the site and to examine the plans with an opportunity to estimate the cost. If the statute or ordinance does not specify the time for advertisement, a reasonable time must be observed; a period of eight days before the opening of bids has been considered as a reasonable time. (*Case* v. *Fowler,* 65 Ind. 29, 1878.)

One week and twice a week for not less than four weeks: Where a statute called for publication of a notice in a newspaper "at least twice a week for a period of not less than four weeks," the court said: "A week is a definite period of time commencing on Sunday and ending on Saturday. . . . The preposition 'for' means of itself duration, when it is put in connection with time, and as all of us use it in that way in our every day conversation, it cannot be presumed that the legislature, in making this statute did not mean to use it in the same way. Twelve successive weeks is as definite a designation of time according to our division of it, as can be made. When we say that anything may be done in twelve weeks, after the happening of a fact which is to precede it, we mean that it may be done in twelve weeks or eighty-four days, or, as the case may be, that it shall not be done before." (*Leach et al.* v. *Burr,* 188 U.S. 510, 567, 1903.)

Two consecutive weeks: Where the statutory period of publication of a notice to bidders provided that the letting could be made only after advertising "for two consecutive weeks" in a newspaper of general circulation, at least fourteen days, excepting Sundays or other days when the newspaper is not published, must intervene between the first publication and the letting of a contract. (*Jenkins* v. *City of Bowling Green,* 64 S.W. 2d 457, Ky., 1933.)

Once a week for two consecutive weeks before letting: Where a statute provides: "The advertisement shall be published in a newspaper of general bona fide circulation in the city, once a week for two consecutive weeks prior to the day set for the opening of the bids," the advertisement appeared on Friday, June 12, Saturday, June 13, and one week later on Friday, June 19 and Saturday, June 20. More than one week intervened between the last day of advertisement on June 20 and the public letting on Monday, June 29. The complaint to the court concerned a contract for constructing streets and stated that since no advertisement appeared during the week of seven days before June 29, the municipality did not comply with the requirement of "two consecutive weeks prior to the day set for the opening of bids." The court rejected the complaint "because the word 'prior' does not necessarily mean 'next prior to be happening of an event." The provisions of the statute quoted above limit the minimum number of times of the advertisement and not the minimum length of time. (*Meahl* v. *City of Henderson,* 290 S.W. 2d 593, Ky., 1956.)

Not less than once a week for two successive weeks: In *Bechthold et al.* v. *City of Wauwatosa et al.,* 280 N.W. 320, Wis., 1938, the case involved a statutory requirement of advertisement "not less than once a week for two successive weeks." The court ruled that the advertisement could be for as much longer a period as the local board may determine, yet there must be compliance with the statutory requirement for advertisement in order to make a valid contract. As the advertisement was published for less than two weeks before the opening of the bids, the contract was declared to be void.

Thirty days before work is finally let: A city charter provided that notice to bidders "shall be given thirty days before the work is finally let, by advertisement in one or more newspapers." The advertisement was placed in weekly newspapers and first appeared on March 2. The notice stated that bids would be received on March 31, and specifications would be ready for examination by prospective contractors on March 15. The contract was finally let at the end of June. The court stated, in an action to force the city to pay for the work under the contract pursuant to the advertisement, that the "charter does not require that notice shall be given 30 days before the bids are made, but 30 days before the work is finally let. Nor can we read the charter as requiring that the advertisement be published on each of the 30 days. . . . We think a single insertion would suffice under this requirement, and in fact numerous insertions were made in each of several newspapers. . . . The fact that the specifications were not exhibited prior to March 15 does not seem to require an extended discussion. The charter does not contain any express provision on the subject; its implied requirement is merely that specifications be provided a reasonable time in advance of bidding. And the time allowed in this instance seems to us sufficient." (*City of Newport News* v. *Potter* 122 F. 321, Va., 1903.)

Not less than five successive days: A local charter provided that in instances where street improvements are to be made "the recorder shall duly give notice of publication for not less than five (5) successive days in a daily newspaper." An advertisement was published on June 5, 6, 7, 8, 9 for a letting to be held "on or after June 10." The bids were opened on June 10, and an award was directed to be made on June 24. The lawsuit was brought on the grounds that the charter provision for advertising for bids was not observed. The court said that such a provision is mandatory, and added: "The notice for bids must be published for the time and in the manner required by the charter; and since the mode is the measure of the power, a failure to follow the prescribed mode will invalidate an attempted special assessment. *Jones* v. *Salm,* 123 P. 109, Oregon; *Matter of Pennie,* 15 N.E. 611, N.Y.; *Upington* v. *Oviatt,* 24 Ohio St. 232; *Breath* v. *City of Galveston,* 49 S.W. 575, Tex.; *Tifft* v. *City of Buffalo,* 25 App. Div. 376, N.Y.; *Michel* v. *Taylor,* 127 S.W. 949, Mo.; *Polk v. McCartney,* 73 N.W. 1067, Iowa; *Meuser* v. *Risdon,* 36 Cal. 239; *Kretsch* v. *Helm,* 45 Ind. 438." In commenting on the phrase "for not less than five (5) successive days," the court said:

"The term 'for' and the words 'not less than' appear in the quoted provision. When used in the connection in which we now find it, the term 'for' means 'through'; 'throughout'; 'during the continuance of'—*Century Dictionary*. If the charter read that the notice must be published 'for five days', by the overwhelming weight of authority it would be interpreted to mean a publication through, throughout, during the continuance of five full days. *3 Words and Phrases 2858; 2 Words and Phrases, 2nd series, 594; Northrop* v. *Cooper,* 23 Kan. 432; *Bacon* v. *Kennedy,* 22 N.W. 824, Mich.; *Wilson* v. *Thompson,* 3 N.W. 699, Minn.; *State* v. *Cherry County,* 79 N.W. 825, Neb.; *Dever* v. *Corwall,* 86 N.W. 227, No. Dak.; *Wilson* v. *Northwestern Mut. Life Ins. Co.,* 65 F. 38; *Finlayson* v. *Peterson,* 67 N.W. 953, No. Dak. The words 'not less than' like the language 'at least' signify 'in the smallest or lowest degree; at the lowest estimate'; and legislation prescribing 'not less than' or 'at least' a specified number of days is usually construed to mean clear and full days for the specified period of time.... Emphatic as is the word 'for' it is, if possible made still more emphatic by the accompanying language 'not less than;' and when combined these words unmistakably mean that the notice must be published for a period of time which cannot be less than five full successive days. In brief, the notice must be published five full days before the right to submit bids is closed."

The decision of the court declared that the advertisement was not published as required by the charter and the right to offer bids should have been kept open until the close of business on June 10, and the public opening of the bids should not have occurred until June 11. (*Albert* v. *City of Salem,* 164 P. 567, Oregon, 1917.)

Plans on file on a day not earlier than ten days after first publication: Where a local statute expressly provided that plans and specifications must be prepared and placed on file in the office of the Board prior to publication of the notice to bidders, it appeared that they were not on file until September 12 or one week after the first publication of the notice. The statute also provided for a public opening of bids "on a day not earlier than ten (10) days after the first publication" and in the instant case the advertisement set the letting date for September 23, or a period of eleven days after the publication of the first advertisement. The court held that the plans and specifications were on file for examination for more than the ten-day requirement, although not for a period of ten days from the first publication. "There was compliance with the essential purpose of the statute which was to insure competitive bidding and there was sufficient substantial compliance with the provisions of the statute as to notice to bidders to such extent that it would be improper for this court to hold such contracts void." The court added that no charge of fraud or illegality was made and there was no evidence that showed any disregard of an opportunity for free, open competitive bidding. (*Feigel Construction Corp.* v. *The City of Evansville et al.,* 150 N.E. 2d 263, Ind., 1958, and citing *Jenny* v. *City of Des Moines,* 103 Iowa 347, 1897; *Carr* v. *Fenstermacher,* 119 Neb. 172, 1929; *Jenkins* v. *City of Bowling Green,* 251 Ky. 119, 1933)

chapter 2

Guidebook for contract awards

After the proposals have been received, opened, and publicly read by the responsible representative of the government agency, it is customary for that agency to prepare and issue a tabulation of the bids.

2.1 Correct computation of bids

The tabulation is useful in any dispute about the method used by the state or municipality to arrive at a correct computation of the bid. An unusual method was the one included in the specifications for a state highway project, which provided that fourteen buildings were to be disposed of by any of three described methods to be selected by the state after an award had been made to the successful bidder. The contractor was required to furnish unit item prices for each method of the building disposition, and payment would be made for the disposition of each building as directed by the state's engineer. The court held that the computation of the gross bid had to include each unit item for the three separate methods of disposition of the properties "even though it may later eventuate that such item may not be called into performance by virtue of the reservations duly provided for." (*Matter of Ottaviano Inc.* v. *Tallamy et al.,* 277 App. Div. 929, N.Y., 1950.)

In the *Matter of V.J. Costanzi, Inc.* v. *Brandt* (Supreme Court, Albany County Special Term, N.Y., Nov. 8, 1940) the petitioner alleged that his bid was the lowest one read at the public letting, but upon verification of the computation of the unit prices by the Superintendent of Public Works, another bid read at the letting as the second low bid disclosed an error in addition of the total bid price. The Superintendent corrected the error, whereupon the second bid became the lowest one, and an award was made to and was accepted by that contractor at the corrected figure. In the opinion by the court, it is stated:

"The statute provides that the lowest bid shall be deemed that which 'specifically states' the lowest 'gross sum' for which the 'entire work' will be performed. The common interpretation of this language—certainly its literal reading—would mean the lowest total as computed by the bidder for which all the work will be done. The further direction of the statute that the 'gross sum' stated must include 'all the items specified' is not the equivalent of a provision that if all the items are properly added they supersede the gross sum elsewhere specifically stated in the bid." In commenting upon the procedure of the Superintendent of Public Works, the opinion added: "I am apprehensive also, that if it be here judicially determined that the stated gross sum of a bid be disregarded in favor of a computation of the items by administrative officers of the state to reach a lower total upon which to award contracts, the way might be opened to the designing and disingenuous for claims based upon purported errors in the statement of specific items, with the state placed upon notice of such possible claims by the patent inconsistency of the computations appearing on the face of the bid. I confess that I cannot fully and to the final end explore all the possibilities. But such a judicial determination would undoubtedly promote litigation and future controversy. That the situation presented here seemed to the state officers . . . to require some form of release from the bidder is suggestive of a similar apprehension on their part."

As a result of that decision, the Highway Law of the State of New York was amended to provide as follows: "The lowest bid shall be deemed to be that which specifically states the lowest gross sum for which the entire work will be performed, including all the items specified in the estimate thereof. The lowest bid shall be determined by the Superintendent on the basis of the gross sum for which the entire work will be performed, arrived at by a correct computation of all the items specified in the estimate therefor at the unit prices contained in the bid." (Sub-div. 3, section 38, Highway Law, N.Y.; similar provisions are contained in sub-div. 6, section 8, Public Buildings Law, N.Y.)

2.2 Bidder's right to an award

After proposals are publicly opened, the awarding official has a reasonable time in which to examine the bids and to satisfy himself of the responsibility of the lowest bidder. Thereafter, he is empowered to exercise his discretion as to the lowest responsible bidder, and in performing that function he may make an award to the lowest bidder, or, if in his judgment the best interests of the public will be served by an award to another bidder instead of the low bidder he has the authority to so decide, or he may reject all of the bids.

Even after public information that a particular party is the lowest bidder, that disclosure does not constitute an award; he is not absolutely entitled to an award; nor does he have any substantive rights to an award. "The mere arithmetical operation of ascertaining which bid is the lowest does not constitute an award." The discretion of the awarding officer is judicial in its nature and the award is the result of a judicial act. (*Erving* v. *Mayor, etc., of New York,* 131 N.Y. 133, 1892; *Matter of Baitinger Elec. Co.* v. *Fones,* 170 Misc. 599, N.Y., 1939.)

The discretion of an awarding official is not subject to interference by a court, unless there is proof of the abuse of that discretion. The courts will intervene where the discretion is arbitrarily exercised. (*State ex rel. United District Heating Inc.* v. *State Office Bldg. Com.,* 179 N.E. 138, Ohio, 1931.) It is a general principle of law that a court will rarely direct an awarding officer to award a contract to a particular bidder. This position has been adopted by most judicial bodies, because in public bidding the statutory duty of the awarding officer allows him a choice in the performance of the discretionary function of his office. However, if the awarding officer awards a public contract as a result of arbitrary and capricious action with willful disregard of the law, then a bidder who is clearly entitled to the award will find assistance in the courts.

2.3 Single bids

The validity of a public letting is challenged occasionally when only a single bid on a particular project is received by the awarding official. The answer is usually found in the statute that governs the awarding of public contracts in that jurisdiction.

Where the legal requirements for public construction contracts call for "competitive bidding" or for "sealed bids to be opened in public," an informative notice to bidders and the medium of publication are the preliminary, and usually the only mandatory, steps that must be observed by the public official or body.

Competitive bidding means that an opportunity is afforded for everyone to bid, and the notice of this opportunity is, as a rule, announced by advertisement in newspapers in the locality of the work to be done and also in accordance with any statute that governs the procedure. Yet if only one bid is received after compliance by the public agency with relation to notice to propective bidders, an award duly made to that single bidder is valid. There is no implication in the procedure for competitive bidding that more than one bid must be submitted in order to justify an award. (*Blanton et al.* v. *Town of Wallins,* 291 S.W. 372, Ky. 1927.)

The requirement of competitive bidding is to hold it open to competition at all times, thereby allowing more than one party to submit a proposal.

Where bids are asked for public work with separate prices for each of nine or ten materials of different kinds, in order to permit the public officials to select one kind, the filing of a single bid after compliance with statutory notice, and in the absence of fraud or collusion, does not violate the legal requirement of competitive bidding, and an award to that bidder is proper. (*Denton* v. *Carey-Reed Co. et al.,* 183 S.W. 262, Ky., 1916.)

2.4 Tie bids

The power of the awarding officer to decide which of two bidders who have submitted identical or tie bids shall receive the contract is based upon the express public policy as disclosed by the statute under which he performs his functions.

The leading case of *Matter of Dictaphone Corp.* v. *O'Leary et al.,* 287 N.Y. 491, 1942 discussed a specification for proposals to furnish office machines. It was specified that "award shall be made to the lowest responsible bidder conforming to specifications. In the event of tie bids, the decision of the Commissioner of Standards and Purchase shall be final." Two bidders offered to supply articles described in the specifications at the same price. The Commissioner awarded the contract to one of them, and the other challenged his procedure. The statute (section 163, State Finance Law) authorized him to perform various duties and to administer "such other matters as may be necessary to give effect to the foregoing rules and the provisions of this article." The clause in the specifications relative to tie bids, which was inserted by administrative action by the Commissioner, was held by the court to be ineffective, because it is deemed to be "a discretion wider than the discretion vested in him under the statute." The opinion of the Court of Appeals stated:

"Where equally responsible bidders offer to provide articles described in the specifications for exactly the same price, the Commissioner in deciding which of the bidders should be accepted as the lowest bidder, may give consideration to 'the qualities of the articles proposed to be supplied, their conformity with the specifications, the purposes for which required and the terms of delivery,' and upon these factors he may base his decision of which bid should be accepted. He may not decide to award the contract to both bidders and thus relieve himself of the duty of determining which of two similar bids is the lowest. The language of the statute is not open to construction that two or more bidders submitting identical bids may be considered 'the lowest bidder.' Such construction would, indeed, tend to thwart rather than to promote the purpose of the statute. It would tend to make ineffective the statutory barriers against collusive bidding and to make the choice of equipment used by the state depend on sale pressure exercised upon departmental heads rather than upon price and quality which the legislature has decreed shall be the factors which must determine choice . . . Under the provisions of the statute, where two bids are tendered at the same price the Commissioner must decide which of the two bids is lowest or must reject both bids.

"His decision, however, must be based upon the considerations specified in the statute. Here the Commissioner claims the right to base decision upon other considerations. It may be that a broader power of choice should be given to the Commissioner, but such arguments must be addressed to the Legislature. We are told that it may be impossible to choose in this case between the two lowest bidders if only the quality of the articles and their conformity to the specifications may be considered. In that case, new bids must be asked, or the supplies required must be purchased in amounts not exceeding $1,000. We recognize that such purchases may work to the disadvantage of the state. The Commissioner must find the remedy within the powers granted to him by the statute, or ask the legislature for increase of such powers.

"He cannot by establishing rules enlarge upon the powers conferred by the Legislature. He derives his authority to make rules from section 163 of the statute, and that authority is limited to the administrative field defined in the first nine subdivisions of the section and to 'such other matters as may be necessary to give effect to the foregoing rules and the provisions of this article.' The clause which the Commissioner inserted in the 'specifications' upon which he invited bids may not be given effect as a rule established in accordance with section 163, if it is intended to confer upon the Commissioner a discretion wider than the discretion vested in him under the statute. Even if the Legislature could delegate to the Commissioner legislative power to extend the scope of a provision contained in the statute, it has not attempted to do so."

Following the court decision in the *Dictaphone Corporation* case, section 174 of the State Finance Law was amended (Chap. 350, Laws of 1944) by inserting the following: "In cases where two or more bidders submit identical bids as to price, the determination of the Commissioner to let a contract to one or more of such identical bidders shall be final."

The "toss of a coin" is an informal procedure that is used in some jurisdictions in the case of tie bids. The tied bidders are invited to consider this method with the clear understanding that the "loser of the toss" will abide by and accept the decision. This understanding also includes the admonition that objection to the "toss" method by either or both parties will result in the rejection of all bids and advertising for new bids.

2.5 Rejection and postponement of bids

The reservation of the right to reject bids is usually expressed in a charter provision or in a statute on the subject. (*McRoberts* v. *Ammons,* 88 P. 2d 1958, Colo., 1939; *Brown* v. *Houston,* 48 S.W. 760, Tex., 1898; *Stern* v. *Spokane,* 111 P. 231, Wash., 1910.) The courts have recognized the right of public officials to reject all bids whether or not there is a statutory provision therefor. (*Davison-Nicholson Co.* v. *Pound,* 94 S.E. 560, Ga., 1917; *Herman Constr. Co.* v. *Lyon,* 211 S.W. 68, Mo., 1919; *Coward* v. *Bayonne,* 51 At. 490, N.J. 1902; *Straw* v. *Williamsport,* 132 At. 804, Pa., 1926.) This right of rejection has also been recognized by the courts when the advertisement for bids stated the reservation, regardless of the existence or omission of a related statute.

In the exercise of the right to reject all bids or to postpone the opening of sealed bids, the courts will not interfere with such a decision when reasonable notice has been released to all of the affected parties; neither will there be judicial interference when there is no proof of caprice or arbitrary action on the part of the responsible public officials. (*Matter of Yonkers Contr. Co.* v. *Thruway Authority,* 283 App. Div. 749, N.Y., 1954.)

A board of education provided the rules governing its public lettings and it was stated that "all proposals received . . . shall be opened . . . immediately after the expiration of the time limited by the advertisement for the receipt thereof, by the officer advertising for the same, in the presence of the secretary of the board, or in the presence of such clerk as the secretary may designate; all proposals shall be thereafter filed in the office of the secretary, and no proposal shall be received or considered that may be offered or submitted after the time limited, as before specified." In commenting on the objection disclosed in a taxpayer's action relative to the award because the letting was adjourned until the day following the date stated in the advertisement for proposals, the court said the objection was untenable, and added: "The provision in this respect is directory. The main purpose to be accomplished was that the bids should be opened speedily after the time had elapsed for their submission and in the presence of the proper authorities. It prejudiced no one by deferring the opening, as was done in this case. It was to bring the officers together who were charged with the duty of making the examination, and they being all present and acting within a reasonable time after the bids were deposited answers the substantial requirements of the provision. The substance of the matter is that they shall examine the bids within a reasonable time and the officers charged with the duty must be present. (*People ex rel. Rodgers* v. *Coler,* 35 App. Div. 401) This was accomplished in the present case."

The motive for the rejection of all bids is not material, particularly when the right to reject is contained in a statute or in a statement in the notice reserving that right. The public official is not compelled to state a reason other than, for example, that the rejection will serve the best interests of the public. This part of the procedure in the consideration of competitive bidding is subject to the control of the governmental agency and not the contractors; also, the notice to bidders is an offer to receive proposals for a contract and not an offer of a contract. (*Trapp* v. *City of Newport,* 74 S.W. 1109, Ky., 1903; *State ex rel. Shay* v. *McCormack,* 167 App. Div. 854, N.Y., 1915.)

2.6 Bid deposits

Most governmental notices to contractors for proposals on public construction provide that each bid must be accompanied by a deposit in the form of a certified check, a bid bond, or some other evidence of a stated sum or a specified percentage of the amount of the respective bid. The purpose of the deposit is to guarantee that the successful bidder will execute the contract. Usually, the return of the bid deposit is determined by statute, and as a rule, the refund is made upon delivery of the completed contract documents. In some places the bid deposit is retained until a certain percentage of the work has been completed. As a general rule the bid deposit is claimed by the government agency in those cases where the bidder to whom an award is made refuses to execute the contract. If a bid bond has been filed as required in the notice to the contractors, then, if the bidder refuses to execute the contract, demand is made upon both the principal (the contractor) and the surety on the bond for payment of the amount that was secured by the bid bond.

Questions arise as to whether or not the requirement of a bid deposit is in the nature of a penalty or of liquidated damages for failure to execute the contract. The intention is not always expressed by the use of the term "penalty" or "liquidated damages." The word "forfeit" is not conclusive. The circumstances must govern in each case, and the courts will be bound to enforce liquidated damage provisions when the contracting parties choose to make such a contract, particularly if the provisions are not severe or inequitable. (*Little et al.* v. *Banks,* 85 N.Y. 258, 1881.)

Davin v. *City of Syracuse,* 69 Misc. 285, N.Y., 1910, affirmed 145 App. Div. 904 is a leading case on the meaning of "liquidated damages" as related to bid deposits. The advertisement stated that a bid deposit was required with the proposal as a guarantee that the successful bidder would execute a contract or, upon refusal to do so, the deposit would be forfeited "as liquidated damages and not as a penalty for such neglect or refusal." The contractor refused to execute the agreement, and he sued to recover the deposit. "It is competent for the parties to a contract to agree" said the court, "as to the amount of damages to be paid upon a breach, instead of leaving the amount to be ascertained by the court and jury. This agreement will not be interfered with unless the amount is 'so grossly disproportionate to the actual injury that a man would start at the mention of it.' *Clement* v. *Cash,* 21 N.Y. 253, 256, per Wright, J. 'So, also, I think where the language employed in that part of the instrument ascertaining the amount of the damage is clear and plainly indicative of an intention to fix a definite sum to be paid by the party failing to perform, and negatives all inferences of an intent to name the sum as a penalty, the courts are not authorized, by construction, to make a new contract for the parties or unmake the one made for them, and hold, from the nature and circumstances of the case that the parties intended something different from what they have expressed.' "

In *Harper, Inc.* v. *City of Newburgh,* 159 App. Div. 695, N.Y., 1913, Harper intended to bid 90 cents per linear foot for stone curbing and 65 cents per linear foot for concrete curbing. The bidder claimed after the letting and after the award was made to him on the day following the letting that the prices were transposed by mistake in his proposal, and he asked the City to release him from the bid and to return his deposit, which was required by the specifications to guarantee that the bidder would execute a contract for the project. The proposal provided that upon failure of the successful bidder to execute a contract, he would forfeit the bid deposit as liquidated damages. If the contract (the proposal) to agree to enter into a contract for the project is rescinded by the court, "then there is no legal obligation upon the plaintiff (Harper) to contract, and it would seem inequitable that the defendant (the City) should have liquidated damages for breach of an extant obligation. Even though the contract as awarded entailed a greater expense upon the city than if the city could have held the plaintiff to a contract upon its mistaken bid, it must be established first that in the eye of the law the defendant (the City) was entitled to such contract from the plaintiff (the bidder)." The court released the bidder from the proposal and ordered a refund of the bid deposit as an exercise of the equitable and discretionary power of the court in such cases.

The opinion in the *Harper* case distinguished the ruling from that in *City of New York* v. *Seely-Taylor Co.,* 140 App. Div. 98, N.Y., 1910, affirmed 208 N.Y. 548. In the latter case, a mistake in his bid was claimed by the contractor who refused to execute a contract. The city sued the contractor and his surety for the difference between his bid and the next lowest bid for which the contract was finally awarded. The city's claim was dismissed upon the grounds that the damages were limited to the amount of the bid deposit.

Brandese v. *City of Schenectady,* 194 Misc. 150, N.Y., 1950, affirmed 273 App. Div. 831 and 297 N.Y. 965, held that when the bid deposit is not grossly disproportionate to the actual injuries or unreasonable the city may retain the deposit as liquidated damages. When a city advertises for proposals and the bidders accept the offer to bid and do submit bids, the bidders become bound by the city's offer. This procedure is in effect a contract. The other contemplated contract under the bidding procedure is the one to do the work. The court directed the forfeit of the bid deposit because of the delay of the contractor in notifying the municipality of his mistake. But, in *O'Connell Inc.* v. *County of Broome,* 198 Misc. 402, N.Y., 1950, the bid deposit, which was required as a guarantee that the bidder would execute the contract, was ordered returned. The contractor had promptly notified the county of his mistake and then applied to the court for rescission. The county made an award to another bidder without the necessity of advertising again but did not return the plaintiff's bid deposit. The court said: "By reason of this mistake on the part of the plaintiff no actual bid remained 'for there was in the eyes of the law no meeting of minds at all.' (*Harper Inc.* v. *City of Newburgh,* 159 App. Div. 695; *Martens & Co.* v. *City of Syracuse,* 183 App. Div. 622. The court may rescind the apparent bid for the mistake of one party only, without finding fraud or inequitable conduct in the other party. (*Moffett, Hodgkins & Clarke Co.* v. *Rochester,* 178 U.S. 373; *New York* v. *Dowd Lumber Co.,* 140 App. Div. 358.)"

When the instructions to bidders provided for a forfeiture of the bid deposit of $1,000 for the contractor's failure to proceed with an award for the construction of a public sewer, the contractor's willingness to take the contract and to proceed only upon acquisition of the necessary right-of-way by the municipality did not relieve him of his obligation under the proposal. The municipality had the right to retain the bid deposit as its damages. While the municipality was required to furnish the right-of-way as a duty to the contractor it was not bound to do so in advance of entering into the contract. The right of eminent domain was available for that purpose under the charter provisions. In *Coonan* v. *City of Cape Girardeau,* 129 S.W. 745, Mo., 1910, the court said: "The case comes down to this, as far as the rule of law governing it is concerned: Plaintiff believed, and perhaps had reasonable cause to believe, the city would be unable to carry out its part of the contract; that is to say, would be unable promptly to furnish a right-of-way for the sewer system and in consequence he might sustain loss; but this was by no means certain, and the law does not relieve a man from a contractual obligation because he believes with good cause the person with whom he has contracted will not be able to perform. *3 Page, contracts sec. 1449; Hathaway* v. *Sabin,* 63 Vt. 527; *Southern Lumber Co.* v. *Supply Co.,* 89 Mo. App. 141; *Hobbs* v. *Brick Co.,* 157 Mass. 109; *Jewett Pub. Co.* v. *Butler,* 159 Mass. 517; *Plummer* v. *Kelly,* 7 No. Dak. 88; *In Matter of Carter,* 21 App. Div. 118; *New England Ins. Co.* v. *Railroad,* 91 N.Y. 153."

The acceptance of a bid provides every element to constitute a contract, notwithstanding the fact that it is intended between the parties that a formal contract is to be subsequently written. It is not material that the bid deposit, being a guarantee that the bidder will enter into a contract, is either a forfeiture or liquidated damage, and it may be used to apply as a credit against any damages actually sustained by the municipality. A proper measure of such damages is the difference between the contract cost to the municipality and the amount of the bid which was defaulted. (*Middleton* v. *City of Emporia,* 186 P. 981, Kan., 1920.

In *Wheaton Bldg. & Lumber Co.* v. *City of Boston,* 90 N.E. 598, Mass., 1910, the three lowest bidders, in turn, declined to accept a contract and their bid deposits were retained by the city. Before the end of the twenty-day period for the acceptance of proposals by the city, an award was made to the plaintiff who refused to take the award on the ground that, in effect, the previous bids were rejected and the award to the plaintiff was therefore, ineffective. The bid deposit was also retained by the city. It was held that the city retained all of its rights under the bids during the stated period of their duration for that purpose. The basis of the lawsuit was an allegation by the plaintiff that the awarding body had misconstrued the specifications to the extent that a higher cost would have been incurred in doing the work. "Against such mistake of law" the court said, "the courts afford no remedy.... The erroneous interpretation of the language of the specifications was not induced by anything said or done by any agent of the defendant. This is the typical case of misunderstanding the legal effect of language used in any instrument fully signed. *Rice* v. *Dwight Manuf. Co.,* 2 Cush. 80; *Taylor* v. *Buttrick,* 165 Mass. 547; *Boyden* v. *Hill,* 188 Mass. 477." The amount of the bid check was less than the actual loss sustained by the city, and by the expressed intent of the parties the deposit could be retained as liquidated damages.

Where there is a provision in a statute or in the specification that no bid can be withdrawn until after a contract has been awarded, the courts will preclude a recovery of the bid deposit even though the bidder may be relieved from a material and honest mistake in his bid. (*Daddario* v. *Wilford,* 296 Mass. 92, 1926; *Ference* v. *State of New York,* 251 App. Div. 13, 1937.)

2.7 Proposals—signatures and forms

The undersigned and unsigned bids: The proposal of a contractor for a public construction project is the offer that usually initiates and leads to the formulation of a complete contract. The bid must be in such form as to be binding upon the bidder when it is accepted by the public agency seeking the bids; the predominant practice is to have the necessary blank forms of proposals furnished by the governmental body. The cautious bidder will use that particular form when submitting his bid, although a proposal form that will merit acceptance by the public official is one that states substantially all of the provisions and representations that are found in the official form. The signature of the bidder is of major importance; it must honestly disclose the name of the bidder, together with such supporting information that may be necessary when the bidder is a partnership, corporation, or are joint adventurers.

Whereas the language of proposals must, of course, conform with the applicable laws, rules, and policies of each jurisdiction where the work is to be done, the provisions shown in the following specimen (taken from the New York State Public Works Specifications) are comprehensive and are recommended for use with appropriate adaptation to meet local requirements:

To: (*name of public department seeking bids*):

In submitting this bid, the undersigned declares that he is or they are the only person or persons interested in the said bid; that it is made without any connection with any person making another bid for the same contract; that the bid is in all respects fair and without collusion, fraud or mental reservation; and that no official of the State (or other governmental unit, as the case may be), or any person in the employ of the State (or other governmental unit, as the case may be) is directly or indirectly interested in said bid or in the supplies or work to which it relates, or in any portion of the profits thereof.

The undersigned also hereby declares that he has or they have carefully examined the plans, specifications and form of contract, and that he has or they have personally inspected the actual location of the work together with the local sources of supply, has or have satisfied himself or themselves as to all the quantities and conditions, and understand that in signing this proposal he or they waive all right to plead any misunderstanding regarding the same.

The undersigned further understands and agrees that he is or they are to furnish and provide for the respective item price bid all the necessary material, machinery, implements, tools, labor, services, etc., and to do and perform all the work necessary under the aforesaid conditions, to complete the improvement of the aforementioned (highway or other comparable project of public work) in accordance with the plans and specifications for said improvement, which plans and specifications it is agreed are a part of this proposal, and to accept in full compensation therefor the amount of the summation of the products of the approximate quantities multiplied by the unit prices bid. This summation will hereinafter be referred to as the gross sum bid.

The undersigned further agrees to accept the aforesaid "unit bid" prices in compensation for any additions or deductions caused by variation in quantities due to more accurate measurement, or by any changes or alterations in the plans or specifications of the work, and for use in the computation of the value of the work performed for monthly estimates.

The term "undersigned," used in the above provisions, is the usual reference to the bidder in the proposal forms, which are provided by government agencies for public construction contracts. The term is defined to mean "the person whose name is signed or the persons whose names are signed at the end of a document; the subscriber or subscribers" (*43 Words and Phrases 105*).

Recognizing the significance of the word "undersigned" in a proposal form, the careful preparation of a firm bid contemplates that it must be signed "by writing one's name at the foot or end of" a document ("subscribe" from Webster's *Universal Dictionary*). Only by so "writing one's name" can the subscriber show that he actually assumes the obligation of the document. Haste in completing the proposal form can produce calamitous results if a signature is inadvertently omitted or if an improper signature is affixed to the proposal. Such an informality with relation to proper signature may justify a rejection of the bid, even if it is the lowest one for the contract, unless the local rules or instructions to bidders provide otherwise.

Form of proposal and signature guide: Some public agencies and officials have special preferences about the manner in which a proposal should be subscribed by the bidder; the concluding paragraphs in the proposal form given here, together with the illustrations of forms of signatures, should prove helpful in meeting the general requirements:

> Accompanying this proposal is a certified check, draft (or bid bond) for $. . . . In case this proposal shall be accepted by your Department (or other public agency) and the undersigned shall fail to execute the contract and in all respects comply with the provisions of (refer to the law governing the particular contract) as amended, the moneys represented by such certified check, draft (or bid bond) shall be regarded as liquidated damages and shall be forfeited and become the property of (name of the public department or agency); otherwise to be returned to the depositor in accordance with the provisions of (name of applicable statute or rule) as amended.
>
> On acceptance of this proposal for said work the undersigned does or do hereby bind himself or themselves to enter into written contract, within ten days of date of notice of award, with the said (name of public department or agency), and to comply in all respects with (name of applicable law or rules), as amended, in relation to security for the faithful performance of the terms of said contract.

	(Signature) . . . . Legal name of bidder
The P.O. address of the bidder is: 55 Pearl St. Arvil, N.Y.	
	(or if bidder conducts business as a sole proprietor and under an assumed name:)
	Construction Service Co. . . . . Name of firm
(Add P.O. address as above.)	(Signature) By . . . . Sole proprietor, d/b/a/ above named

Proposal form subscribed by individual bidder

The P.O. address of the bidder is:
100 Main St.
Municipal Town, N.Y.

Manning and Olds Construction Co.
....
Legal name of firm

By (Signature)
....
Partner

Name of members	Address
George Manning	65 Residence St., Wickston, N.Y.
Arthur Olds	65 Residence St., Wickston, N.Y.

Proposal form subscribed for partnership

State-wide Construction Corp.
....
Legal name of corporation

The P.O. address of the bidder is:
44 High Terrace
Hightstown, N.J.

By (Signature)
....
President

Officers of the corporation

Name		*Address*
Martin Smith	President	55 Park Place, Hightstown, N.J.
Anna Brown	Secretary	50 Avenue A, Hightstown, N.J.
George Gray	Treasurer	57 Dall Drive, Hightstown, N.J.

Proposal form subscribed for corporation

A joint adventure by the undersigned:

(Signature) .... Name of individual	and	State-wide Construction Corp. .... Name of corporation
	By	(Signature) .... President

The P.O. address of each of the joint venturers is:

Jonathan Doe, 56 Pearl St., Arvil, N.Y.
State-wide Construction Corp., 44 High Terrace, Hightstown, N.J.

Send all notices to State-wide Construction Corp.

Officers of the corporation

Name		*Address*
Martin Smith	President	55 Park Place, Hightstown, N.J. ....
Anna Brown	Secretary	50 Avenue A, Hightstown, N.J. ....
George Gray	Treasurer	57 Dall Drive, Hightstown, N.J. ....

Proposal form subscribed by individual and corporation for joint adventure

When a bidder pasted the printed bid form on a validily executed bond but did not fill in or sign the printed bid form, the court, in an action to test the validity of the proposal, said: "It is not claimed that the blank by itself would not constitute a bid; but it is argued that being attached to the bond which was (properly) signed, and identified by reference in the bond, it must be held sufficient. . . . The reference (in the bond) is to a bid, which the blank paper is not." The reference in the bond to the bid does not give life and validity to the bid which was incomplete and not signed. The bidder in this case could not be held as having made a valid proposal, and he could refuse to accept and to execute the contract if it were offered to him for such purposes. (*Williams et al.* v. *Bergin,* 62 P. 59, Cal., 1900.)

A joint adventure by the undersigned:

(Signature) .... and Manning & Olds Construction Co. ....

Name of individual — Name of firm

By (Signature) ....

Partner

The P.O. address of each of the joint venturers is:

Joseph Roe, 55 Pearl St., Arvil, N.Y.
Manning & Olds Construction Co., 100 Main St.,
Municipal Town, N.Y.

Send all notices to Joseph Roe, 55 Pearl St., Arvil, N.Y.

Names of partners	*Address*
George Manning	65 Residence St., Wickston, N.Y.
Arthur Olds	63 Residence St., Wickston, N.Y.

Proposal form subscribed by individual and partnership for joint adventure

In the case where the proposal or bid proper stated "The undersigned bidder hereby agrees . . .," the bidder did not sign as required, although other pages forming specific parts of the proposal were signed, the municipality's right to make an award to this bidder was challenged in the court. The decision held: "It is clear from the instrument itself that the essential part of the proposal was not signed. The (municipality) had the right to waive such defect, but they (the council) had the right to insist upon the due execution of the written proposal by a proper signature. This is true, notwithstanding the required check of $2,000.00 or 5 per cent of the bid accompanied the instrument. The (municipality) was not required to take a chance of possible litigation because of the defective instrument, and they had a legal right to give no consideration to the bid." (*Interstate Power Co.* v. *Incorp. Town of McGregor et al.,* 296 N.W. 770, Iowa, 1941.) As a matter of administrative procedure it is desirable, without question, to insist upon a signature by the bidder in the place on the bid form which validates

A joint adventure by the undersigned:

Manning & Olds Construction Co.		State-wide Construction Corp.
....	and	....
By (Signature)		By (Signature) ....
Attorney-in-fact for partners		Attorney-in-fact for President

Certified copy of power of attorney is attached hereto.

The P.O. address of each of the joint venturers is:

State-wide Construction Corp., 44 High Terrace, Hightstown, N.J.
Manning & Olds Construction Co., 100 Main St., Municipal Town, N.Y.

Send all notices to State-wide Construction Corp. at above address.

The members of the partnership are:
George Manning, 65 Residence St., Wickston, N.Y.
Arthur Olds, 63 Residence St., Wickston, N.Y.

The officers of the corporation are:
Martin Smith, President, 55 Park Place, Hightstown, N.J.
Anna Brown, Secretary, 50 Avenue A, Hightstown, N.J.
George Gray, Treasurer, 57 Dall Ave., Hightstown, N.J.

Proposal form subscribed by partnership and corporation for joint venture

and formalizes the first step in the bidding procedure. A waiver of such an informality strikes at the fundamental purpose of the signature on a contract form.

In a lawsuit that involved a signature of a corporation in a proposal it appeared that the name of the corporation was signed, but the signature of its authorized officer was missing. The awarding officer did not question the method of signing and its effect on the validity of the bid. As a result, an award was made on the bid and the corporation had substantially performed the work by the time the lawsuit was instituted. The decision stated that the principal reason for denying the court action to set the contract aside was the fact that the contract for the work had already been undertaken. (*Prendergast* v. *City of St. Louis et al.,* 167 S.W. 970, Mo., 1914.) It seems that the courts in most jurisdictions will not approve as valid a proposal which is not properly signed, but when the procedure had already advanced through the performance of the work under the contract, it is reasonable to expect the courts to accept the accomplished fact, as in the case here cited.

The great importance that is attached to the proper signature by a corporation on a document is pointed out in the case where a promissory note was filled out in the name of the corporation on the signature line with the corporate seal impressed thereon, but the word "by" was not written before the signed names of the individuals. In commenting on the proper subscription by a corporation, the court said: "A corporation, as such, cannot write a word or do any other physical act. When business necessity requires its name to be subscribed on paper, it can only be placed thereby and through its officers or representatives. It is a custom of universal practice in the present day for those who sign the name of a corporation to any business or other document, with the intent to bind the corporation only, to append to the corporate signature their own names, or initials, with their official titles, intending thereby to characterize such signing as an act of the corporation only. Such is the common understanding throughout the business and commercial world, and it would seem reasonable for a judicial tribunal, when called upon to interpret a note or other written document so signed, to recognize such custom and give it full force and effect." (*New England Electric Co.* v. *Shook,* 145 P. 1002, Colo., 1915.)

A proposal signed:

Buckeye Meter Company
a Division of & owned by
The Buckeye Incubator Co.
By (s/s) Geo. H. Partin
Sales Manager.

has the legal effect of constituting "The Buckeye Incubator Co." the actual bidder and is not the bid of Buckeye Meter Company, which is a division of the parent company for bookkeeping purposes. (*Hines* v. *City of Bellefontaine,* 57 N.E. 2d 164, Ohio, 1943.)

2.8 Firm bids

A firm bid is one that conforms substantially to and without material variance from the advertised terms, plans, and specifications and does not permit an advantage or benefit that is not enjoyed by the other bidders. The bid or proposal is, in effect, the offer of the contractor to make a contract with the state or municipality to perform the work required by and in accordance with the plans, drawings, and specifications. When the awarding officer determines after a public letting that the bidder meets the requirements of an award as well as the standards of responsibility and that the contractor is qualified in all respects to perform the work satisfactorily, the award by the governmental agency is its acceptance of the contractor's proposal or offer. Basically, this offer and the acceptance constitute a contract between the bidder and the governmental body to enter into a contract for the performance of the work and to obligate the public body to pay for such work in accordance with the prices stated in the bidder's proposal.

The test of a firm bid is found in the unqualified and inviolable promise and commitment by the bidder to accept the award as and when it is made to him. Usually there is a statutory or other specified period of time during which the proposal is binding and effective and within which the government body is required to act, unless the bidder extends such effective date. The proposal cannot be withdrawn by the act of the bidder alone, unless he thereby assumes the risk of forfeiting his bid deposit by such act. If there is a qualification or reservation set up by the bidder in his proposal which gives him the option of accepting, modifying, or rejecting the contract, then it cannot be said that he has submitted a firm bid.

A public officer charged with determining a successful bidder in competitive bidding also has the responsibility of determining whether the proposal is in all respects a firm bid or whether the prospective contractor has included any statement that reserves to him the option of accepting or rejecting the contract when it is offered to him. The categories of proposals most frequently disclosed at public lettings are discussed in this chapter.

2.9 Informal bids and irregular bids

When the awarding official makes a decision as to the intention of the bidder as disclosed by his proposal, the official must also determine whether or not any variance or discrepancy will cause an important or major change in the accomplishment of the work under the contract. However, if the procedure or the result will not produce the thing required because of the variance, then the proposal may be deemed to be informal and invalid.

An informal bid is one that fails to comply with the express requirements that permit a bidder to submit a proposal, or is one that discloses a variance that does not comply with the specifications.

The weight or value attributed to "informality" varies from case to case because some omissions or variances have no material effect upon the proposal, whereas others are of importance. If an informality or irregularity does not materially affect the proposal, then the awarding official may waive the omission or variance. Every variation in a bid does not invalidate it; the variation must be of a substantial nature to warrant that conclusion. A variation is deemed to be of a substantial nature when it affords the bidder an advantage over the other bidders and affects one or more of the elements that were contemplated or considered in reaching a price figure so that the result affects the amount of the bid.

The lowest responsible bidder should not lose his right to an award because of mere irregularity in a proposal. The public official who has the duty of awarding a contract to the lowest responsible bidder will find it difficult, if not impossible, to justify the rejection of the lowest responsible bidder because of such an irregularity. The form of a proposal is usually prescribed by the governmental agency, and a substantial compliance with the requirements of the form is enough to overcome a failure to comply technically with the required form. The irregularity that does not mislead or injure and that can be corrected without changing the substance of the bid, should be waived. (*Faist* v. *Mayor, etc., of City of Hoboken,* 60 At. 1120, N.J., 1905.) Inconsequential variances or minor omissions may be waived without jeopardizing the standards of the bidding procedure.

There is an important distinction between the *power* and the *duty* of a public official who exercises his discretion to accept or reject an informal bid. There is usually a *mandatory duty* upon that official to make an award to the lowest responsible bidder. However, one who has not complied with all of the reasonable and necessary requirements of the governmental agency is not a *bidder,* and in that sense, the public official has no mandatory duty of making an award to him. In the reasonable exercise of the power of determining to waive an inconsequential informality and to accept the bid, or to reject it because of the informality, the public official who is clothed with that power, must first explore the importance of the informality, and then give consideration to the rights of the public whom he represents, and the effect of his action upon the other bidders. Experience has disclosed that sound administrative policy is implemented in a practical manner, by reliance upon inflexible standards in the bidding procedure that are invoked with uniformity and assurance.

The local standards of the bidding procedure are, of course, the guide in the performance of the official duty of deciding the importance or the materiality of an informality or irregularity in a proposal. Applicable statutes establish the basis for such a decision, and these legislative expressions of public policy usually admonish the public official to recognize the "best interests of the public" as an integral part of the bidding procedure. Those interests are best served when a comprehensive examination of the particular situation permits the application of equitable and reasonable rules to the facts at hand. For example, a public officer is justified, as serving the best interest of the state or municipality, by waiving an informality due to a variance in a bid deposit that is slightly less than the amount required in the notice to bidders.

Where a proposal called for a unit bid price for installing 4,000 tons of bituminous material for a section of a public way, the bidder was also required, by a separate unit item, to bid for installing 90 tons of the same type of material for private driveways upon adjacent property. The specifications stated that the unit price of the driveway work "is not to exceed" the unit price of the material used on the public way, by 25 per cent." This kind of a statement of price limitation is a commonly used administrative provision that is accepted in most jurisdictions as an implied power of the public official who is responsible for public contract procedure. Such a provision is deemed to be a reasonable inclusion in a requirement in the specification under the general authority of the public officer in formulating the specification and need not be expressed in a related statute.

When the proposals for the project described in the preceding paragraph were opened, it was found that the first three low bidders disregarded the 25 per cent limitation and the fourth low bidder, who did comply with the limitation, offered to do the entire work of the contract for a total price that was $40,000 higher than the lowest bid. The lowest bidder exceeded the 25 per cent limitation by $2 per ton or a total of $180. The public officials were confronted with a choice of one of three courses of action: (1) they could exercise the right of rejection of all bids and advertise for new proposals; (2) they could waive the irregularity in the lowest bid; or (3) they could reject the first three low bids and make an award to the fourth bidder. Here was an instance where public policy demanded cautious application of local standards. This responsibility was met when it was decided that the variance by the lowest bidder was of a relatively minor nature, and that the public interest will be served by an award to him, even at a total bid that included the sum of $180 which had created the situation. The public officers who made the decision felt justified in waiving the variance as an illustration of their recognition of the interest of the public in this case.

In *Cohen* v. *City of New York et al.*, 205 Misc. 105, N.Y., 1953, the issue before the court was the propriety of three requirements stated in the specification that had to be met in order to qualify as a bidder. The requirements were: (1) he must be regularly engaged in the business covered by the proposal; (2) he must be possessed of satisfactory ability, equipment, and organization to insure speedy and satisfactory service; and (3) he must have a place of business or a warehouse in the City of New York. The court held these requirements to be of a fundamental nature, and a waiver of any of them would not serve the best interests of the City. The court said: "Failure to meet or comply with basic requirements is not a mere informality; such a failure is fundamental. Informality does not reach matters of essence and substance. Informality is want of regularity as to that which is not basically essential. Informality is the antithesis of formality and regularity and its distinguishing feature is that it does not affect essence and substance. (*Phoenix Bldg. and Homestead Assn.* v. *Meraux,* 189 La. 819)".

The interest of the public is served when the awarding officer seeks, within the scope of his authority, to preserve the lowest bid insofar as the legislative policy and administrative rules will permit. For example, in most jurisdictions, the requirement of a certified check is deemed to be complied with when the bid deposit is in the form of a United States Post Office money order or a telegraphed money order. Also, there is no substantial distinction between a cashier's check and a certified check, for the purpose of qualifying a bidder. (*Hornung et al.* v. *Town of West New York et al.,* 81 At. 1116, N.J., 1911.) Another issue in the same court decision was the failure of the lowest bidder to state the number of working days in which the project would be completed, such statement being an express requirement of the specifications. The court ruled that the omission is material to the bid and, therefore, the bid was declared to be informal.

An illustration of adherence to the principles of competitive bidding as an aid to the efficacy of the statute relating to the procedure therefor, is found in the case when a bidder omitted to deposit a certified check, which was required by the local statute and which was called for by the information in the advertisement for proposals. The court ruled that the term "lowest bidder" is not restricted to the amount of the bid; it means also that the bid conforms with the specifications. Such a requirement is material, mandatory, and jurisdictional, and the informality should not be waived. (*City of Hartford* v. *King,* 249 S.W. 2d 13, Ky., 1952; *Hillside Twp.* v. *Sternin,* 136 At. 2d 265, N.J., 1957.)

A variation in the requirement of bid deposit in the form of a certified check, is disclosed in *P. Michelotti & Son, Inc.* v. *Fairlawn,* 152 At. 2d 369, N.J., 1959. In this case the local statute did not require a bid deposit, but the court ruled that the governmental agency had the power to require a certified check for that purpose. The charter of the municipality also provided that there is reserved to its representatives the right to waive any informalities in the bids received for a public project. The low bidder submitted a check for the correct amount, but it was not certified; an award was made pursuant to the waiver of that informality, and the check was later certified. The court said: "This was not a case of a 'material departure,' 'substantial variance' or a 'substantial non-compliance' which would stand in the way of a valid contract and which could not be waived by the municipality. . . ." Among public officials, there is not unanimity of opinion as to this judicial holding. The certification of the check indicates the financial position of the bidder's bank account, and the waiver of the requirement of a certified check when one is expressly called for in the notice to bidders, is, in a sense, discriminatory to the extent that it confuses prospective bidders as to the enforcement of the requirement in future bidding and creates uncertainty where confidence should prevail.

2.10 Variance by bidder and stipulations

As a general rule the instructions to the bidder for preparing a proposal for a public agency warn him to avoid any change in the form that may make his bid ineffective. Discrepancies are created despite the warning, however, and it then becomes the duty of the awarding officer to determine the bidder's intention. Each situation, whether deliberate or inadvertent, calls for particular consideration and must be decided upon the contents of the proposal itself, without any extrinsic or explanatory information or stipulation. The proposal must be complete and adequate in form to afford the bidder full and equal opportunity to submit a firm bid.

An administrative problem in ascertaining the bidder's intention was disclosed in one case when the bidder reduced the quantities of paint as stated in the proposal form. The bidder wrote his own estimate of the required number of gallons of paint, in the space provided for the unit item price in words. This estimate was lower than that expressly called for in the proposal form. He used this lesser gallonage and his stated price therefor as the factors of the product in the total unit price, and then he contended that his estimate of the required gallonage was more accurate than that called for in the proposal form. He admitted that his method was motivated by a desire for an advantage over other bidders who computed their prices and total bid price pursuant to the stated gallonage as estimated by the public department and indicated in its proposal form. Sound bidding practice does not condone variances which implement selfish advantage and which create controversy between the contracting parties. Bidders will feel assured of equal opportunity when the governmental agency and the awarding officer reject a bid which discloses a material change in the specifications, and does thereby threaten uniformity of procedure.

The filing of a stipulation of intention or a supplemental statement to support and explain a proposal is a challenge to the validity of the bid. (*Knickenberg* v. *State of New York,* Claim No. 25,452, Court of Claims. Opinion was filed but not reported.) The acceptance by the governmental agency of such explanatory and supporting writing stamps the proposal as lacking the firmness of a bid and permits an opportunity for inequitable and unfair advantage over other bidders.

In the *Knickenberg* case the contractor bid $152,192.90 for a highway contract, which was the lowest total gross bid. One unit item called for 630 cubic yards of two-course cement concrete pavement, and in preparing the proposal, the contractor inadvertently wrote "six cents" in the column headed "items with unit bid price written in words," and $6.00 in the column for numerals. The total amount of this item was $3,780, which was the product of the numerals and the specified quantity. Before awarding the contract to Knickenberg, the commissioner of highways, maintaining that the written word must govern, gave the contractor the option of withdrawing his bid or of agreeing to a price of 6¢ per cubic yard for the unit item, which would reduce his total bid to $149,182.70. The contractor executed and delivered his agreement for the reduced figure, and the contract was awarded to him. The contractor was paid at the rate of 6¢ per cubic yard for the amount of work done under the item, and, after due notice to the state, he sued to recover at the figure of $6.00 per cubic yard. The state rested its defense on the contractor's agreement to accept the 6¢ rate. The court of claims upheld the contractor and admonished the state that "Such an agreement is in direct contravention of the state's policy to award contracts to the 'lowest responsible bidder.' " It was also ruled judicially that "the commissioner should have exercised his option to reject or to award the bid as it was, and not have insisted on an agreement which did not express the intention of the bidder and was grossly inadequate. Nowhere in the Highway Law is there any specific grant of authority to the commissioner to modify or change a submitted proposal before award, nor can such power be implied."

2.11 Modification of bid by telegram

Most statutes that govern the bidding procedure of public construction contracts require sealed bids, which are to be opened at a public letting. After a sealed bid has been delivered to the awarding officer, the bidder may find, prior to the time of the public opening of the sealed bids that it is important and necessary for him to change or modify his proposal, principally as to the bid price. The practical method is to arrange for withdrawal of the sealed bid and substitution of the modified bid before the time stated for the public letting. It is usual to find in the instructions to bidders, a prohibition against withdrawing a bid after the time set for the public letting.

Administrative problems are created when telephone calls or telegrams are used by the bidder to communicate to the awarding officer the modification or change of the bid, which has been sealed as required by the related statute and which has already been deposited in the public office in contemplation of the scheduled public letting.

In *People* v. *Caldwell-Garvan and Bertini, Inc. et ano.,* 161 Misc. 864, N.Y., 1937, a telegram was received by the awarding officer before the date and hour fixed for the opening of sealed bids. This telegram directed a reduction of the price in the sealed bid by $1,500. Upon opening the sealed bids, the one from the defendant-contractor was the lowest without regard to the telegram. On the question of the validity of the telegram, the court held that the awarding officer "could not in any event accept any bid except a sealed bid, and this telegram could not constitute a sealed bid or any part thereof."

In later court decisions, significant rulings were made that disclosed a liberal construction of the statutory requirement of sealed bids for public construction contracts. In *Pearlman* v. *State of New York,* 18 Misc. 2d 494, N.Y., 1959, a public letting was scheduled for March 22. The contractor mailed his sealed bid the day before the letting, after which he received by mail an addendum from the State that required bidding on additional work and materials. At noon on March 22, he telegraphed to the awarding officer a direction to increase the price in the sealed bid by $3,400. The State rejected the telegram, and, at the public letting, the sealed bid of the contractor was the lowest, whereupon an award was made to him at the price in the sealed bid. Upon a suit to recover the additional $3,400, the court held: "Until the State had accepted the proposer's bid there was no valid contract." The decision also stated: "In the first place claimant had the right previous to the opening of bids, to modify his proposal by letter or telegram. When the telegram was received there was as yet no valid or binding contract. If the State's officers were of the opinion that a telegram did not comply with the conditions for bidding they could have regarded the bid as informal. Proper advertisement requires reasonable notice. Opening of bids may be postponed without abuse of discretion."

In another lawsuit under somewhat similar conditions, the judicial answer to the validity of a telegraphic direction for modification of a sealed bid already filed was stated as follows: "So that at the time of the plaintiff's telegram and at the time its letter was written there was no valid or binding contract, and so long as there was not a valid contract the plaintiff was free to withdraw its proposition, without sacrificing the deposit which it had been compelled to make as a condition of bidding." (*North-Eastern Const. Co.* v. *Town of North Hempstead,* 121 *App. Div.* 187, N.Y., 1907.)

In trying to determine the effect, if any, of a modification of a bid by telegram or letter when the requirement is for a sealed bid and the modification is directed at a proposal already properly filed, the awarding officer must weigh his responsibility to the other bidders who have complied with the instructions and who have the right to expect that consideration will be given to sealed bids only. The administration of this responsibility has a direct effect on the standards of the bidding procedure, and to administer those standards in various ways as opposed to uniform practice could create confusion and opportunity for fraud. It seems that relatively recent judicial opinions leave the awarding officer but one course of action, and that is to authorize the withdrawal of a proposal when a letter or telegram making such a request is received from the bidder and also to declare a bid to be informal when the bidder directs a modification in the price or other material item by telegram or letter. Equitable and uniform practice is highly desirable to maintain the integrity of the bidding procedure, whether the telegram or letter directs an increase or even a decrease in the price. Equal attention and treatment are mandatory to preserve the status and reputation of the public agency with respect to awards for public construction projects by the established and accepted method of sealed bids in competitive bidding.

2.12 Words, figures, and "no charge"

Generally the instructions to bidders include the statement that a bid price must be written in words and in figures, and if they are not in agreement, the written words "shall" be considered binding. It is preferable to construe the word "shall" as expressing futurity and not a mandate. The reason for this preference is integrated with the duty and function of the awarding officer to seek and to consider the intention of the bidder as disclosed by the contents and the context of the proposal.

Sometimes proposals are submitted in which the bidder has plainly indicated that he does not intend to charge for a particular item. Such intention may be shown by the words "no charge" or by "0." The total bid price should reflect that intention. However, when a bidder omits to write in the price of unit item, but leaves the space provided for writing in such price, blank, then the intention must be ascertained by considering all of the related data in the proposal. When the total bid price is greater than the correct computation of the unit prices, it is fair to assume that the bidder had intended to charge for the item, but due to oversight the omission was created. A reasonable determination by the awarding officer would be that the omission to indicate a price or to write in "no charge" or "0," resulted in an incomplete bid, not a firm bid. Sound business principles do not lend themselves to the production of costly work without charge therefor, and even though the bidder who omitted the unit price from his proposal offers to accept an award without compensation for work of a substantial nature and value, proper standards of bidding procedure do not permit the acceptance of such a proposal. The omission of the unit price may have been deliberate as a means to have this bidder assume control of the procedure to the point where he, and not the awarding officer, could decide whether this bid is formal or informal, depending upon his own evalution of the other bids.

If a bidder omits to state a price upon one or more of the important items, the bid is informal. In *State ex rel. Eberhardt* v. *Cincinnati,* 1 Ohio Nisi Prius Reports 337, 1895, the lowest bidder failed to fill in the blank space left for the price of the particular item. In the list of requirements, it was stated that "Bids must be made on every item. . . ." Although the bidder wrote to the awarding officer after the public letting that he intended no charge for the item, the court ruled that the board had the right to reject the bid as being informal, and that the offer to perform the work without charge was a proposition of a new bid by the contractor.

2.13 Unbalanced bids

An "unbalanced bid" is one that is usually related to a proposal composed of unit items. Such a proposal form states the estimated quantities of the various classes of work to be performed, and the bidder is required to show in the same proposal form the unit price for each class of work. An unbalanced bid usually shows a low or even a nominal price for the performance of certain of the estimated quantities disclosed in the proposal and a high or enhanced price for other estimated quantities of work. The result is that, as a rule, a computation based upon true quantities known to the bidder would provide a high bid for the project.

Another instance of an "unbalanced bid" is found when a contractor alleges a material variance in the engineer's estimate of quantity of a certain item as compared with the actual quantity that is discovered during the contract operations. Generally, when a contractor accepts the estimated quantity in the engineer's estimate as being reasonable, the computation of the contractor's unit price for the item comprises the factors that make up the contractor's cost of the work itself plus overhead and profit. A substantial reduction or increase in the quantity when the contract operations are already under way could well cause a distortion of the factors of the contractor's cost so that the unit bid price in the proposal would be thrown out of balance with a resultant loss to the contractor.

A significant discussion of the corelated terms "unbalanced bid," "unit price bid," and "lump sum bid" is found in the following excerpts from *Armaniaco* v. *Borough of Cresskill,* 163 At. 2d 379, 1960:

"Where the types of construction are largely standardized and where a variety of operations is required which make it impractical to break down the work required under a construction contract into units, it is generally customary to award contracts on a lump sum bid. On the other hand, where the work requires large quantities of relatively few types of construction and the volume of the work cannot be determined in advance, resort is had to the unit price form of bid. The advantage of the unit type bid under the circumstances referred to is that a contractor and municipality are not obliged to gamble on uncertain conditions, and know in advance the price they will, respectively, receive and be obligated to pay for various extra items.

"An unbalanced unit price bid is one where one or more of the items bid does not carry its share of the cost of the work and the contractor's profit. Such bids are admittedly susceptible of fraud and collusion and carry the additional danger of placing an irresponsible bidder in a position of bidding higher on the earlier work to be done under the contract and lower on the later work. Such a bidder could, after having taken his profit out of his early payments on the job, fail to complete the work called for. One control of such danger is by including a provision in the specifications, as here, for the permissive rejection of unbalanced bids.

"The justification of the unbalanced bid lies in the fact that the expenses of mobilizing the construction plant, bringing equipment and materials to the site, and the general costs of getting work started are appreciable. These items usually do not appear in the bid and, therefore, are liquidated only as the work of the bid items progresses. This causes a hardship to the contractor in that his working capital is unnecessarily tied up in the work, without compensation, and may result in a failure to bid, thus reducing competition. . . .

"In the absence of fraud, corruption, or abuse of discretion, the determination of the proper officers in making an award will not be disturbed."

The specifications should provide that an obviously unbalanced bid may be rejected by the awarding official. The intention of the provision is to discourage that type of bidding, but it is not a prohibition against unbalanced bids. The proposal as submitted by the bidder establishes a condition that affords the government agencies an opportunity to determine whether or not the bid would actually raise the cost of the project.

When a proposal contains a lump sum as the final price for a substantial part of the work or for an important item, the statutory requirement for competitive bidding is disregarded. The object of the requirement is to "enforce a submission of every important item for competition, naming the quantity so far as it could reasonably be ascertained" (*Matter of Merriam,* 84 N.Y. 596, 1881).

A frequently cited court decision in lawsuits involving unbalanced bids is *Matter of Anderson,* 109 N.Y. 554, 1888. Anderson, as the owner of real property, appealed to the court for a reduction in a street assessment that was levied by the city for grading a part of the street, and he claimed that the assessment proceedings were fraudulent or erroneous. The city charter provided that, in contracts for work costing more than $1,000, public, competitive bids are required after competent engineers have surveyed and planned the project, and, also, that the quantities should be estimated as closely as possible to accomplish the work. The published estimates included 10,000 cubic yards of earth excavation and 20,000 cubic yards of rock excavation. A contractor named Kane was the successful bidder at a total price of $17,100, and his unit bid price was $1.625 per cubic yard for earth excavation and 2 cents per cubic yard for rock excavation. In the performance of the work more than 21,000 cubic yards of earth excavation, but only 9,200 cubic yards of rock excavation, were required. The contractor was paid $33,620 or about twice the amount of his total bid price. The court ruled that no bona fide effort was made to comply with the charter provisions as to the careful estimating of the quantities of the work to be done, but, instead, the estimates were the result of careless and inadequate examination by the city's engineers as to the quantities of earth and of rock excavation.

In the *Anderson* case the court stated that every unbalanced bid is not per se (in itself) fraudulent, nor is it evidence of substantial error, but as to this particular case, the court added:

"We think there was substantial error in letting this contract to Kane, who was in fact nearly the highest instead of the lowest bidder for the work. . . . The quantities of rock and earth, respectively, as stated in the advertisement, were mere estimates or random guesses, without any basis whatever to rest upon, and the ordinance which required the quantity of each kind of material, and its nature, to be stated as near as possible, was in no sense complied with, and therefore there was no basis for a valid contract. . . . We are also of opinion that, upon the facts of the case, it is a just inference that the contract was the result of fraud and collusion. . . . The bid on its face is suggestive of fraud. The officers of the city, when they saw in the bid 2 cents per cubic yard for rock excavation and $1.625 for earth excavation, ought to have known, and must have known, if they read it, that some fraud was contemplated; . . . It is not needful, however, for us to find that there was actual fraud, but it is sufficient that all the facts of the case were such as justified an inference of fraud in the court below." The judicial opinion also pointed out: "An unbalanced bid that does not materially enhance the aggregate cost of the work cannot be complained of. If there is no deception or mistake as to the quantities, and if the ordinances have fairly been complied with, and the quantity and quality of the work has been estimated as nearly as practical, there is no ground for alleging substantial error merely because of an unbalanced bid under which the contract was let, and if the cost of the work has not thereby been enhanced, there is no ground for alleging fraud."

The court granted Anderson's claim, and the assessment was reduced because the contract for the construction work was not prepared in accordance with the charter requirements.

The above quoted *Anderson* case was cited in *Pearlman* v. *City of Pittsburgh,* 155 At. 118, Pa., 1931, as support for a judicial decision with relation to a public contract for water meters. The information to bidders named several types of meters of which any one type would be acceptable to the city. The Neptune Meter Company was about 5 per cent lower in its bid than any other bidder, but the awarding officer rejected the Neptune proposal for several reasons including a charge that the bid was unbalanced. This charge was based on the alleged fact that for certain specified repair parts, the Neptune Company bid was much below their cost and unfair. The contract was for five years, and the successful bidder would supply the needed repair parts. The court ruled in favor of the Neptune Company and said: "It was the policy of the Neptune Company, in connection with furnishing water meters, to contract for supplying repair parts at nominal cost, based doubtless largely on the experience that the original parts did not wear out. The bid would not have been invalid even had it agreed to furnish the needed repair parts free of charge. It is a novel proposition that the bid of a perfectly responsible party can be ignored because it is in whole or in part too low. There was no suggestion either by the director or by the trial court that the Neptune Company's bid as to any item was too high; hence in a legal sense, the bid was not unbalanced."

2.14 Alternate bids

When preparing proposal forms many public works officials include several different materials or procedures to accomplish the same general purpose, and a choice of the alternative is to be made by the awarding officer after the proposals have been received and opened by the governmental unit. This provision in a proposal or specification, when required by a local law or when clearly and definitely stated, is deemed to be within the requirements of competitive bidding and awarding to the lowest bidder. (*Barber Asphalt Paving Co.* v. *Garr,* 73 S.W. 1106, Ky., 1903; *Trapp* v. *City of Newport,* 74 S.W. 1109, Ky., 1903; *Baltimore* v. *Flack,* 64 At. 702, Md., 1906; *Attorney-General ex rel. Cook et al* v. *City of Detroit,* 26 Mich. 263; *Muff* v. *Cameron,* 114 S.W. 1125, Mo., 1909; *Dixey* v. *Atl. City & D. River Quarry & Constr. Co.,* 58 At. 370, N.J., 1904; *Gilmore* v. *Utica,* 29 N.E. 841, N.Y., 1892; *Ampt ex rel. Cincinnati* v. *Cincinnati,* 54 N.E. 1097, Ohio, 1899.

When a committee of taxpayers in a municipality asked a court to restrain the city engineer from entering into a contract for street paving with asphalt concrete, they complained of the policy of the public works department which was to require either asphalt concrete or Portland cement concrete by specifying only one kind of such material for a given project, in an attempt to maintain equality between each industry. The local laws directed the public works engineer to prepare alternate plans and specifications and to use all standard materials for high type paving, unless, in his judgment, conditions require the use of a particular type of pavement. The local laws also provided that only asphalt concrete and Portland cement concrete were to be used for projects defined in the law as "high type paving." It appeared that a decision at the outset to use Portland cement concrete was later changed to asphalt concrete, principally because the Portland cement concrete was more expensive than the asphalt product. Yet, it was admitted by the public engineer that, disregarding cost, a satisfactory highway could be made with Portland cement concrete. The court decided that the public officer was guilty of an abuse of discretion in failing to call for alternate bids. In the judicial opinion, it was pointed out that: "It was undoubtedly the intention of the Legislature to eliminate favoritism, prejudice and artificially maintained equality between producers, and to make the choice of pavement a result of free, competitive bidding between two recognized superior materials. The primary object is of course economy in road construction, but the Legislature has determined that such economy is to be achieved by means of competition between the producers of two standard materials. If the department substitutes its own judgment on costs, the benefits of such competition are lost. The judgment of the department may be an entirely honest one, and in the present proceeding no bad faith is charged. But, under the statute, an honest opinion that one type of pavement will cost more than another cannot be the basis of the department's action. Physical conditions alone are committed to the discretion of the (public officers)." (*Landsborough* v. *Kelly,* 37 P. 2d 293, Cal., 1930.)

A distinction in statutory requirements is shown in *Stocking* v. *Warren Bros. Co.,* 114 N.W. 789, Wis., 1908. The local laws required the public works board to specify the material to be used before the board advertised for bids. Four sets of plans and specifications were prepared for a paving job on one street; each set called for a different kind of pavement and no decision was to be made by the board as to the material to be used until after the letting. The court held that the action by the board was wrong because the local charter required competitive bids on a predetermined pavement material.

In most localities, applicable statutes require only that an award for public construction work must be made to the lowest responsible bidder after advertising for competitive bids. Under such a local statute, a public works official (*Mayor etc. of City of Baltimore* v. *Flack,* 64 At. 702, Md., 1906) prepared separate sets of specifications, each one calling for a distinct kind of material – one for asphalt block pavement, the second for vitrified brick pavement, and the third for bitulithic pavement. The bitulithic pavement is a patented process, and the specification for that material included an agreement by the owner of the process to provide all of the material to any bidder for a uniform price of $1.45 per square yard of pavement. After advertising, two bids were received award was authorized upon the bitulithic material, to the lowest on vitrified brick, one on asphalt block, and three on bitulithic. An responsible bidder, although one bid on vitrified brick was lower in price than the material upon which the award was made. The discretion as exercised by the awarding officer was challenged as being invalid. The court discussed the basis of alternate bidding and indicated that the exercise of judgment in selecting a paving material is founded on comparison or observation whereby a suitable material can be decided upon. The court said that there "are two kinds of competition – the one, competition between different things which will equally answer the same general purpose (e.g., alternate materials included in the same proposal), and the other, competition between the prices bid respectively upon each of those distinct things (e.g., separate proposals for a stated material)." The court stated that there is merit in getting bids on alternate materials and pointed out that seeking proposals on a predetermined material necessarily excluded everything else that might have been an adequate substitute for the material specified. The judicial opinion stated: "That the commissioners for opening streets had the power to put asphalt block, bitulithic and vitrified brick pavements in competition with each other, and after the bids had been opened by the Board of Awards, to select the one of three pavements to be used, and that the Board of Awards had the power to award the contract to the lowest responsible bidder upon the kind or character of the pavement so selected, even though a bid had been filed on some other material which was lower than the lowest responsible bid on the selected material." The opinion added that "no judicial tribunal is clothed with jurisdiction to rejudge its conclusions" when the awarding body or officer exercises the power of such selection in good faith and in the absence of fraud.

In the *City of Baltimore* case, the court's opinion stated that *Attorney-General ex rel. Cook et al.* v. *City of Detroit,* 26 Mich. 263 is the leading case on the subject of issuing separate proposals, each for different materials and for the paving of a certain street. Such separate proposals were prepared for several kinds of wood and stone pavement. The local charter required public advertisement and an award to the lowest responsible bidder. After competitive bids were received, a Ballard pavement was selected although other responsible bidders had submitted lower bids on other kinds of specified pavement. The Attorney-General contended that the public awarding officer should first decide upon the kind of pavement material and should then advertise for proposals for that kind alone. He insisted that wood pavement cannot be put in competition with stone pavement for the paving of one street. The Attorney-General argued that the law intended that the right to an award is based on price, but if all bids are not based on the same specifications, the awarding officer could reject the lowest bid on a pretense that the material is inferior in quality. The court said that the local statute did not require the public officials to determine the material to be used before advertising for bids. All persons had equal opportunity to enter into the competition; therefore, the judicial decision upheld the awarding officer.

The judicial answer in the *City of Detroit* case described above was approved in *Holmes* v. *Council Detroit,* 120 Mich. 226, 1899; *Fones Hardware Co.* v. *Erb,* 54 Ark. 645, 1891; and in *Trapp* v. *City of Newport,* 74 S.W. 1109, Ky., 1903. In *Trowbridge* v. *Hudson,* 24 Ohio C.C. 76, 1873, it was decided that a requirement in a local or state law for an award to the lowest bidder does not demand the selection of the lowest bid given when alternate or different kinds of materials or procedures are called for in the advertisement for bids.

2.15 Failure to bid on alternates

The general rule with relation to a requirement in a proposal form that all spaces must be filled in and all alternates and unit bid prices must be bid upon is that a bidder's failure to comply with the requirement imposes upon him the risk of losing the contract if the alternate is selected or if the bid is rejected because of its irregularity. In considering the acceptance or the rejection of a bid, it must be remembered that a variation in a proposal must be substantial and must be material so as to provide an advantage for a particular bidder. In other words, the test to be applied seems to be whether or not the waiver of an irregularity will adversely affect fair and competitive bidding. Thus, the awarding officials have the authority to decide whether or not the failure to comply with the requirement to bid on alternates failed to meet that test. The duties "of officers intrusted with the letting of contracts for works of public improvements to the lowest bidder are not duties of a strictly ministerial nature but involve the exercise of such a degree of official discretion as to place them beyond the control of courts by mandamus." (*Devin* v. *Belt,* 17 At. 375, Md., 1889; *City of Baltimore* v. *Clack,* 64 At. 702, Md., 1906.)

In the case of *George A. Fuller Co.* v. *Elderkin,* 154 At. 548, Md., 1931, the Fuller Company proposed to install a heating control system, and the "Instruction to Bidders" stated that all the information called for by the proposal form must be given, all spaces filled in, and all alternates and unit prices must be bid upon. Seven different sets of alternates were called for, and the Board of Awards reserved the discretionary power that, by provisions in the city charter, authorized the Board to modify the original specifications by adding alternates that were described in the specifications. Twelve bidders submitted proposals, and an award of the contract was recommended by the chief engineer of the Board to the Fuller Company on its base bid of $2,044,000, which was reduced to $2,003,000 upon selection of certain alternates. The next bid, including certain alternates, was $2,017,000. The second bidder protested the award for several stated reasons as to omissions relative to certain alternates and also because the Fuller Company did not include a guaranty in saving of steam consumption, as called for in the proposal; the omitted quantity was for the alternate that was adopted by the Board. The second bidder claimed that his guaranty in the proposal of a 25 per cent saving constituted his bid as the lowest one.

The outcome of the case was that the court upheld the Board in awarding the contract to the Fuller Company, and ruled that the conclusion of the Board is not subject to review by the court, because the

Board acted within its discretionary powers and there was no proof of fraud, arbitrary conduct, or collusion. The decision stated that failure of the Fuller Company to bid on certain alternates, which were not in the plan as finally adopted, had no effect on other bidders. Also, the compliance with the "Instruction to Bidders" was not made a condition for considering the proposal as submitted, but the instruction only reserved the right to reject any bid not in compliance with the instruction. (*Pascoe* v. *Barlum,* 225 N.W. 506, Mich., 1929; *Andrews* v. *City of Detroit,* 206 N.W. 514, Mich., 1925.)

2.16 Joint adventures and joint ventures

The terms "joint adventure" and "joint venture" are synonymous. There is no fixed definition of a "joint adventure," but in each case in which the question arises, the determination as to the creation or existence of that relationship depends upon the particular facts and the agreement or the intention of the parties upon which the joint relationship of "joint adventurers" is created. (*Myers et al.* v. *Lillard et al.,* 220 S.W. 2d 608, Ark., 1949; *Wyoming-Indiana Oil and Gas Co.* v. *Weston et al.,* 7P. 2d 206, Wyoming, 1932.) "The now widely recognized legal concept of joint adventure is of modern origin. It has been said to be purely the creature of American courts. The early common law did not recognize the relationship of coadventurers unless the elements of partnership were disclosed and proved, but it is now generally understood that two or more persons may, by combining their property or labor in a joint venture, create a status, which while having some or many of the characteristics of a partnership, is not identical therewith." (*30 Am. Jur. 676.*) The principal distinction between a joint venture and a partnership is that the former relates to a single transaction, which may require a period of years for accomplishment. Where there is a dispute between partners, the remedy is usually to sue for an accounting, whereas one joint adventurer may sue the other or others for breach of contract or for contribution to any losses. (*Goss* v. *Lannin,* 152 N.W. 43, Iowa, 1915.)

Two or more persons who join in a particular business enterprise for profit create a joint adventure or venture. The relationship need not be that of a partnership or an agency and may be inferred from the conduct of the parties without express agreement. (*Libby* v. *L. J. Corporation,* 247 F. 2d 78, District of Columbia, 1957.)

Broadly speaking, a joint adventure may be characterized as a quasi-partnership in a single adventure undertaken for mutual gain. . . . If the venture be for pleasure rather than profit, it is sometimes called a joint enterprise. (*Bradbury* v. *Nagelhus,* 319 P. 2d 503, Mont., 1959.)

In most states a corporation may join with an individual or a partnership in a joint adventure, providing the transaction is within the authorized scope of the purposes for which the corporation was formed.

It is considered as being against public policy when bidders for a public construction contract agree that one or more shall refrain from bidding in order to give another bidder an undue advantage. However, joint adventures are permissible as a cooperative effort, even though this arrangement does, in a sense, reduce competition. The size of the contract, the financial ability, the "know-how," and the experience to perform certain portions of the work, together with the ownership of proper equipment, are reasonable and honest factors to support a voluntary combination of rival contractors. A bid for a joint adventure that is made honestly and in good faith is acceptable as a reasonable, disclosed assurance of capability in all respects to perform the work of the contract.

Secret agreements, whereby several bidders have a common interest to share in a contract and separate bids are submitted by each of the group, by mutual arrangement, are not proper and will not be enforced or condoned in any court in case of a litigated dispute between the parties to the secret agreement.

A striking illustration of an improper situation is found in a lawsuit where one contractor asked a court to make his associate pay his share of the profits from a contract to construct a water pipeline for a municipality. (*Hoffman* v. *McMullen,* 83 F. 372, Oregon, 1897.) The decision declares the approval of the court of a joint adventure when the parties thereto act in good faith in obtaining the public construction contract, and the court also expresses judicial contempt of an illegal agreement for an unconscionable combination of persons. During the trial of the case between the contracting parties it was disclosed that in pursuance of their verbal agreement, Hoffman submitted a bid for the work in the name of "Hoffman & Bates." McMullen, with the knowledge and concurrence of Hoffman, submitted a separate bid in the name of "San Francisco Bridge Co." The latter bid was $49,000 higher than the Hoffman bid. Hoffman received the award of the contract for the construction project, whereupon he and McMullen entered into a written agreement which recited the procedure they followed in bidding for the contract and also provided for an equal sharing of the profits and the losses under the contract with the municipality. The profit was nearly $140,000, but Hoffman refused to recognize McMullen's claim under the agreement that was made between them. Hoffman's position was based upon the ground that the bidding as they participated in it, reduced the competition and operated as a fraud on the municipality. The court decided that it would not attempt to determine which one of the parties was more faithless to the other and that neither party had a standing in the court in an attempt to enforce the illegal agreement. In commenting on joint enterprises or joint adventures, the decision stated: "There is no valid objection to such voluntary combinations if the joint action of the parties is done honestly and in good faith." When a joint proposal or a joint venture is undertaken as the result of honest cooperation, it is not prohibited by public policy even though it lessens competition. The association is public and not secret, and the risk as well as the profit, is joint and openly assumed. The public may well benefit by the open disclosure of joint responsibility and of the joint ability to do the work. The public officials having such honest information in the proposal can evaluate the bid without searching for motives of the bidders.

In the absence of words or the provisions of a statute making a written contract several or joint and several, an obligation under a contract by more than one person is presumed to be joint. The pronoun "we" in a written instrument indicates a joint obligation, unless a different intention is expressed. The use of the pronoun "I" in a contract where the obligation applies to more than one person is joint and several.

"The word 'joint' means united or coupled together in interest or liability, opposed to several." It "connotes that each and all parties are responsible for the entire obligation." (23 Words and Phrases 43.)

chapter 3

Mistakes in bids and specifications

When a bidder discovers that his proposal contains a mistake, and when the proposal has already been filed with a government agency and perhaps even been opened and read publicly, it is imperative that he notify the awarding officer promptly. If such a claim of mistake should be made by the apparent lowest bidder, steps must be immediately taken to safeguard the interest of both the public and the bidder. Usually, the public official offers the bidder an opportunity to explain the basis of his claim in person and to submit his worksheets for inspection. The mistake is generally a careless act, an error in computation, the result of incorrect judgment, or an incorrect statutory procedure.

The practical considerations of such a conference on a claim of mistake suggest that the meeting be conducted in an informal manner and without a stenographic record. The exploratory questions and the inspection of the worksheets are important in determining whether or not the claim is based upon an honest mistake and if its presence in the proposal is in good faith. An informal discussion permits inquiring about the bidder's experience in other contracts with the same or other government agencies, as well as to the financial condition of the bidder and his capability to perform the work. Representatives of the public agency who arrange the conference will support their own record by a list of the conferees and by their notes concerning the questions and answers.

3.1 Conditions for release from bid

As a rule, it is difficult to solve the bidder's claim of mistake, but a practical criterion is found in certain important factors. The first point to be considered is the seriousness of the mistake and the unconscionable advantage of enforcing the contract that would be based upon the proposal. The mistake that the bidder claims is ground for his being released from the bid must relate to a material feature of the work. A showing of reasonable care and diligence in preparing the proposal is important; furthermore the public official must be certain that by granting the bidder's request, the public, for whom the bid was received, will incur no injustice or serious loss. In any event, timely communication of the discovery of an error in a bid is of first importance to prevent an acceptance of the proposal, which could have the effect of a completed offer and result in an enforceable contract. The relief to which a bidder is entitled when a valid claim of mistake in a bid meets the general criteria suggested here, is not affected by a statutory prohibition or restriction against withdrawals of public bids. (*Moffett, Hodgkins & Clarke Co.* v. *Rochester,* 178 U.S. 373, 1900; *Baltimore* v. *DeLuca-Davis Const. Co.,* 124 At. 2d 557, Md., 1956.) Under conditions as described above, the bidder is also entitled to recover his bid desposit. (*Rushlight Automatic Sprinkler Co.* v. *Portland,* 219 P. 2d 732, Oregon, 1950.) If the mistake is an obvious, but honest error and is likely to cause the contractor irreparable damage, then a reasonable policy is to relieve the bidder of his proposal and to return his bid deposit.

Some public officials insist upon a "gentlemen's agreement," when a claim of mistake is granted to a bidder, that if the project is again advertised for proposals, the bidder will not submit a proposal for the new letting. By this procedure the public official makes an attempt to avoid a situation at the second letting where the "mistaken" bidder will be the lowest bidder again, but at a higher price than the first bid. The implication attributed to such a situation is that, by collusion, the first bid is returned because of the claimed mistake, and the bidder succeeds in getting a better price for the same project than that which he originally intended. The difficulty with such a "gentlemen's agreement" is the challenge that it is restrictive of competitive bidding for public work.

In some instances it is the practice to reject all bids when the lowest bidder is relieved of his bid because of a claim of mistake. The theory behind this practice is that an award to the second lowest bidder subjects the awarding official to public and private attack by uninformed and unscrupulous people. These people allege that the awarding official openly and deliberately disregarded the legal requirement of making an award to the lowest bidder; meanwhile, the awarding official does not have an opportunity to refute such unfounded charges. Justifiable complaints are usually registered by the bidders when all proposals are rejected solely because the lowest bid was submitted in error. The cost of inspecting the site of the proposed work, the charge for the plans and specifications, the services and expenses of the estimating personnel, and other disbursements connected with submitting a proposal are relatively expensive. Of even greater importance is the fact that the bidders who did not claim any mistake in their respective proposals have disclosed their prices, thus substantially committing themselves to a fixed pattern for the new letting. If the public official is satisfied that the second bid meets all the requirements of the specifications and the award of a contract, there is merit to a determination to make such an award, particularly when the public official declares the mistaken bid to be informal. Therefore, it is not a valid proposal and also is not the "lowest bid" in law or in fact. This situation would indicate that the first valid bid that meets all requirements is the "lowest bid."

When a bidder claims a mistake in the preparation of his proposal and seeks to withdraw the bid or to be released from it there should be proof of good faith and real cause for the request in order to avoid loose procedure in competitive bidding. The objective of the contractor's claim of mistake is a rescission of the proposal, and in this endeavor to annul the offer of a contract, the notice of the claim must be given to the awarding official within a reasonable time, the mistake must be with regard to material facts, and, it must be recognized that consent for a rescission will place both parties – the contractor and the governmental agency – in the same situation as existed before the proposal or offer of the contractor was made or submitted.

3.2 Stating claim of mistake

If a bidder notifies the public official by letter before the contract is awarded that the bid contains a mistake, does not disclose its character in the letter, and indicates that he wants to withdraw the proposal, the consent of the public official is governed by the circumstances in the particular case.

In *City of Hattiesburg* v. *Cobb Bros. Const. Co.,* 184 So. 630, Miss., 1938, the contractor claimed that his letter to the Mayor stated that he had "found some error;" he asked that he be allowed to withdraw his bid and that the bid deposit be returned to him. The public officials rejected the request because the particulars of the mistake were not disclosed in the contractor's letter. The court to which the contractor appealed upheld the public officials and pointed out:

"The letting of public contracts by competitive bidding is for the protection of the public, and the public authorities are without the right to permit a bid for the contract to be withdrawn in the absence of circumstance that would render it inequitable not to permit its withdrawal. The inequitable circumstance here claimed is an honest mistake in determining the amount of the bid. Unless the mistake was in fact made, and honestly made, no right of withdrawal would appear. In determining whether to permit the withdrawal of this bid, the Mayor and the Commissioners were under the duty to the public to ascertain whether a mistake affecting the amount of the bid had in fact been made. In order to do this, it was necessary for them to be advised of the character of the claimed mistake, so that they might consider it in connection with the bid and the advertisement therefor.

"The mere claim that a bidder has 'made a mistake' or 'found some error' in his bid neither gives him the right to withdraw his bid nor imposes on the public authorities any duty to examine the bid in order to ascertain whether or not a mistake appears therein. Another reason for requiring a statement of the character of the mistake made to be set forth in a notice from the bidder of withdrawal of a bid, is that in an action to rescind the contract made by the acceptance of the bid by the awarding officer and to recover a benefit conferred by the bidder on the other party to the contract, the bidder may be confined to the particular mistake claimed to have been made when the notice of withdrawal was given."

In a leading case on the subject of mistakes by bidders, *Moffett, Hodgkins & Clarke Company* v. *Rochester,* 178 U.S. 373, 1900, the city invited proposals, but the forms to be used by the bidders for filing at the letting on December 23 were not available until about December 15. The contractor, who was "extremely nearsighted," became nervous and confused because of the insufficient time to prepare the proposal and in transcribing the figures he prepared, he accidently made certain clerical errors. In writing the proposal, he hastily wrote 50 cents per cubic yard for open trench excavation when he intended to charge 70 cents per cubic yard, and he also wrote $1.50 per cubic yard for excavation in tunnel, when he intended that the price should be $15 per cubic yard. The estimated quantities of the items disclosed a bid price of $63,800 lower than the contractor intended. At the public letting, the contractor's representative heard the erroneous prices in the proposal read by the clerk of the awarding unit, and the representative informed the board of the errors. Nevertheless the board accepted the proposal and offered the contractor as an alternative to executing the contract, that his bid deposit of $90,000 be forfeited, because the city charter provided that no bid deposit or surety bond could be withdrawn until the board awarded a contract. In supporting the contractor's claim, the court said that the forfeit of the bid deposit was not commanded by the charter and was not proper in this particular. The decision of the court said: "The rule between individuals is that until a proposal be accepted it may be withdrawn, and if this principle cannot be applied in the pending case, on account of the charter of the city, there is certainly nothing in the charter which forbids or excuses the existence of the necessary elements of a contract."

Continuing the preceding judicial comment on the specific subject of the charter provision as to the restricted right to withdraw the bid deposit or surety bond, the court ruled:

"The complainant (contractor) is not endeavoring to withdraw or cancel a bid or bond. The bill (complaint) proceeds upon the theory that the bid upon which the defendants (city) acted was not the complainant's bid; that the complainant was no more responsible for it than if it had been the result of agraphia or the mistake of a copyist or printer. In other words, that the proposal read at the meeting of the board was one which the complainant never intended to make and that the minds of the parties never met upon a contract based thereon. If the defendants (city) are correct in their contention there is absolutely no redress for a bidder for public work, no matter how aggravated or palpable his blunder. The moment his proposal is opened by the executive board he is held as in a grasp of steel. There is no remedy, no escape. If, through an error of his clerk, he has agreed to do work worth a million dollars for ten dollars, he must be held to the strict letter of his contract, while equity stands by with folded hands and sees him driven to bankruptcy. The defendants (city) position admits of no compromise, no exception, and no middle ground." The decision emphasized the pertinence of these statements by adding that: "The transactions had not reached the degree of a contract – a proposal and acceptance. Nor was the bid withdrawn or cancelled against a provision of the charter. A clerical error was discovered in it and declared, and no question of the error was then made or of the good faith of the complainant (contractor)."

3.3 Honest mistake in a bid

In *Colella* v. *County of Allegheny,* 137 At. 2d 265, Pa., 1958, the contractor submitted a bid of $144,900 to the County of Allegheny for the construction of a sewage disposal plant at a municipal airport and deposited a bid bond of $17,500. The next two lowest bids were

$218,995 and $233,000, respectively. Before an award was made to him, the contractor notified the County that a clerical mistake had been made in preparing his proposal, and he requested that he be relieved of his bid and that the bid bond be returned. The request was denied by the County, and after readvertising, the County made an award to another contractor for $214,000. The contractor sued the County for the return of the bid bond and the County counterclaimed for $69,100, which was the difference between the contract price as awarded and the bid of the plaintiff-contractor in this lawsuit. The court held against the contractor, because the mistake was unilateral and was due to his negligence; the County claimed that no error was apparent and there was no knowledge on the part of the County of any clerical mistake in the bid. However, the damages allowed by the court were limited to $17,500, the amount of the bid bond, which was declared to be the liquidated damages for refusing to execute the contract. Upon appeal to a higher court, it was held: "If a person, firm or corporation submits a sealed bid on public works, the principle contended for by the contractor, namely, that after all bids are opened he can withdraw his bid under the plea of a clerical mistake, would seriously undermine and make the requirement or system of sealed bids a mockery; it would likewise open wide the door to fraud and collusion between contractors and or between the contractors and the public authority. What is the use or purpose of a sealed bid if the bidder does not have to be bound by what he submits under seal? What is the use or purpose of requiring a surety bond as further protection for the public, i.e. the municipality, if a bidder can withdraw his bid under plea of clerical mistake, whenever he sees that his bid is low, and that he must have made an error in judgment?"

Ordinarily, a party to an executory contract (that is, one that is to be performed in the future) will not be granted relief in case of a unilateral mistake. (*Saligman et al.* v. *United States,* 56 F. Supp. 505, 1944.) If, however, the awarding officer receives a proposal that discloses a mistake because of the amount of the bid or otherwise, and he knows or has reason to know of the error, the bid is voidable by the bidder. (*2 Restatement of Contracts, sec. 503.*)

3.4 Right of rescission

In *Harper, Inc.* v. *City of Newburgh,* 159 App. Div. 695, N.Y., 1913, the contractor alleged that he transposed two items by mistake in bidding for pavement work. The court accepted the contractor's version of the mistake and declared that the mistake was not due to the plaintiff's negligence, but gave judgment for the defendants. On the plaintiff's appeal, the higher court reversed the judgment and granted a new trial on the ground that there is "no legal reason, then, why a court of equity, in the exercise of its discretion, could not have afforded rescission and a refund of the deposit within the principles of *City of New York* v. *Doud Lumber Co.* (140 App. Div. 358) and *Moffett, Hodgkins, and Clarke Co.* v. *Rochester* (178 U.S. 373). *Board of School Commissioners of City of Indianapolis* v. *Bender* (72 N.E. 154) is quite in point."

The appellate court also pointed out:

"There was neither mutual mistake nor fraud, deceit or bad faith on the part of the defendants; the mistake was not apparent on its face, nor was the defendant's attention called to it until after the bid had been accepted and the resolution to award the contract had been passed by the common council; that under such circumstances the awarding of the work made a complete contract binding on both parties and neither could escape therefrom except upon proof of fraud or of bad faith or of mutual mistake.

"Yet, there can be rescission (i.e., annulling or making void) of a contract for unilateral mistake. The rule stated by the learned Special Term applies to reformation (i.e., making anew). (*Moran* v. *McLarty,* 75 N.Y. 25.) For reformation affords a contract. And consequently when reformation is sought for the mistake of one party only it is essential that fraud or inequitable conduct be found in the other, else the court in determining that there is a contract at the instance of one might be doing right to that one and equal wrong to the other when without legal fault."

The opinion added that because of the mistake of one party only, without fraud or inequitable conduct in the other, a contract may be rescinded. Rescission is available even though there does not exist a completed contract. If the courts, by rescission, cannot place the parties in status quo, then that relief is afforded "only when the clearest and strongest equity imperatively demands it (*Grymes* v. *Sanders,* 93 U.S. 62)."

3.5 Obvious error in bid

An illustration of the principles that are recognized by the courts as an inequitable advantage for the government in a claim of a mistake by a bidder was disclosed in the case of *C. N. Monroe Mfg. Co.* v. *United States,* 143 F. Supp. 449, Mich., 1956. The three lowest bidders for certain types of equipment offered unit prices of $3.91, $6.40, and $18.00, respectively, whereas the next six bidders' prices ranged up to $39.50. The lowest bidder asked the court to decide that the government should have recognized the price differences, and by the great discrepancies between the two lowest bidders and the others, the government was on notice of the bidder's unilateral mistake. The court ruled: "It is well settled that if a unilateral mistake is known to the other party or because of accompanying circumstances the other party had reason to know of it, the party making the mistake has a right of rescission. *Moffett, Hodgkins & Clarke Co.* v. *Rochester,* 178 U.S. 373; *Armour & Co.* v. *Rinaker,* 202 F. 901; *United States* v. *Jones,* 176 F. 2d 278; *Saligman et al.* v. *United States,* 56 F. Supp. 505; *State of Connecticut* v. *F. H. McGraw & Co.,* 41 F. Supp. 369; *Kemp* v. *United States,* 38 F. Supp. 568." The awarding officer admitted that even a layman could tell that the items could not be manufactured for $3.91 or $6.40. The court pointed out: "It is well known that the vast operations of the Government require the awarding of a tremendous number of contracts on competitive bids, and as stated in *Saligman et al.* v. *United States,* 56 F. Supp. 505, the government should not be obliged to act as a nursemaid for careless bidders. It is true that in this case the plaintiff (bidder) made an honest but careless mistake. The defendant (government) did not have actual knowledge of the error and no inference of bad faith exists on the part of the government in awarding the contract. However, there is justice in plaintiff's complaint that to hold it to the severe consequences of its unilateral mistake would be harsh and inequitable. It is not believed that an honest, fair-minded and reasonable man in good conscience would seek to maintain such a gross and inequitable advantage of a mistaken bidder, when as in this case he had reason to know of the mistake, and the government is required to maintain the same standard of fairness with those who contract with it. (*Kemp* v. *United States,* 38 F. Supp. 568.) Under all the circumstances it must be held that the government should have known at the time it awarded the contract that a gross mistake had been made by the plaintiff and therefore it is inequitable to hold the plaintiff to the terms of the contract. . . . The government therefore could not be prejudiced in giving the relief the plaintiff is seeking other than its loss of a bargain that was brought about through a gross and obvious mistake."

3.6 Principles of bidder's claim of mistake

A comprehensive review of the subject of mistakes by bidders and a statement of the general principles that govern the disposition of their requests for release from the proposal is contained in *Kemper* v. *City of Los Angeles,* 235 P. 2d 7, Cal., 1951. The plaintiff inadvertently left an item of $301,769 out of his final worksheet, and it was not included in the total amount of his bid. Several hours after the letting, the plaintiff notified the city of its mistake. An informal hearing was held, but the city refused to release the plaintiff from its bid. The company refused to execute the contract, and after making an award to the next bidder for $1,049,592, the city sought a forfeiture of the plaintiff's bid bond of 10 per cent of the sum of the bid, which was "as a guarantee that the bidder will enter into the proposed contract if it is awarded to him," but upon failure to execute the contract the proceeds of the bond "become the property of the city." The trial court found that the mistake was honest and excusable, without negligence of the bidder, and there was prompt notice to the city. The appellate court held:

"Once opened and declared, the company's bid was in the nature of an irrevocable option, a contract right of which the city could not be deprived without its consent unless the requirements of rescission were satisfied. . . . The company seeks to enforce rescission of its bid on the ground of mistake. . . . The city contends that a party is entitled to relief on that ground only where the mistake is mutual, and it points to the fact that the mistake in the bid submitted was wholly unilateral. . . . However, the city had actual notice of the error in the estimates before it attempted to accept the bid, and knowledge by one party that the other is acting under mistake is treated as equivalent to mutual mistake for purposes of rescission. . . . Relief from mistaken bids is consistently allowed where one party knows or has reason to know of the other's error, and the requirements for rescission are fulfilled. . . .

"Rescission may be had for mistake of fact if the mistake is material to the contract and was not the result of neglect of a legal duty, if enforcement of the contract as made would be unconscionable, and if the other party can be placed in status quo. . . . In addition, the party seeking relief must give prompt notice of his election to rescind and must restore or offer to restore to the other party everything of value which he has received under the contract. . . .

"Omission of the $301,769 item from the company's bid was, of course, a material mistake. The city claims that the company is barred from relief because it was negligent in preparing the estimates, but even if we assume that the error was due to some carelessness, it does not follow that the company is without remedy. Civil Code sec. 1577, which defines mistake for which relief may be allowed, describes it as one not caused by 'the neglect of a legal duty' on the part of the person making the mistake. It has been recognized numerous times that not all carelessness constitutes a 'neglect of a legal duty' within the meaning of the section. . . . On facts very similar to those in the present case, courts of other jurisdictions have stated that there was no culpable negligence and have granted relief from erroneous bids. The type of error here involved is one which will sometimes occur in the conduct of reasonable and cautious businessmen, and under all the circumstances, we cannot say as a matter of law that it constituted a neglect of legal duty such as would bar the right to equitable relief."

The opinion cited, in support of the court's position, leading decisions in California and in other states, as follows: *Conduit & Foundation Corpn.* v. *Atlantic City,* 2 N.J. Super. 433; *School District of Scottsbluff* v. *Olson Const. Co.,* 153 Neb. 451; *McCall* v. *Superior Court,* 1 Cal. 2d 527, 535, 536; *Seeger* v. *Odell,* 18 Cal. 2d 409, 417, 418; *Los Angeles R. R. Co.* v. *New Liverpool Salt Co.,* 150 Cal. 21, 28; *Mills* v. *Schulba,* 95 Cal. App. 2d 559, 565; *Burt* v. *Los Angeles Olive Growers' Assn.,* 175 Cal. 668; *Board of Regents* v. *Cole,* 209 Ky. 761; *Geremia* v. *Boyarsky,* 107 Conn. 387; *Barlow* v. *Jones,* 87 At. 649; *Martens & Co.* v. *City of Syracuse,* 183 App. Div. 622, N.Y.; *R. O. Brumagin & Co.* v. *City of Bloomington,* 234 Ill. 114; *Board of School Comrs.* v. *Bender,* 36 Ind. App. 164; *Moffett, Hodgkins & Clarke Co.* v. *Rochester,* 178 U.S. 373, and *Steinmeyer* v. *Schoppel,* 226 Ill. 9.

In the *Kemper* case, the opinion stated also that:

"The evidence clearly supports the conclusion that it would be unconscionable to hold the company to its bid at the mistaken figure. The city had knowledge before the bid was accepted that the company had made a clerical error which resulted in the omission of an item amounting to nearly one-third of the amount intended to be bid, and under all the circumstances, it appears that it would be unjust and unfair to permit the city to take advantage of the company's mistake. There is no reason for denying relief on the ground that the city cannot be restored to status quo. It had ample time in which to award the contract without readvertising, the contract was actually awarded to the next lowest bidder, and the city will not be heard to complain that it cannot be placed in status quo because it will not have the benefit of an inequitable bargain. *Union & Peoples Nat. Bank* v. *Anderson-Campbell Co.,* 256 Mich. 674. . . . Finally, the company gave notice promptly upon discovering the facts entitling it to rescind, and no offer of restoration was necessary because it had received nothing of value which it could restore. See *Rosemead Co.* v. *Shepley Co.,* 207 Cal. 414, 420 to 422. We are satisfied that all the requirements for rescission have been met."

The city contended that the notice to bidders in the invitation and in the proposal form that they "will not be released on account of errors" bars the relief sought by the company. The court replied:

"There is a difference between mere mechanical or clerical errors made in tabulating or transcribing figures and errors in judgment, as for example, underestimating the cost of labor or materials. The distinction between the two types of errors is recognized in the cases allowing rescission and in the procedure provided by the state and federal governments for relieving contractors from mistakes in bids on public work. See *School District of Scottsbluff* v. *Olson Const. Co.,* 153 Neb. 451.... Generally, relief is refused for error in judgment and allowed only for clerical or mathematical mistakes.... When a person is denied relief because of an error in judgment, the agreement which is enforced is the one he intended to make, whereas if he is denied relief from a clerical error, he is forced to perform an agreement he had no intention of making."

The city also claimed that its charter provision, which prohibits withdrawal of bids after the letting, requires a literal interpretation to comply with the public interest. The opinion holds that "a bid is in the nature of an irrevocable offer or option, but the offer is subject to rescission upon proper equitable grounds, and the cases recognize no distinction between public and private contracts with regard to the right of equitable relief."

As to the right of the city to have forfeiture of the bid bond, the court said that no such right is available "when the bidder has a legal excuse for refusing to enter into a formal written contract, because the contingency in favor of a forfeiture did not arise." The decisions in other states are in line with this conclusion.

In *City of N.Y.* v. *Doud Lumber Co.,* 140 App. Div. 359, N.Y., 1910, errors were noted in extensions of unit prices for materials by the city's representative immediately after the bid opening and before the award was made to the bidder. The proposal showed, for example, 3,000 pieces of lumber at 35 cents per piece for $105, instead of $1,050, and the total of erroneous extensions was $543, whereas the correct total of the items would be $5,430. The bidder promptly notified the city of the error and asked to be relieved of the bid. The city refused the request, bought the lumber elsewhere, and sued to recover the excess cost. In its finding for the contractor, the court said:

"This action, in substance, is one to recover damages from the lumber company for the breach of the agreement, contained in its bid, to execute and carry out a contract upon the terms stated in the bid. The answer of the company is, in effect, that it never consciously or intentionally entered into such an agreement, its apparent agreement being the result of an honest mistake in transcribing the bid. Therefore, it says it would be unconscionable to hold it in damages. This is not an action to reform a written contract wherein it is usually necessary to show mutual mistake, or mistake on one part and fraud on the other. It is a demand that that which, on its face, appears to be a contract, shall be held to be no contract at all. It was not necessary to bring an independent action for the rescission of the contract. It was sufficient to state the facts as a defense. (*Born* v. *Schrenkeisen,* 110 N.Y. 55.) There are numerous cases to be found in the books wherein the courts have relieved parties from the consequences of mistakes similar to these proven in this case. A notable one is *Moffett, Hodgkins & Clarke Co.,* v. *Rochester,* 178 U.S. 373, where the plaintiff had made a mistake in its bid, which, however, was not patent upon the face of the bid itself. In the course of its opinion the court remarked that where there is a mistake on one side only it may be ground for rescinding a contract, but not for reforming it; for where the minds of the parties have not met there is no contract, and therefore, nothing to be reformed. To the same effect are *Hyde* v. *Tanner,* 1 Barb. 75, N.Y.; *Smith* v. *Mackin,* 4 Lans. 41, N.Y.; and *Flynn* v. *Smith,* 111 App. Div. 873, 874, N.Y. The rule in such cases was well stated by Lamar, Jr., in *Singer* v. *Grand Rapids Match Co.,* 117 Ga. 94, as follows: 'A slip of the pen or a slip of the tongue ought not to be treated as a deliberate contract, unless the other party has acted thereon to his injury. . . . But, if by reason of ambiguity in the terms of the contract, or some peculiar circumstances attending the transaction, it appears that one of the parties has, without gross fault or delay on his part, made a mistake, that this mistake was known, or ought to have been known to the opposite party, and that the mistake can be relieved against without injustice, the court will afford relief, either by refusing to decree specific performance, by cancellation or by refusing to give damages. . . . But, where the mistake is patent, where the opposite party knew or should have known of it, no contract has been made, the minds of the parties have not met, and they will be left where the mistake places them.' "

In an action by the State Road Commission (*State of Utah* v. *Union Construction Company and ano.,* 339 P. 2d 241, Utah, 1959) against a contractor and the surety on his bid bond, it appeared that the contractor had submitted a proposal for construction of five miles of a bituminous-surfaced road. He refused to accept an award and claimed that, upon visiting the site two days after the letting, he found he had been misled by some old stakes, which had been placed by the Commission's agents to show the route, and in particular by a red flag placed on a gully, which the State's engineer admitted should have been pulled. The contractor notified the Commission of the mistake immediately. The contractor's reliance upon the stakes led him to believe the roadway was going through an area of loose soil, but the plans actually showed rock, which had to be excavated. The rock excavation would have cost the contractor about $29,000 more than he figured for removing the loose soil. In allowing the claim of the contractor, the court cited the case of *Puget Sound Painters Inc.,* v. *State of Washington,* 278 P. 2d 302, and quoted from that case the following criteria for determining the claim of mistake and the disposition of the bid deposit: "... that equity will relieve against forfeiture of a bid bond, (a) if the bidder acted in good faith, and (b) without gross negligence, (c) if he was reasonably prompt in giving notice of the error in the bid to the other party, (d) if the bidder will suffer substantial detriment by forfeiture, and (e) if the other party's status has not greatly changed, and relief from forfeiture will work no substantial hardship on him...."

As for a contention by a government agency that a release of the lowest bid would cost the public a higher amount when an award is made to the next bidder, the answer to that type of claim is suggested in *Kutsche* v. *Ford,* 192 N.W. 714, Mich., 1923, as follows: "It does not appear that plaintiff's (contractor) mistake has made the school building cost more than it otherwise would have cost. The school district, if placed back where it was before the bid, loses nothing except what it seeks to gain out of the plaintiff's mistake. To compel plaintiff to forfeit his deposit because of his mistake would permit the school district to lessen the proper cost of the school building at the expense of the plaintiff, and that, in equity, is no reason for refusing plaintiff relief."

In a case involving an unusual statute that expressly provided for correction of a mistake in a bid for public contracts, the contractor claimed that he was entitled to the return of his bid deposit because of his error in misreading the adding machine tape. The faint figure "6" in the thousand column of the total was read as "0" and was carried into the total, resulting in an error of $6,000. The contractor complied with the statute by giving the awarding officer immediate notice of the mistake and then submitted clear evidence to show that it was not the result of carelessness in examining the plans and specifications. The court ruled in favor of the contractor on the ground that he had complied with the statute governing the procedure in such cases. (*Krasin* v. *Almond,* 290 N.W. 152, Wis., 1940.)

3.7 Misinterpretation of specifications

Claims of mistake are sometimes alleged by a bidder who asserts that his error is due to a misinterpretation of the specifications. The case of *Greene* v. *City of New York et al.,* 283 App. Div. 485, N.Y., 1954 related to specifications that permitted the contractor to exercise his own option in the use of plasterboard partitions in bathrooms, but there was also a general provision that purported to bind the contractor to State rules affecting this kind of work. Because the State rules required a more expensive method of construction, the city refused to pay for it on the theory that the State rules prevailed by virtue of the general provision in the specifications. The court said: "The city prepared the plans and specifications and asked for competitive bidding on the basis thereof. It could not be expected that the bidder would examine the various laws and building codes as to each item specified to see if the codes required some different method of construction. Such procedure would make the bidder's interpretation of the law rather than the specifications controlling. It would make the specifications so indefinite and uncertain as to destroy the validity of any contract awarded pursuant thereto."

3.8 Mistakes by the government agency

Cases have reached the courts in which the public agency seeks to be relieved of a contract because of a mistake it made preparing the proposal form. For example, the State of New York once advertised for bids for 600 rolls of wire fencing and, after the letting, awarded a contract for that number which anticipated a need for 60,000 feet. It developed that the requirement by the State was really 60 rolls or 6,000 feet of fencing. (*Wholesale Serv. Sup. Corp.* v. *State of New York,* 201 Misc. 56, N.Y., 1951.) The court said:

"We do not believe the present contract falls within the line of cases cases stemming from *Brawley* v. *United States,* 96 U.S. 168, which holds that where a specific quantity if qualified by such words as 'about,' 'more or less,' etc., and such qualified words are followed by other stipulations or conditions, the contract is to be governed by such conditions. The underlying reason for the development of such a theory of contract law in connection with government contracts arises from the fact that in procuring commodities for military establishments and like agencies, it is often impossible, because of changing conditions, to specify an exact amount, and to protect the public and the taxpayer from the needless dissipation of public funds an approximate quantity is denoted. (*Field* v. *United States,* 16 Ct. Cl. U.S. 434; *Gemsco, Inc.* v. *United States,* 115 Ct. Cl. U.S. 209.) In the present instance, the quantity found in the proposal and award was not qualified by the words 'more or less' or words of similar import, and, in addition under the conditions herein involved, there was no reason why the agency requiring the fencing would not be in a position to determine the exact quantity needed."

In the contract relations between a state or a municipality and its contractor, the courts have consistently applied equitable principles when there has been negligence, incompetence, or carelessness, for example, a material understatement of the work required to be done. In *Palmberg* v. *City of Astoria (Oregon),* 199 P. 630, 1921, the defendant's engineer made a mistake as to the amount of embankment required to complete a street improvement. The quantity specified was 17,087 cubic yards, but actually the work involved 28,567 cubic yards, and upon discovering the error, the contractor notified the city that he would complete the project under protest and expected the city to pay damages for extra cost to him. The bid was for a lump sum for the entire contract and did not provide for payment at a unit price for the quantity of the embankment, and there was no provision for payment for extra work. The court commented that in a similar situation between private parties, the negligent and misleading computation by an employee of the owner would make the owner liable for damages to the other party who suffered by the mistake.

"But", said the court, "municipal corporations of the present day are so hedged about with provisions restricting their liability that it becomes a matter of extreme nicety to determine whether or not such liability exists in a case like the present.

"While the same honesty is to be expected from a municipal corporation as from a private individual, peculiar provisions are sometimes found in statutes relating to governmental bodies, which exempt a state or municipality from acts or omissions for which a private party would be held responsible. In the *Palmberg* v. *Astoria* decision, the court stated the rule in its jurisdiction by the following: 'It is settled, however, in this State (Oregon), that a municipal corporation cannot escape liability for an ordinary tort arising from its negligent acts or omissions, unless its charter or ordinance provides an equivalent remedy against the officer through whose agency or negligence the wrong was committed.' "

Whereas a municipality has no responsibility for the mistake by an employee without proof of negligence, carelessness, or incompetence, a contractor is entitled to be paid for extra work because the city's engineer made a mistake in the depth for excavation of a sewer trench. (*McCann* v. *Albany,* 158 N.Y. 634, 1896.) A municipality was held responsible for extra brick masonry to align a tunnel because the negligence of the engineer resulted in a line that was not straight. (*Chicago* v. *Duffy,* 218 Ill. 242, 1905, 75 N.E. 912.)

In *Leary* v. *City of Watervliet,* 222 N.Y. 337, 1918, it was specified that the estimated quantities were not intended as the actual quantities, but were reasonable approximations, and payment or deductions would be made for any increase or decrease. The contractor had inspected the site before he bid for the work, which the court said placed all of the risk on the contractor as to the extent and amount of rock he would encounter. Whereas there was no claim of fraud or concealment by the municipality, the court held: "A variation in the estimate by the commission of the work to be done arising by mistake or inadvertence or by changes of work contemplated, was at the risk of the contractor."

Becker v. *City of N.Y.,* 176 N.Y. 441, 1903 disclosed a contract provision for a street improvement project that "a city surveyor will be employed by the parties of the first part (the city) to see that the work is completed in conformity with the profile, and to ascertain and certify the quantity of the work done. Said surveyor at the request of the contractor, will be directed to designate and fix grades for his guidance during the progress of the work, without charge, provided that the said parties of the first part shall not be liable for any delay or for any errors of said surveyor in giving such grades, and said surveyor shall be considered as the agent of the contractor as far as giving such grades is concerned, and not the agent of the city of New York." A profile was attached to the contract and the street was to be graded in accordance with the profile. The city surveyor set the stakes and marked the erroneous grades. Although the contractor protested the surveying inaccuracies to the municipality, he was directed by the superintendent to follow the grades. Therefore, he excavated to a greater depth than was necessary, and he had to refill the excavation, for which he sought damages from the city. The court excused the city from responsibility because of the mistake in the grade, as it was the duty of the contractor to grade the street according to the profile attached to the contract, and any direction by the superintendent of streets to follow the grades set by the city surveyor was an unauthorized modification or variance of the contract.

chapter 4

Lowest responsible bidder

A municipal charter provision requiring awards on public contracts to be made to the "lowest responsible bidder" has been declared by the courts as "based upon motives of public economy, and originated, perhaps, in some degree of distrust of the officers to whom the duty of making contracts for the public service was committed. If executed according to its intention, it will preclude favoritism and jobbing, and such was its obvious purpose. It does not require any argument to show that a contract made in violation of its requirements is null and void" (*Brady* v. *New York,* 20 N.Y. 312, 1859).

4.1 Definition of lowest responsible bidder

The expression "lowest responsible bidder" as it is used in statutes and local charters means the lowest bidder whose offer best responds in quality, fitness, and capacity to the particular requirements of the proposed work. The meaning of the term is not confined to the lowest bidder whose pecuniary ability is the best, but rather to the bidder who is most likely with regard to skill, judgment, and integrity as well as sufficient financial resources, to do faithful, conscientious work and promptly fulfill the contract according to its letter and spirit. The awarding officer, who has the function of determining the "lowest responsible bidder," also has the judicial duty and discretion to investigate the integrity and moral worth of various bidders and to examine their background. In the absence of any factual showing of dishonesty, fraud, collusion, corruption, or bad faith, any determination he makes will not be overturned by the courts. (*Koich* v. *Cvar,* 110 P. 2d

964, Mont., 1941; *Hodgeman* v. *City of San Diego,* 128 P. 2d 412, Cal., 1942; *Application of Limitone,* 189 N.Y.S. 2d 738, 1959.)

In defining the "lowest responsible bidder," the word "responsible" is as important as the word "lowest." The term "lowest bidder" is not to be construed literally; in determining who is the lowest bidder, the quality and utility of the thing offered and its adaptability to the purpose for which it is required should be considered. (*Clayton* v. *Taylor,* 49 Mo. App. 117, 1871; *Cleveland Fire Alarm Teleg. Co.* v. *Bd. of Fire Comrs.,* 55 Barb. 288, N.Y., 1869; *Porterfield* v. *City of Oakland,* 159 P. 202, Cal., 1916; *Commonwealth* v. *Mitchell,* 82 Pa. 343, 1876; *Findley* v. *City of Pittsburgh,* 82 Pa. 351, 1876.)

Where an awarding officer in a municipality rejected the lowest bid for an asphalt pavement project, the affected contractor asked the court to command the public official to observe the provisions of the local charter, which provided that an award is to be made, after advertising, to the "lowest responsible bidder." The charter also empowered the Commissioner of Public Works to make awards of contracts in projects of this type. The ordinance of the Common Council approving the project called for asphaltum equal in quality to that found in Pitch Lake, Trinidad, and the Commissioner of Public Works decided that the material offered by the lowest bidder had not yet been adequately tested, and that in the Commissioner's opinion the offered material did not measure up to the standards of a durable pavement. Information had also been received by the municipality that the contractor's work in other places disclosed poor workmanship and unsatisfactory materials, which were attributed principally to lack of related experience. The contractor challenged the language of the ordinance, which called for an award to the "lowest reliable and responsible bidder," whereas the charter provision directed an award to the "lowest responsible bidder." The contractor claimed that he was "responsible" because of his financial condition, and the quoted word refers only to that financial or pecuniary responsibility of the bidder to do the work. Upon his compliance with this standard, the contractor insisted that he was entitled to an award under the provisions of the local charter.

In the case described in the preceding paragraph, the court stated that the term "responsible" must not be construed in a situation of this kind that is narrow and limited. "In one sense the term is synonymous with 'accountable' and means answerable legally or morally for the discharge of a duty, trust, debt, service or other obligation. In that sense, a bidder entering into a contract, becomes responsible for its fulfillment; but this was not the sense in which the term is employed. In the statute, it was evidently used in its other sense, of ability to respond.

In that sense, 'responsible' means to be able to answer or respond in accordance with what is expected or demanded, and to discharge a claim or duty. The scope of the term depends upon the subject matter to which it is applied." In deciding against the contractor's claim to the award, the court added: "As ('responsibility') applied to a bidder who proposes to undertake the performance of the stipulations and conditions of such a contract as this, we regard the term as including the ability to respond by the discharge of his obligations in accordance with what may be expected or demanded by the terms of the contract. It was not unreasonable to suppose that the intent of the statute was to limit competition to parties able to perform the conditions of the contract. Such ability is of great importance to the city and the parties assessed to pay for the improvement. It is only by letting the contract to a bidder capable of performing it that the full benefit and advantage sought to be derived from the improvement can be realized" (*People ex rel. Assyrian Asphalt Co.* v. *Kent, Commissioner,* 43 N.E. 760, Ill., 1896).

In the *Matter of Kaelber* v. *Sahm et al.,* 305 N.Y. 858, 1953, the Town Board sought bids for a mechanical stoker and asked the bidders to submit statements showing experience in similar work. Whereas there was no dispute about the financial responsibility of the lowest bidder, he did not have any experience in constructing a mechanical stoker, whereas the bidder to whom an award was made had patented that type of equipment and had installed a number of them. Although the Town Board was required to award to the lowest responsible bidder, its duty required that the members of the board consider the purpose of the installation of the equipment and the qualifications of the bidders to perform the work. The bidders had notice of the standard of qualifications demanded of them, and in the absence of fraud, the action of the Town Board in rejecting the low bid was deemed by the court to be not arbitrary but reasonable and plausible.

Following this theory of bidding procedure, the award to the lowest responsible bidder must be in accordance with the specifications that were made available to the prospective contractors. Competition for public construction contracts assures equal opportunity, and it is reasonable for bidders on public work to expect that a basis has been established in advance to permit an exact comparison of the proposals; this can be accomplished only when the same work or thing to be done is disclosed to all of the bidders. They may be expected to submit intelligent bids on a common basis when the plans, drawings, and specifications are definite and explicit in their contents and are available in sufficient time for prospective contractors to consider the information provided. If the work to be done, the manner in which it is required, and the materials to be used are not specified, the lowest responsible bidder cannot be chosen. The point is that whereas there may be no statutory requirement to furnish plans and specifications in some states and municipalities, the requirement of an award to the lowest responsible bidder presupposes that the same, necessary information must be given to all bidders, and, consequently, the proposals will meet the reasonable standards of comparison of bids. (*Detroit* v. *Hosmer,* 79 Mich. 384, 1890; *Wells* v. *Burnham,* 20 Wis. 112, 1865.)

In most jurisdictions, it is contrary to the theory of open competitive bidding on public construction work if the public body does not provide plans and specifications but instead, directs contractors to furnish their own plans and specifications as part of their sealed bids. (*Boren* v. *Darke County Comrs.,* 21 Ohio St. 311, 1871; *State* v. *Barlow,* 48 Mo. 17, 1871; *Re Eager,* 46 N.Y. 100, 1871; *People* v. *Buffalo County Comrs.,* 4 Neb. 150, 1875.)

A leading case in which the public body did not provide plans and specifications in advance of a public letting is *Fones Hardware Co.,* v. *Erb.,* 54 Ark. 645, 1891. The Board of Commissioners advertised for and received bids with competitive plans and specifications for a bridge. The law authorizing public construction required competitive bidding and directed that awards are to be made to the lowest responsible bidder. The court granted an injunction against the board who were about to enter into a contract for the construction of the bridge, and the decision stated: "When a contract to build a bridge is to be let, there are two kinds of competition that may arise: first, that between persons desiring to build different kinds of bridges; and, second, that between those desiring to build the same kind. And as was said by Judge Chrisiancy in discussing a provision similar to that under consideration, the bidding which it contemplates is of the latter kind—bidding for the same particular thing, to be done according to the same specifications. For, says he, no bids for different kinds of work, and referring to different specifications, could be recognized as coming in competition with each other for the purpose of determining the lowest bid within the requirement of this section, without opening the door to the same corrupt combinations, and furnishing facilities for the same fraudulent practices, which it was the purpose of this provision to prevent." (*Attorney-General ex rel. Cook et al.* v. *City of Detroit,* 26 Mich. 263.) Also see *Ertle* v. *Leary* 46 P. 1, Cal., 1896; *Andrews* v. *Ada County,* 63 P. 592, Idaho, 1900; *Littler* v. *Jayne,* 16 N.E. 374, Ill., 1888; *Bluffton* v. *Miller,* 70 N.E. 989, Ind., 1904; *Packard* v. *Hayes,* 51 At. 32, Md., 1902; *Louisiana* v. *Shaffner,* 78 S.W. 287, Mo., 1904; *Browning* v. *Bergen County,* 76 At. 1054, N.J., 1910.

A variation in the requirement is found in *Yaryan* v. *Toledo,* 81 N.E. 1199, Ohio, 1907, which holds that when the necessary information is contained in the written specifications, it is not a drastic omission if the sketches and drawings are not prepared.

It is also important to remember in this regard that no material contingencies can be imposed upon the award that were not disclosed before the letting date. In *North-Eastern Const. Co.* v. *Town of North Hempstead,* 121 App. Div. 187, N.Y., 1907, the Town Board adopted a resolution "that the secretary notify the North-Eastern Construction Co. that their bid for construction of the bridge across Udall's Pond, Great Neck, be accepted, conditioned upon leave being granted by the Board of Supervisors to issue bonds of the Town in the sum of $20,000." The court ruled that the town "did not absolutely accept the plaintiff's bid; the acceptance depended upon a contingency which the plaintiff (contractor) was not asked to consider in making its bid. This requirement was contrary to an elementary principle in contract law that both parties must be bound."

4.2 Determination of irresponsibility

A determination by a public official or agency that a bidder is irresponsible should be supported by a record of both the investigation that disclosed the facts and the opportunity that was offered to the bidder to present his qualifications. Unless there is a statute or specification that calls for a formal hearing, the preferred method, whether required or not, is an informal conference without technical rules of evidence. The determination is of a quasi-judicial nature and not a mere administrative act; the issues should be considered after reasonable notice and a fair opportunity has been afforded to the apparent low bidder before a determination is made that actually denies him an award. A reasonable practice is to notify the bidder by letter containing sufficient information as to the reason, scope, and purpose of the meetting so that he will be able to assemble the data that he believes entitle him to a favorable determination. Courts do not, as a rule, direct a public administrative official to award a contract to a certain bidder, except where there has been a clear violation of law or where the official failed to perform an act that is required by law. (*Matter of Albro Contracting Corp.* v. *Dept. of Public Works, etc.,* 13 Misc. 2d 845, N.Y., 1958.)

In *Matter of Ward LaFrance Truck Corp.* v. *City of New York,* 7 Misc. 2d 739, N.Y., 1957, the awarding officials decided that the bidder who had submitted the lowest bid was not responsible because 143 pieces of fire-fighting equipment previously purchased from the bidder required 467 repairs in seven categories over a period of several years. Because of the necessary repairs the equipment was unavailable for use by the Fire Department for long periods of time. The city charter (section 343, sub. b.) and a regulation of the Board of Estimate called for a filed decision and a stated reason for eliminating the lowest bidder. The court ruled: "Upon the facts here presented, the determination of the petitioner's (the bidder) nonresponsibility and the refusal to award it the contract appear to have been the result of hearings at which the essential elements of a fair trial were not observed. It appears that respondents have not pursued their authority in the mode required by law and that fundamental rules of law have been violated to the prejudice of petitioner." The court did not direct an award to the lowest bidder but returned the matter to the responsible city officials for reconsideration of the determination after following the provisions of the statute on hearings and a determination.

4.3 Judgment and discretion of awarding officer

Another illustration of the application of the term "responsible" in its relation to the "lowest bidder" is found in the case where officials of the City of New York rejected the lowest bid for sewer construction work because the bidder was actually a "front" for a person who over a period of years had been in frequent conflict with the criminal law relative to public construction contracts. (*Picone* v. *City of N.Y.*, 176 Misc. 967, N.Y., 1941.) Although the city charter required an award to the "lowest responsible bidder," the court ruled that the discretion used in rejecting the bid "was exercised with an honest desire to award the contract to the lowest responsible bidder and is therefore not subject to review by the courts (cases cited)." When there is no proof of fraud or bad faith on the part of a public official, the courts will not interfere with administrative discretion or try to substitute a decision by the bench in place of administrative discretion. (*Campbell* v. *City of N.Y.*, 244 N.Y. 317, 1927.)

In the case of *Syracuse Intercepting Sewer Board* v. *Deposit Co. of Maryland*, 255 N.Y. 288, 1931, the surety company brought suit to recover the excess cost of completing a contract after the original contractor had abandoned the project. The surety company challenged the Sewer Board's rejection of the lowest bidder for the completion of the unfinished work. After rejecting the lowest bidder, the Sewer Board had made an award to a contractor whose bid was higher. At the trial the Sewer Board contended that the experience with the contractor who abandoned the project justified a careful inquiry about the finances and equipment of another contractor for the completion of the job. The court did not pass upon the ability of the lowest bidder on the completion job to perform the work, but the Court of Appeals did uphold the action of the Sewer Board on the proof that the lowest bidder refused to furnish a financial statement, depending upon borrowed money to do the job. The evidence also included information from Dun and Bradstreet and from another surety company that was unsatisfactory. The high court held that the Sewer Board had exercised "honest judgment and discretion" and that it had the right to reject the unqualified bidder.

Quite often, speed and promptness of performance of a contract are factors in establishing the responsibility of a contractor. In this connection, the bid by a contractor who maintained the policy of "open shop" in the employment of workmen could be a test of the rule of an award to the "lowest responsible bidder." It is generally known that union mechanics will strike a job when nonunion workmen are also employed on the same contract. In the case of *Taylor* v. *Board of Education,* 253 App. Div. 653, N.Y., 1938, affirmed 278 N.Y. 641, a taxpayer sued to restrain the Board from rejecting bids from "open shop" contractors. The court dismissed the complaint, stating that "the rejection of bids was done in good faith by responsible officials vested with and in the exercise of a proper discretion and . . . to avoid increased burdens and threatened dangers to the municipality."

In the case of *People ex rel. Martin* v. *Dorsheimer et al., Commissioners of the New State Capitol in Albany,* 55 How. Pr. 118, N.Y., 1871, the relator's (that is, the party on whose behalf the case was brought) bid for certain carpenter work was the lowest that was received. The bid was rejected because the contractor was not financially responsible, although he offered a good bond for faithful performance. The law governing the project declared that "all contracts shall be awarded to the lowest bona fide responsible bidder. . . ." The same statute also provided: "The said new capitol commissioners are hereby directed to take such measures as shall insure the completion and finishing of that portion of the new capitol containing the assembly chamber for occupation on the 1st day of January, 1879, by the senate and assembly." The court held that the commissioners had a large discretion, and, in order to meet the deadline, they could discriminate the men to be employed. The conclusion stated in the opinion was as follows: "As efficiency and promptness in doing any work must depend largely upon the man who does it, having reference to his integrity, ability and responsibility, the board was authorized to discriminate between bidders."

4.4 Administrative decision and court judgment

In 1892 the New York Court of Appeals explained the significance of the steps that precede the making of a formal public contract. In *Erving* v. *Mayor etc. of New York,* 131 N.Y. 133, 1892, it was stated: "The mere fact that a party who has made proposals for public work in the city of New York is the lowest bidder and knows that fact, does not necessarily entitle him to the contract and does not constitute an award to him of such contract within the meaning of the law regulating the letting of work upon competitive bids. The officer charged with the duty of awarding the contract may, and in some cases would very properly, determine that the interests of the city would be best served by rejecting proposals to enter into a contract with a party who, though he might be the lowest bidder, yet did not possess such advantages and facilities for the performance of the contract as to render his proposal, on the whole, for the best interests of the city. The commissioner may reject all the bids and readvertise the work, if in his judgment and discretion such course is for the best interest of the municipality. The awarding of the contract on the part of the officer to one of several bidders requires the exercise on his part of judgment and discretion and the award itself should be manifested by some formal official act on his part, and ordinarily reduced to writing and made a part of the records in his department. In no other way can the rights of the parties be preserved, at least prior to the actual execution of the contract. The mere arithmetical operation of ascertaining which bid is the lowest does not constitute an award. The duty of the commissioner is to examine the proposals and the award of the contract is judicial in its nature and character (*East River Gas Light Co.* v. *Donnelly,* 93 N.Y. 55; *People ex rel. Coughlin* v. *Gleason,* 121 N.Y. 631), and the award is the result of a judicial act."

Although the court will not substitute its own judgment for that of a public official or agency who is responsible for public bidding functions, the application of this rule was modified because of questionable procedure in *Arensmeyer et al.* v. *Wray et al.,* 118 Misc. 619, N.Y., 1922. The related statute required an award to the "lowest responsible bidder furnishing security as required," and the court said that although the awarding agency could reject the bids because of some informality, no informality was disclosed, even though the lowest bidder failed to state the names of each member of the firm. This omission was not the type of informality to justify the board to reject the bids. Good faith is required of the awarding officer and his action must be above "the suspicion of favoritism or corruption. Under these circumstances the courts will control the action of the board and require an observance of the statutes." An individual has latitude in the selection of a contractor, but the public agency cannot reject all of the bids unless "there is some substantial reason for that course." The court directed an award to the "lowest responsible bidder furnishing security as required."

4.5 Influence of surety bond

Lawyers and public officials have long been considering the weight or influence of a surety bond in determining the lowest responsible bidder. In some quarters, particularly among awarding officials, it is believed that the lowest bidder has responsibility if he can provide a surety bond. The surety companies have urged an entirely different approach to this subject. They have clearly indicated that "a contract bond might reasonably enough be issued in favor of a contractor to whom it would yet be undesirable, on the whole, to award the given contract. If, for example, an applicant for a contract bond is a man of good character or a concern of high standing, and is known to be abundantly responsible in a financial way, a bonding company is hardly in a position to withhold its suretyship even if the man or the concern has apparently bid too little or is for some other reason not likely to complete the contract in a satisfactory way or in any event with profit." (*How Can Contract-Bond Conditions Be Improved?* issued by The Surety Association of America, 1925.)

The surety bond companies prefer to have a contract awarded to one who is able to complete the contract or to pay damages to the owner in case of failure of performance under the contract. The companies have the primary function of guaranteeing indemnity in case the contractor defaults. The awarding officer, however, has the broader duty of insuring that the award is made to a bidder who has the ability to complete the contract as well as to respond in damages in case he defaults.

In *Williams* v. *City of Topeka,* 118 P. 864, Kan., 1911, the Supreme Court of Kansas said: "We conclude that the word 'responsible' in the phrase 'lowest responsible bidder' was used by the legislature in the sense in which it had long been interpreted by the court and text writers and must be held to imply skill, judgment and integrity necessary to the faithful performance of the contract as well as sufficient financial resources and ability." The function of the awarding official is to evaluate the bidder's responsibility on the basis of these qualifications without regard to a surety bond to indemnify the public in case of default.

4.6 Prequalification

Many contractors and public officials believe that it is good practice for a governmental agency to qualify bidders before the issuance of a set of plans and specifications or before a bidder has submitted his proposal. In the usual method, a contractor qualifies for an award after he has submitted his proposal. The first method is called "prequalification" and the latter method is called "postqualification." The merits and drawbacks of each process have been debated for several decades, and notwithstanding the fact that the number of states requiring prequalification has grown over a period of years, there are other states that have employed postqualification for many years, with no indication that they will change.

Its proponents have pointed out that although prequalification before the issuance or sale of a set of plans and specifications or the acceptance of a proposal might appear to restrict the number of responsible bidders, experience in localities where this procedure is in effect shows increases rather than decreases in that number. One particular argument that has been emphasized by some of the proponents of prequalification is that it eliminates the irresponsible bidder whose sharp business policy is deliberately to bid below cost because he does not have serious intentions of completing the job. They claimed, therefore, that competitive bidding without prequalification does, in effect, result in forcing the responsible bidder out of the bidding for the particular contract.

Some reasons in favor of prequalification have been issued by representatives of a major national organization of construction contractors:

> 1. It saves an enormous amount of money expended in the preparation of proposals. To perform investigations, locate sources of material supply, compute quantities, secure subbids, and devise a plan of construction, all of which is necessary for an intelligent proposal on a government building may cost $1,000 or $1,500. Naturally, the contractor dislikes to waste the money by being disqualified after bids are received.
>
> 2. It insures that all bidders who submit proposals are competent to do the work, and that the award will be made without protracted delay, legal battle or subsequent difficulty with the accounting department.
>
> 3. It removes the temptation of awarding officials to make awards to irresponsible bidders who would not deliver the quality of work that must be produced by a concern of integrity.

In connection with the statements quoted above, the following minimum requirements have been suggested (*The American City*, June, 1928) for a basic measure in judging a contractor's responsibility:

> (a) Under ordinary circumstances, a contractor probably should not bid on a project more than double or treble the value of the largest of similar nature that he has previously constructed.
>
> (b) In building work, he should probably not have on hand projects which, all totaled, exceed twenty times the amount of available liquid assets.
>
> (c) In highway and engineering work, requiring a heavy equipment outlay, he should not have on hand projects valued at more than ten times his available liquid assets. This presumes that his equipment is paid for.

A fundamental purpose of competitive bidding is to stimulate competition so that as many contractors as possible will submit proposals for the project. Those who oppose prequalification insist that the time to evaluate the responsibility of a bidder is after the bids have been received. They contend that the postqualification method places no barrier against one who has enough interest in the contract to examine the plans and to determine whether or not his experience and capabilities are good enough to submit a bid. In the opinion of those who oppose its use, prequalification shuts out new contractors, particularly because they do not meet the set of rules as to "experience" and thus are rejected without an equal opportunity to expand their operations and to grow in their field of endeavor. Another point against prequalification is that a public official must determine the responsibility of a contractor by the answers to a questionnaire that the contractor must submit before he is eligible to receive a set of plans or before he may submit a proposal to the public official. This circumstance introduces the human element of honest mistake, the possibility of fraudulent application of unequal and inaccurate measures in the official determination, and the pressure of political influence. Prequalification, it is charged by those who do not accept its use, results in higher costs, because the restriction reduces the number of bidders and places them in a position to impose their demands upon material suppliers and equipment distributors. In applying prequalification, opponents feel, the public official is trying to safeguard the contractor's business from the contractor, whereas the time and effort of the public official should be devoted to a method that will conserve the public dollar by free and open competition, subject to postqualification and an informative tabulation of bids.

In relation to the cost of federal aid public construction projects, it is of interest to note the express policy of the Bureau of Public Roads of the U.S. Department of Commerce, on prequalification in the administration of federal aid for highways, as follows (Sec. 1.16—Regulations effective May 11, 1960):

> With respect to federal aid projects, no procedure or requirement for prequalification, qualification or licensing of contractors shall be approved which, in the judgment of the Administrator, may operate to restrict competition to prevent submission of a bid by, or to prohibit the consideration of a bid submitted by, any responsible contractor, whether resident or nonresident of the State wherein the work is to be performed. No contractor shall be required by law, regulation or practice to obtain a license before he may submit a bid or before his bid may be considered for award of a contract. This, however, is not intended to preclude requirements for the licensing of a contractor upon or subsequent to the award of the contract if such requirements are consistent with competitive bidding. Prequalification of contractors may be required as a condition for submission of a bid or award of contract only if the period between the date of issuing a call for bids and the date of opening of bids affords sufficient time to enable a bidder to obtain the required prequalification rating. Requirements for the prequalification, qualification or licensing of contractors, that operate to govern the amount of work that may be bid upon by or may be awarded to a contractor, shall be approved only if based upon a full and appropriate evaluation of the contractor's experience, personnel, equipment, financial resources, and performance record.

In a case where a statute required that a construction contract of more than $1,000 could only be awarded after advertising and competitive bidding, a statement in the published notice required each application for a set of drawings to be accompanied by financial and experience statements upon forms to be obtained from the architect. The public agency rejected a proposal because, by its rules, the bidder did not meet the standards of the prequalification procedure. The bidder carried his claim against the rejection to the courts and alleged that the action that limited the number of bidders was illegal. Although the court recognized the objective of the public agency as being an act of good faith, it decided, however, that the statutory power of the public body did not authorize the attempt to limit competition by prequalification. In other states where prequalification is in effect, there is express statutory power to proceed by that method. (*Matter of Weinstein Building Corp.* v. *Scoville,* 141 Misc. 902, N.Y., 1931, citing *Harris* v. *City of Philadelphia,* 149 At. 722, Pa., 1930.)

The following significant comment is quoted from the *Weinstein* case, which is discussed in the previous paragraph:

"Although not intending to do so, the requirement that applicants for the plans and specifications must file financial statements and experience questionnaires before being allowed to bid, and the fact that their applications for the plans and specifications may be rejected, discourages and prevents the competitive bidding which the statute requires.

"To enter into the competition, a prospective bidder must first apply for the plans and specifications and must file a financial statement and experience questionnaire. If his application for the plans and specifications is rejected, he must litigate his right to receive the plans and specifications, and his right to be allowed to bid. This involves expense. If successful in the litigation, he may then submit his bid, only to find that his was not the lowest bid, so that the expense of litigating his right to bid has been lost. Confronted with an expenditure of money to enforce his right to bid, before he knows whether his bid is the lowest, it can scarcely be doubted that many possible competitors might retire from the field of competition, to the injury of the taxpayers of the city of Mount Vernon."

The court ruled that the authorization for prequalification must be found in an applicable statute.

It is important to distinguish between prequalification and a provision in an advertisement for bids that prescribes the qualifications of contractors who will be considered for an award. For example, bridge commissioners included a statement in an advertisement that steel bids would be accepted only from those bidders whose plants were adequate to do the work and who were in successful operation for at least one year. They also prescribed that steel containing more than a specified percentage of foreign elements would be excluded from a consideration of an award. As there was no fraud alleged or involved in the award, the court ruled that the restrictions were not illegal. (*Knowles* v. *City of New York,* 176 N.Y. 430, 1903.)

Where special or expert workmanship is required, a governmental body has the right to protect itself from inexperienced, irresponsible, or incompetent persons, and a provision as to experience and ability that is fair and reasonable is not an illegal restriction of competitive bidding. (*Heninger* v. *City of Akron,* 112 N.E. 77, Ohio, 1951.)

4.7 Contractor's statement

A noteworthy example of the "Contractor's Statement of Experience and Financial Condition," used by the Department of Public Works of the State of California, is illustrated here. It shows a typical set of representations by the prospective bidder in accordance with a statutory authorization for the requirement of the form, "for the purpose of inducing the Department of Public Works to supply the submitter with plans and specifications." The prequalification law is summarized in the form along with the instructions for preparing and filing the necessary information.

4.8 Licenses, taxes, and fees

The different classifications of contractors, as well as the general business of contracting, have been the subject of occupation, license, or privilege taxes pursuant to state statutes or local ordinances that impose those particular taxes. In each jurisdiction having such a statute or ordinance, it is, of course, imperative for a prospective contractor to inquire about and to be familiar with the requirements, the safeguards, and the penalties under the provisions. In some states the license statute has an effect upon the right of a contractor or subcontractor to enforce payment of his bill for performance of the work.

"The statutory requirement to obtain a license before engaging in the trade is a police regulation for the protection of the public, *Smith* v. *American Packing & Provision Co.,* 130 P. 2d 951; a penalty is provided for the violation of the statutory exaction; and it is the well settled general rule that in ordinary circumstances, a contract entered into by an unlicensed person in contravention of the statutory provisions of this kind will not be enforced. *Wedgewood* v. *Jorgens,* 190 Mich. 620; *Hickey* v. *Sutton,* 191 Wis. 313; *Sherwood* v. *Wise,* 132 Wash. 295; *Lund* v. *Bruflat,* 159 Wash. 89; *American Store Equipt. & Constr. Corpn.* v. *Jack Dempsey's Punch Bowl, Inc.,* 283 N.Y. 601; *Massie* v. *Dudley,* 173 Va. 42; *Board of Education* v. *Elliott,* 276 Ky. 790.

"But that general rule does not have application in a case of this kind in which an unlicensed member of a profession or trade seeks to recover from a licensed member for services rendered or labor performed pursuant to a contract entered into by them." (*Dow* v. *U.S.,* 157 F. 2d 707, Ut., 1946, and also citing *Martindale* v. *Shaba,* 51 Okla. 670; *White* v. *Little,* 131 Okla. 132; *Ferris* v. *Snively,* 172 Wash. 167; *Cf. John E. Rosasco Creameries* v. *Cohen,* 276 N.Y. 274.)

If the license tax was enacted to secure revenue without restrictions upon the class of persons who would be liable for the tax, then the statute usually imposes a penalty for failure to pay the tax in due time, and the right of an unlicensed person to collect his bill under a contract is not affected. However, if the intention of the statute is to protect "the public from imposition by ignorance or dishonesty," then an unlicensed person will find that he does not have that right. (*Draper* v. *Miller,* 140 P. 890, Kan., 1914; *Alvarado* v. *Davis, et al.,* 6 P. 2d 121, Cal., 1931.)

A striking example of the modern concept of governmental control and regulation of professional and vocational standards is contained in the California Contractors' Law, Chapter 9 of Division 3 of the Business and Professions Code. The duty of administering the law is vested in the Contractors' State License Board, which is in the Department of Professional and Vocational Standards and consists of seven members who are

STATE OF CALIFORNIA
DEPARTMENT OF PUBLIC WORKS
SACRAMENTO, CALIFORNIA

CONTRACTOR'S STATEMENT OF EXPERIENCE AND FINANCIAL CONDITION

DIVISION OF HIGHWAYS
HIGHWAY CONSTRUCTION

(1) *If you are a General Contractor interested in bidding on all types of highway construction, mark cross after "All Classes of Construction" only.*

(2) *If you are interested in contracting direct with the State for certain types of work only, mark cross in the column provided after the particular types of work on which you wish to bid.*

TYPE OF WORK	MARK WITH (x)
1 All Classes of Construction	
2 Grading and Paving	
3 Clearing and Grubbing	
4 Grading	
5 Culverts and Small Structures	
6 Paving, Portland Cement Concrete	
7 Paving, Asphalt Concrete	
8 Bridges	
9 Painting	
10 Traffic Signals and Lighting	
11 Erosion Control and Roadside Development	
12 Seal Coats	
13 Joint Sealing	
14 Other (List and mark with X)	

DIVISION OF ARCHITECTURE
BUILDING CONSTRUCTION

(1) *If you are a General Contractor interested in bidding on all types of building construction, mark cross after "All Classes of Construction" only.*

(2) *If you are interested in contracting direct with the State for certain types of work only, mark cross in the column provided after the particular types of work on which you wish to bid.*

TYPE OF WORK	MARK WITH (x)
1 All Classes of Construction	
2 Excavating and Grading	
3 Concrete	
4 Masonry	
5 Structural Steel	
6 Steel and Miscellaneous Iron	
7 Sheet Metal	
8 Roofing	
9 Lathing and Plastering	
10 Painting and Decorating	
11 Floor Covering	
12 Elevator	
13 Plumbing	
14 Heating and Ventilating	
15 Air Conditioning	
16 Boiler and Equipment	
17 Electrical	
18 Laboratory Equipment	
19 Well Drilling	
20 Landscaping	

[1]

Contractor's statement of experience and financial condition appointed by the Governor. All members of the Board are actively engaged in the business of contracting and must have at least five years experience. One member is chosen from the group of general engineering contractors, three members from the general building contracting business, and three members are specialty contractors.

An applicant for a contractor's license is required to select the clas-

CONTRACTOR'S STATEMENT OF EXPERIENCE

NAME ______________________________ ☐ A Corporation ☐ A Co-partnership ☐ An Individual ☐ Combination*

PRINCIPAL OFFICE ______________________________

The signatory of this questionnaire guarantees the truth and accuracy of all statements and of all answers to interrogatories hereinafter made

1. Are you licensed as a Contractor to do business in California? ________ License No. ____________
2. How many years has your organization been in business as a contractor under your present business name? ____________
3. How many years experience in ____________ (Type) construction work has your organization had:
   (a) As a general contractor? ____________ (b) As a subcontractor? ____________
4. Show the projects your organization has completed during at least the last five years in the following tabulation:

YEAR	TYPE OF WORK (See page 1) (Give kind of pavement)	VALUE OF WORK PERFORMED	LOCATION OF WORK	FOR WHOM PERFORMED

5. Have you or your organization, or any officer or partner thereof, failed to complete a contract? ________ If so, give details ____________

* Where prequalification is based on a combination of several organizations, show the experience, equipment, and financial resources of the combined organizations.

[2]

Contractor's statement of experience and financial condition (cont.)

sification "that best suits the contracting business in which he intends to engage," from the following list of contractors' classifications and the related definitions of the respective classifications:

General Engineering Contractor—A

A general engineering contractor is a contractor whose principal contracting business is in connection with fixed works requiring

6. If you have a controlling interest in any firms presently prequalified with this department, show names thereof

7. In what other lines of business are you financially interested?

8. Name the persons with whom you have been associated in business as partners or business associates in each of the last five years.

9. What is the construction experience of the principal individuals of your present organization?

INDIVIDUAL'S NAME	PRESENT POSITION OR OFFICE IN YOUR ORGANIZATION	YEARS OF CONSTRUCTION EXPERIENCE	MAGNITUDE AND TYPE OF WORK	IN WHAT CAPACITY

10. List construction equipment owned by you in the following tabulation. It is preferred that like items be grouped as a single item but items may be detailed. Also, a separate schedule may be attached if desired.

QUANTITY	DESCRIPTION, NAME AND CAPACITY OF ITEMS	AGE OF ITEMS*	PURCHASE PRICE	DEPRECIATION CHARGED OFF	BOOK VALUE

* In this column show range of ages of items of equipment in a group when items are grouped; thus: 1-5 yrs., 2-10 yrs., etc.

[3]

Contractor's statement of experience and financial condition (cont.)

specialized engineering knowledge and skill, including the following divisions or subjects: irrigation, drainage, water power, water supply, flood control, inland waterways, harbors, docks and wharves, shipyards and ports, dams and hydroelectric projects, levees, river control and reclamation works, railroads, highways, streets and roads, tunnels, airports and airways, sewers and sewage disposal plants and systems,

CONTRACTOR'S FINANCIAL STATEMENT

NAME ________________

Condition at close of business ________________ 19___

ASSETS	DETAIL	TOTAL
Current Assets		
1. Cash		
2. Notes receivable		
3. Accounts receivable from completed contracts		
4. Sums earned on incomplete contracts		
5. Other accounts receivable		
6. Advances to construction joint ventures		
7. Materials in stock not included in Item 4		
8. Negotiable securities		
9. Other current assets		
TOTAL		
Fixed and Other Assets		
10. Real estate		
11. Construction plant and equipment		
12. Furniture and fixtures		
13. Investments of a non-current nature		
14. Other non-current assets		
TOTAL		
TOTAL ASSETS		
LIABILITIES AND CAPITAL		
Current Liabilities		
15. Notes payable, exclusive of equipment obligations and real estate encumbrances		
16. Accounts payable		
17. Other current liabilities		
TOTAL		
Other Liabilities and Reserves		
18. Real estate encumbrances		
19. Equipment obligations secured by equipment		
20. Other non-current liabilities		
21. Reserves		
TOTAL		
Capital and Surplus		
22. Capital Stock Paid Up		
23. Surplus (or Net Worth)		
TOTAL		
TOTAL LIABILITIES AND CAPITAL		
CONTINGENT LIABILITIES		
24. Liability on notes receivable, discounted or sold		
25. Liability on accounts receivable, pledged, assigned or sold		
26. Liability as bondsman		
27. Liability as guarantor on contracts or on accounts of others		
28. Other contingent liabilities		
TOTAL CONTINGENT LIABILITIES		

NOTE.—Show details under main headings in first column, extending totals of main headings to second column.

[4]

Contractor's statement of experience and financial condition (cont.)

waste reduction plants, bridges, overpasses, underpasses and other similar works, pipelines and other systems for the transmission of petroleum and other liquid or gaseous substances, parks, playgrounds and other recreational works, refineries, chemical plants and similar industrial plants requiring specialized engineering knowledge and

DETAILS RELATIVE TO ASSETS

1 **Cash:**
(a) On hand ________ $________
(b) Deposited in banks named below ________ $________
(c) Elsewhere—(state where) ________ $________

NAME OF BANK	LOCATION	DEPOSIT IN NAME OF	AMOUNT

2* **Notes Receivable:**
(a) Due within one year ________ $________
(b) Due after one year ________ $________
(c) Past due ________ $________

RECEIVABLE FROM	FOR WHAT	DATE OF MATURITY	HOW SECURED	AMOUNT

Have any of the above been discounted or sold? ________ If so, state amount, to whom, and reason ________

3* **Accounts receivable from completed contracts exclusive of claims not approved for payment $________**

RECEIVABLE FROM	TYPE OF WORK	AMOUNT OF CONTRACT	AMOUNT RECEIVABLE

Have any of the above been assigned, sold or pledged? ________ If so, state amount, to whom, and reason ________

4* **Sums earned on incomplete contracts, as shown by engineers' or architects' estimates ________ $________**

RECEIVABLE FROM	TYPE OF WORK	AMOUNT OF CONTRACT	AMOUNT RECEIVABLE

Have any of the above been assigned, sold or pledged? ________ If so, state amount, to whom, and reason ________

* List separately each item amounting to 10 per cent or more of the total and combine the remainder.

[5]

Contractor's statement of experience and financial condition (cont.)

skill, powerhouses, power plants and other utility plants and installations, mines and metallurgical plants, land leveling and earthmoving projects, excavating, grading, trenching, paving and surfacing work and cement and concrete works in connection with the above mentioned fixed works.

DETAILS RELATIVE TO ASSETS (Continued)

5* **Accounts receivable not from construction contracts** ________ $________

RECEIVABLE FROM	FOR WHAT	WHEN DUE	AMOUNT

What amount, if any, is past due? ________ $________
Assigned, sold, or pledged ________ $________

6 **Advances to construction joint ventures** ________ $________

NAME OF JOINT VENTURE	TYPE OF WORK	AMOUNT

What amount, if any, has been assigned, sold, or pledged? ________ $________

7 **Materials in stock and not included in Item 4**
(a) For use on incomplete contracts (inventory value) ________ $________
(b) For future operations (inventory value) ________ $________
(c) For sale (inventory value) ________ $________

DESCRIPTION	QUANTITY	VALUE		
		FOR INCOMPLETE CONTRACTS	FOR FUTURE OPERATIONS	FOR SALE

What amount, if any, has been assigned, sold, or pledged? ________ $________

8** Negotiable Securities (List non-negotiable items under Item 13)
(a) Listed—present market value ________ $________
(b) Unlisted—present value ________ $________

ISSUING COMPANY	CLASS	QUANTITY	BOOK VALUE		PRESENT VALUE (ACTUAL OR ESTIMATED)	
			UNIT PRICE	AMOUNT	UNIT PRICE	AMOUNT

Who has possession? ________
If any are pledged or in escrow, state for whom and reason ________

Amount pledged or in escrow ________ $________

* List separately each item amounting to 10 per cent or more of the total and combine the remainder.
** IMPORTANT: Items listed under this heading will be given no consideration as working capital unless actual or estimated market value is furnished.

[6]

Contractor's statement of experience and financial condition (cont.)

General Building Contractor—B1

A general building contractor is a contractor whose principal contracting business is in connection with any structure built, being built, or to be built, for the support, shelter and enclosure of persons, animals, chattels or movable property of any kind, requiring in its construction the use of more than two unrelated building trades or crafts, or to do or superintend the whole or any part thereof.

DETAILS RELATIVE TO ASSETS (Continued)

9 **Other current assets**
Bid deposits, prepaid expenses, cash value of life insurance, accrued interest, etc. .... $______

DESCRIPTION	AMOUNT

10* Real estate { (a) Used for business purposes .... $______
Book value { (b) Not used for business purposes .... $______

LOCATION	DESCRIPTION	HELD IN WHOSE NAME	VALUE

11* **Construction plant and equipment** .... $______
(Show details on Page 3, Item 10)

What is your approximate annual income from rental of equipment owned by you, exclusive of such income from associated concerns having same ownership .... $______

12* **Furniture and fixtures** .... $______

13 **Investments of a non-current nature** .... $______

DESCRIPTION	AMOUNT

14 **Other non-current assets** .... $______

DESCRIPTION	AMOUNT

TOTAL ASSETS $______

* Show book value (cost less depreciation) unless an appraisal schedule prepared by an *independent* appraiser is attached; in which case appraised value may be shown.

[7]

Contractor's statement of experience and financial condition (cont.)

Specialty Contractor—B2

A specialty contractor is a contractor whose operations as such are the performance of construction work requiring special skill and whose principal contracting business involves the use of specialized building trades or crafts.

DETAILS RELATIVE TO LIABILITIES

15 Notes Payable, exclusive of equipment obligations and real estate obligations $

TO WHOM PAYABLE	WHAT SECURITY	WHEN DUE	AMOUNT

16* Accounts Payable: (a) Not past due $
(b) Past due $

TO WHOM PAYABLE	FOR WHAT	WHEN DUE	AMOUNT

17 Other current liabilities $
Accrued interest, taxes, insurance, payrolls, etc.

DESCRIPTION	AMOUNT

18 Real estate encumbrances $

19 Equipment obligations secured by equipment: (a) Total payments due within six months $
(b) Total payments due after six months $

TO WHOM PAYABLE	HOW PAYABLE**	AMOUNT

20 Other non-current liabilities $

DESCRIPTION	FOR WHAT	WHEN DUE	AMOUNT

21 Reserves $

DESCRIPTION	AMOUNT

22 Capital stock paid up: (a) Common $
(b) Preferred $

23 Surplus (or Net Worth) $

TOTAL LIABILITIES AND CAPITAL $

** In this space show amount and frequency of installment payments.

* List separately each item amounting to 10 per cent or more of the total and combine the remainder.

[8]

Contractor's statement of experience and financial condition (cont.)

The following are the classifications of specialty contractors:

Boiler, Hot Water Heating and Steam Fitting	C-4
Cabinet and Mill Work	C-6
Cement and Concrete	C-8
Electrical (general)	C-10
Electrical signs	C-45

If a corporation, answer this:

Capital paid in cash, $________

When incorporated________

In what State________

President's name________

Vice President's name________

Secretary's name________

Treasurer's name________

If a copartnership, answer this:

Date of organization________

State whether partnership is general, limited or association

Name and address of each partner:

Where prequalification is based on a combination of organizations, the appropriate affidavits below must be executed for each member of such combination.

AFFIDAVIT FOR INDIVIDUAL

________, *certifies and says: That he is the person submitting the foregoing statement of experience and financial condition; that he has read the same, and that the same is true of his own knowledge; that the statement is for the purpose of inducing the Department of Public Works to supply the submittor with plans and specifications, and that any depository, vendor, or other agency therein named is hereby authorized to supply said Department of Public Works with any information necessary to verify the statement; and that furthermore, should the foregoing statement at any time cease to properly and truly represent his financial condition in any substantial respect, he will refrain from further bidding on State work until he shall have submitted a revised and corrected statement.*

I certify and declare under penalty of perjury that the foregoing is true and correct.

*Subscribed on*________ (date) *at*________ (city), ________ (county), *State of*________

________ (Applicant must sign here)

AFFIDAVIT FOR CO-PARTNERSHIP

________, *certifies and says: That he is a partner of the partnership of*________; *that said partnership submitted the foregoing statement of experience and financial condition; that he has read the same and that the same is true of his own knowledge; that the statement is for the purpose of inducing the Department of Public Works to supply the submittor with plans and specifications, and that any depository, vendor, or other agency therein named is hereby authorized to supply said Department of Public Works with any information necessary to verify the statement; and that furthermore, should the foregoing statement at any time cease to properly and truly represent the financial condition of said firm in any substantial respect, they will refrain from further bidding on State work until they shall have submitted a revised and corrected statement.*

I certify and declare under penalty of perjury that the foregoing is true and correct.

*Subscribed on*________ (date) *at*________ (city), ________ (county), *State of*________

________ (Member of firm must sign here)

________ (Name of firm)

The foregoing statement and affidavit are hereby affirmed.

____

________ (REMAINING MEMBERS OF FIRM SIGN HERE)

[9]

Contractor's statement of experience and financial condition (cont.)

Elevator Installation	C-11
Excavating, Grading, Trenching, Paving, Surfacing	C-12
Fire Protection Engineering	C-16
Flooring (wood)	C-15
Glazing	C-17
House and Building Moving, Wrecking	C-21

AFFIDAVIT FOR CORPORATION

________________, certifies and says: That he is ________________
(Official capacity)

of the ________________, the corporation submitting the foregoing statement of experience and financial condition; that he has read the same, and that the same is true of his own knowledge; that the statement is for the purpose of inducing the Department of Public Works to supply the submittor with plans and specifications, and that any depository, vendor, or other agency therein named is hereby authorized to supply said Department of Public Works with any information necessary to verify the statement; and that furthermore, should the foregoing statement at any time cease to properly and truly represent the financial condition of said corporation in any substantial respect, it will refrain from further bidding on State work until it shall have submitted a revised and corrected statement.

I certify and declare under penalty of perjury that the foregoing is true and correct.

Subscribed on ________ (date) at ________ (city), ________ (county), State of ________

____
(Officer must sign here)

Note.—Use full corporate name and attach corporate seal.

CERTIFICATE OF ACCOUNTANT

STATE OF ________________

We have examined the Financial Statement of ________________ as of ________.
Our examination was made in accordance with generally accepted auditing standards, and accordingly included such tests of the accounting records and such other auditing procedures as we considered necessary in the circumstances.

In our opinion, the accompanying financial statement included on pages ____ to ____, inclusive, sets forth fairly the financial condition of ________________ as of ________________, in conformity with generally accepted accounting principles.

____
(Certified Public Accountant) (Public Accountant)

License No. ________

Special Note to Accountant:
Review instructions on inside of front cover carefully before completing statement.

The above Certificate of Accountant must not be made by any individual who is in the regular employ of the individual, co-partnership or corporation submitting this statement, nor by any individual who is a member of the concern unless he discloses his financial interest therein.

The Certificate of Accountant will be required in all cases where the contractor desires to qualify for work in excess of $50,000. If the contractor does not desire to qualify for work in excess of $50,000, the accountant's certificate need not be filled out.

[10] [illegible] ① SPO

Contractor's statement of experience and financial condition (cont.)

Insulation	C-2
Landscaping	C-27
Lathing	C-26
Masonry	C-29
Ornamental Metals	C-23
Painting and Decorating	C-33
Pipeline	C-34
Plastering	C-35
Plumbing	C-36
Refrigeration	C-38
Roofing	C-39
Sewer, Sewage Disposal, Drain, Cement Pipe Laying	C-42
Sheet Metal	C-43
Steel, Reinforcing	C-50
Steel, Structural	C-51
Swimming Pool	C-53
Tile (ceramic and mosaic)	C-54
Warm Air Heating, Ventilating and Air Conditioning	C-20
Welding	C-60
Well Drilling (water)	C-57
Classified Specialists	C-61

C-61, Classified Specialists, is for a specialty contractor whose operations as such are the performance of construction work requiring a special skill not included in this list.

The general policies of enforcement are stated in the License Board's "Rules and Regulations," as follows:

700. Authority for Law in Police Powers. Recognizing that the source of the authority vested in it by Legislature to adopt rules and regulations to enforce the Contractors' License Law is derived from the police powers granted by the state and federal constitutions, the board declares its general policies in the enforcement of said law as follows:

701. Purpose of Law—Protection, Health and Safety of Public. The board interprets the primary intent of the Legislature in enacting Chapter 9, Division 3 of the Business and Professions Code to be the protection of the health, safety, and general welfare of all those persons dealing with persons engaged in the building contracting vocation, and the affording to such persons of an effective and practical protection against the incompetent, inexperienced, unlawful and fraudulent acts of building contractors with whom they contract. All rules, regulations or orders adopted by the Board or its registrar shall be interpreted and construed in light of the policies announced herein.

The Contractors' State License Board furnishes a form known as "Application for State Contractor's License," which is a questionnaire concerning the applicant's integrity, experience in the classification of contracting work he expects to engage in, and requires a financial statement of assets and liabilities. The "Rules and Regulations" contain a declaration of the policy of the Board "that experience of each applicant for license be sufficient to protect the health, safety and welfare of the public as in these rules provided," and under paragraph 724, the experience requirement is stated as follows:

> Every applicant for a contractor's license must have had, within the last 10 years immediately preceding the filing of the application, not less than four years experience as a journeyman, foreman, supervising employee or contractor in the particular class within which the applicant intends to engage as a contractor.
>
> An applicant who was a former license holder or who was a member of a licensed entity actively engaged in conducting the business of such licensee in the same classification applied for may compute his experience as a journeyman, foreman, supervising employee or contractor in the classification applied for without regard to the ten-year limitation.
>
> An applicant shall not be jeopardized in computing time by his service in the armed forces of the United States during a National Emergency and such length of service may be added to the 10 years mentioned above.
>
> Acceptable technical training in an accredited school will be counted as experience, but in no case will technical training count for more than three years experience.
>
> The required experience shall be possessed by one member of the applicant entity or by a responsible managing employee therefor and he shall be required to take the examination.

The Registrar of the License Board is authorized to adopt procedures for examination of applicants, but in his discretion, the Registrar may waive the written examination for an original license and determine the qualifications of an applicant by an investigation as he may require, provided the applicant meets the following conditions and requirements:

> (a) For a total of four (4) years within the four and one-half (4½) years immediately preceding the date of filing an application the qualifying member of the applicant entity shall have been on record in the official license files of this board as a member of the personnel of a licensee, licensed in good standing in the same classification within which the applicant applies for license.

(b) The qualifying member of the applicant entity shows by such means as the registrar may require that during the period of time prescribed in subsection (a) and as a member of the personnel of a licensee as prescribed in subsection (a) he has been actively engaged in the actual construction phase of the contracting business in the same classification within which the applicant applies for license. (Paragraph 775–*Rules and Regulations*)

The Registrar is also authorized to exercise his discretion of waiving the written examination for an original license, under the following conditions:

(a) When an applicant is an immediate member of the family of a licensee whose license is in good standing, the applicant shall show by such means as the registrar may order, that:

(1) The applicant has been actively engaged in the licensee's business;

(2) The license is required to continue the business in the event of the absence or death of the licensee;

(3) The application is for a license in the same classification or classifications within which the licensee is or was licensed.

(b) When the qualifying member of an applicant entity was, within a period of fifteen (15) months immediately preceding the date of filing an application, the qualifying member of the personnel of a licensee, licensed in good standing in the same classification within which the applicant applies for license. (Paragraph 775.1–*Rules and Regulations*)

A "Business and Professions Code" is published and issued by the License Board and it provides, among other things, a disciplinary proceeding against any contractor for wrongs or omissions.

4.9 Summary of rules and requirements in fifty states

Appendix A presents a summary of the requirements of each state of the United States for prequalification, postqualification, licenses, taxes, and fees that apply to the construction business. These provisions are offered as a guide to such requirements but with the firm recommendation that the contractor or subcontractor investigate the subject himself if he is to avoid pitfalls and penalties in this field, in which only the specialist can be fully informed. The source of the summary is the Bureau of Contract Information, Inc., Washington, D.C. The Bureau states the following as a word of caution with relation to the subject-matter:

"Information contained herein is based on State legislation and regulations in effect at time of publication. While believed to be substantially correct no warranty as to accuracy or completeness is given or implied.

"Municipal and County requirements are *not* covered.

"Before taking any action contractors should check carefully with all administrative agencies concerned, in order to be certain of having complete and up-to-date information on these matters."

chapter 5

Warranty of plans and extra and additional work

Government units are frequently engaged in many contracts for public work, some of which are complicated and conducted on widely separated areas. The many different phases of the contracts require an army of employees, and because of the involved procedures the responsible officials are always seeking ways to protect the public interest. The history of claims in the courts discloses the trouble and losses that are experienced in public construction work due to conditions at the site or in the nature of the work, both above ground and subsurface.

Although there are sometimes cases of negligence in the preparation of the specifications, the conditions caused by the peculiarities of certain projects, areas, and locations provide an incidence of errors that cannot always be foreseen and are unavoidable.

5.1 Contractor's obligations and warranty of plans

Modern specifications for public construction usually include an express agreement by the contractor that he is informed of the conditions relating to the site and to other matters affecting the work to be done; that in preparing his proposal, he relies upon his own personal inspection and investigation, and does not depend upon information that has been furnished or has been made available by the state or the municipality. Frequently the contractor also agrees by the appropriate language in the same specifications that he will make no claim against the state or municipality by reason of the information he received from those sources.

In many jurisdictions the rule has been established that a contractor who bids for public work has the right to depend upon the plans and specifications to indicate the conditions of the job, and, if the actual conditions are found to be otherwise, necessitating extra work and expense, he is entitled to proper compensation. (*United States* v. *Gibbons,* 109 U.S. 200, 1883; *Hollerbach* v. *United States,* 233 U.S. 165, 1914; *Maney* v. *Oklahoma City,* 150 Okla. 77 and 300 P. 642, 1931.) Stated in another way, the courts have said: "Where one party furnishes specifications and plans for a contractor to follow in a construction job, he thereby impliedly warrants their sufficiency for the purpose in view. *Montrose Const. Co.* v. *County of Westchester,* 80 F. 2d 841, 1936; *United States* v. *Spearin,* 248 U.S. 132; *Penn Bridge Co.* v. *City of New Orleans,* 222 F. 737; *MacKnight & Flintic Stone Co.* v. *Mayor of New York,* 160 N.Y. 72."

When the information to the contractor by the state or the municipality is affirmative and definite and there is included in such data a record of subsurface conditions, those representations to the contractor are in the nature of a warranty. The right to compensation is for a breach of contract analogous "to the right of a vendee to elect to retain goods which are not as warranted, and to recover damages for the breach of warranty." (*Hollerbach* v. *United States,* 233 U.S. 165, 1914; *Christie* v. *United States,* 237 U.S. 234, 1915; *United States* v. *L.P. and J. A. Smith,* 256 U.S. 12, 1921; *United States* v. *Atlantic Dredging Co.* 253 U.S. 1, 1920.) These were actions against the government for damages incurred because the character of the material found in making excavations was different from that described in the specifications referred to in the contract. It was held by the courts that the statements in the specifications as to the nature of the material to be excavated, though made honestly and without intent to deceive, must be taken as true and binding on the government, and that loss resulting from a mistaken representation of an essential condition should fall on the government rather than on the contractor. (*E. & F. Const. Co.* v. *Town of Stamford,* 158 At. 551, Conn., 1932.) When a party makes representations about the subject matter of a contract, and about which it is reasonable to suppose he has information, the other party has the right to rely upon those representations. This situation is not necessarily a fraudulent one, but even as an innocent representation it is accepted and acted upon by the party to whom it was furnished as part of the facts and conditions of a business transaction.

A leading decision that is cited frequently in support of applying the rule of representation of existing conditions is *Hollerbach* v. *United States,* 233 U.S. 165, 1914. The contractor agreed to repair a dam, and the specifications stated that the quantities given were approximate only and that each bidder should visit the site to determine the nature of the work. The specifications also set forth: "The dam is now backed for about 50 feet with broken stone, sawdust and sediment to a height within 2 or 3 feet of the crest." The excavations behind the dam were to be made to the bottom. The contractor proceeded with his work and then found that the dam was not backed with the material stated in the specifications, and, also, that from 7 feet below the top to the bottom there was a backing of cribbing of an average height of 4.3 feet of sound logs filled with stone. This condition increased the contractor's cost. The court ruled in favor of the contractor, and pointed out that the specifications spoke with certainty as to a part of the conditions to be encountered by the contractor. "True, the claimants (the contractor) might have penetrated the 7 feet of soft slushy sediment by means which would have discovered the log crib work filled with stones which was concealed below, but the specifications assured them of the character of the material—a matter concerning which the government might be presumed to speak with knowledge and authority. We think the positive statement of the specifications must be taken as true and binding upon the government, and that upon it rather than upon the claimants, must fall the loss resulting from such mistaken representations. We think it would be going quite too far to interpret the general language of the other paragraphs as requiring independent investigation of facts which the specifications furnished by the government as a basis of the contract left no doubt. If the government wished to leave the matter open to the independent investigation of the claimants, it might easily have omitted the specification as to the character of the filling back of the dam. In its positive assertion of the nature of this much of the work it made a representation upon which the claimants had a right to rely without an investigation to prove its falsity. See *United States* v. *Utah, N. & C. Stage Co.,* 199 U.S. 414."

Another long-established decision that has been used consistently as a precedent on this subject is *Christie* v. *United States,* 237 U.S. 234, 1915. The specifications for the construction of three locks and dams provided: "The material to be excavated, as far as known, is shown by borings, drawings of which may be seen at this office, but bidders must inform and satisfy themselves as to the nature of the material." The contractor examined the drawings, which showed "gravel, sand and clay of various descriptions," and showed no other materials. The actual material to be excavated consisted largely of stumps below the surface of the earth, buried logs of cemented sand and gravel—not of the kind described in the drawings—which was more difficult to penetrate and excavate. Borings had been made by the government engineers; obstructions were discovered by the particles broken off and floating to the surface, and most of the obstructions were not shown on the drawings. The boring apparatus was moved elsewhere to a place where the drill could penetrate when an obstruction was encountered, and the later result was shown as if it had been taken at the stake-out place. The court decided in favor of the contractor, and said: "It makes no difference to the legal aspects of the case that the omissions from the records of the results of the borings did not have sinister purpose. There were representations made which were relied upon by the claimants, and properly relied upon by them, as they were positive."

Contractors must determine how much reliance they should place on representations of conditions at the site, especially subsurface conditions. If the information furnished by the governmental agency is upon an express opinion, there is no basis for recovery of damages. A jury verdict may be necessary when a representation by a state or municipality is either one of opinion or of fact. (*McClung Constr. Co.* v. *Muncy,* 65 S.W. 2d 786, Tex., 1933.) In *Emmerson* v. *Hutchinson,* 63 Ill. App. 203, 1896, it was stated: "Indefinite statements, expressions of opinion, conjectural views of cost, expense or value, cannot be made the basis of an action for deceit."

In *Gisel* v. *City of Buffalo,* 48 Hun 615, N.Y., 1888, the contract for constructing a brick and tile sewer provided that the city had estimated the quantities and work, that no claim would be made by the contractor for extra work beyond that enumerated in the contract, and that the contractor would examine the site and assume all risk of variance in the work. In a place where the testing-rod had been sunk, the excavators actually found rock and had to perform extra work. The only representations made by the city were on the profile on file in the city engineer's office; the tests had actually been made, but no rock had been found at the affected places. No compensation was awarded by the court because there was no misrepresentation or fraud.

When the underwater elevations on the contour map that is part of the contract documents are material representations on which the contractor relies, the contractor is entitled to damages for breach of contract, even though there was a provision in the contract by which he agreed that he was fully informed of the conditions and would make no claim by reason of representations made by the state. In *Young Fehlhaber Pile Co. Inc.* v. *State of New York,* 177 Misc. 204, N.Y., 1941, affirmed 265 App. Div. 61, that rule was applied when it was learned after the contract was signed, that the data was from a survey made six years previously and that in the meantime the creek had been dredged. Because of the greater depth, more costly work became necessary, and the contractor sued for damages. The state's defense was based upon the purported waiver of claim in the contract, but the courts rejected it on the ground that the state had created a situation upon which the contractor had relied. The court pointed out: "In the face of such facts, it becomes immaterial whether the bidder had time to take soundings or not, and its omission to do so does not deprive it of its remedy for the wrong committed by the state. *Christie* v. *United States,* 237 U.S. 234; *Cauldwell-Wingate Co.* v. *State of New York,* 276 N.Y. 365. It is only where uncertainty is indicated that the contractor is bound to satisfy himself as to existing conditions or suffer the penalty imposed by the contract terms for such failure. *Foundation Co.* v. *State of New York,* 233 N.Y. 177; *Faber* v. *City of New York,* 222 N.Y. 255; *Weston* v. *State of New York,* 262 N.Y. 46."

The case of *Foundation Co.* v. *State of New York,* 233 N.Y. 177, 1922 illustrates the contractor's obligation when the state's "boring sheet" was not a part of the plans but was furnished to the contractor at his request. The specifications included a waiver of any claim against the state because of erroneous tests or representations concerning the work. The excavation to bedrock exceeded the estimated depth, and the contractor sued the state for the cost of the additional work. The claim was disallowed because the "boring sheet" was not prepared or used as a part of the plans; it was independent information that the state had assembled, and the bidder assumed the risk of its accuracy when he relied upon it.

An instance where the state was held responsible for misleading and imperfect foundation plans is found in the case of *Cauldwell-Wingate Co.* v. *State of New York,* 276 N.Y. 365, 1938. The contract was for the construction of a bridge superstructure that was to be started after another contractor had completed the foundations, and the document provided that the contractor would be held to his own appraisal of site conditions made after visits to the premises. The State described the foundation work in the plans. The superstructure contractor assumed the risk of delay in the work of the foundation contractor, but did not assume any loss due to the State's misleading plans to both contractors. The whole scheme of the foundation construction had to be changed because of an underground pond and other obstacles that were uncovered by the excavating. The Court of Appeals said that the superstructure contractor was not obligated to make borings and was not bound to awareness of the conditions, because it would be unreasonable "to hold that a contractor, whose only work commenced when the foundations were finished, was obliged to make soundings and borings to discover whether the plans and specifications of the State regarding such foundations were true or false. Whether this was McDonald's (the foundation contractor) duty, we need not now determine." The State did pay McDonald the sum of $168,580.47 voluntarily because of its mistake and he also received $29,622.86 through the Court of Claims. The court referred to the *Foundation Co.* case and said: "The contract in the *Foundation* case required foundations to rest on bedrock. The contractor apparently relied upon boring sheets which the court said were not part of the plans and specifications or representations. Even this court was divided. In the case now before us the plans and specifications were an integral part of the contract, furnished 'for information pertaining to work included in the foundation contract'."

5.2 Contractor's procedure

When a contractor discovers that the government unit has misrepresented the conditions or by "constructive fraud" has induced him to enter into the contract, he must consider the application of the rules in such cases and decide what procedure to follow to protect his rights.

The elements of the legal concept of fraud against a public construction contractor are as follows:

1. The government agency must have made a misrepresentation of fact.
2. The fact must have been a material one.
3. The representative of the state or the municipality must have made the representation with knowledge of its falsity, or as being of his own knowledge when he only believed it to be true or was recklessly indifferent as to whether it was true or false.
4. The government agency must have perpetrated the fraud to induce the contractor to rely upon it.
5. The contractor must have believed the fraudulent conditions and relied upon them.
6. The contractor must have been warranted in doing so.
7. It must have been such as to operate to his damage. (Summarized from *166 American Law Reports 938.*)

In the case of *Benjamin Foster Co.* v. *Com.,* 61 N.E. 147, Mass., 1945, the contract was entered into for work on a water supply system, and the lawsuit related to the construction of the embankment for the main dam of the reservoir. A breach of contract was charged by the contractor on the ground that the engineer in charge wrongly interpreted, but within the express authorization in the contract, the "theoretical core line" as marking the limit to which material not suitable for the core should be permitted to penetrate in the normal process of sluicing. Certain markings on the plans supported the interpretation. The contractor contended that he was induced to make the contract by reason of a false misrepresentation or of representations of partial truth which amounted to falsehood, by entries upon the plans and by statements of one of the subordinate engineers, which, in themselves were literally true, to believe that he could build the entire embankment of the modified drift found in the so-called lower borrow area, which was convenient and easily worked. It appeared that the engineer in charge had already decided against that operation and insisted upon other material at a less convenient place and harder to work. The court held that according to the evidence, the contractor did not rely upon the nonexistence of the undisclosed decision as an inducement to enter into the contract. Mere failure to communicate the engineer's decision, the court said, "is not fraud justifying rescission. . . ."

How reasonable is the extent the contractor's reliance upon the plans and specifications is measured largely by the definiteness of the representations in the plans and specifications, the purpose intended by such representations, the opportunity that the contractor had to find out the true conditions for himself, whether the facts were specially known to the contractee, and whether the contract provisions contemplate the positiveness of such representations. (*United States* v. *L. P. and J. A. Smith,* 256 U.S. 12, 1921; *Jahn Contracting Co.* v. *Seattle,* 170 P. 549, Wash., 1918; *Christie* v. *United States,* 237 U.S. 234, 1915; *Maney* v. *Oklahoma City,* 150 Okla. 77 and 300 P. 642, Okla., 1931; *Richmond* v. *I. J. Smith & Co.,* 89 S.E. 123, Va., 1916; *Semper* v. *Duffey,* 227 N.Y. 151, 1919; *Lentilhon* v. *New York,* 102 App. Div. 548, N.Y., 1905.)

When a contractor reasonably relies upon the plans as being sufficient because of representations made by the governmental unit and then discovers a misrepresentation before the contract is completed, he has two choices. He may refuse to continue the work (*United States* v. *Spearin,* 248 U.S. 132, 1918; *Passaic Valley Street Comrs.* v. *Tierney,* 1 F. 2d 304, 1924), or he may complete the project under the contract and then sue to recover the added cost. (*McConnell* v. *Corona City Water Co.,* 85 P. 929, Cal., 1906; *Lentilhon* v. *New York,* 185 N.Y. 548, 1906.) If he does complete the work, the contractor has the right to sue for damages for breach of the contract or he may seek to recover the reasonable value of the extra work. (*Carroll* v. *O'Connor,* 35 N.E. 1006, Ind., 1893.)

5.3 Responsibility for defects in specifications

The general rule in almost all states is that a contractor, when he decides the plans and specifications are guaranteed to be sufficient by the owner, will not be liable, in the absence of his own negligence or warranty, for the sufficiency of the plans and specifications. "A contractor is required to follow the plans and specifications and when he does so, he cannot be held to guarantee that the work performed as required by his contract will be free from defects, or withstand the action of the elements, or that the completed job will accomplish the purpose intended. He is only responsible for improper workmanship or other faults, or defects resulting from his failure to perform . . ." (*Puget Sound Nat. Bank of Tacoma* v. *C. B. Lauch Const. Co.,* 245 P. 2d 800, Idaho, 1952).

An exception to the general rule is disclosed in *Lonergan* v. *San Antonio Loan & T. Co.,* 104 S.W. 1061, Tex., 1907, which involved an uncompleted project. The contractor entered into an agreement with the loan company to erect a building strictly in accordance with the plans and specifications that were prepared by an architect whose "decision is to be final and conclusive on all points." As the construction work neared completion, the building fell down and the contractor abandoned the project. The loan company sued him to recover damages for the breach of the contract. The contractor claimed that he had no obligation under the contract to rebuild the structure and contended that the building did not fall because of any defect in the material or faulty workmanship. He defended his position on the grounds that there were defects and imperfections in the plans and specifications furnished to him and that the architect was not skilled in his work. The court decided against the contractor upon the ground that there was no provision in the contract that the loan company must guarantee the sufficiency of the specifications, even though the architect was to have supervision and control of all the work. These functions of the architect were, in the opinion of the court, "simply for the protection of the owner, who was represented in the execution of the work by the architect, while the builder represented himself. We are of the opinion that (the contractor) having failed to comply with their agreement to construct and complete the building in accordance with the contract and specifications, must be held responsible for the loss, notwithstanding the fact that the house fell by reason of its weakness arising out of the defects in the specifications and without any fault on the part of the builder." The court pointed out that as a matter of law, the specifications are not guaranteed by either party to the other, and if the building falls or is destroyed after the contract is finished, the contractor is not liable for the loss. The decision added: "It has been just as uniformly held, however, that whenever the building or structure has been destroyed by reason of any defect in the work done, or by any accident or any means whatever before the contract has been completed, then the contractor must bear the loss, no matter what might be the occasion thereof, unless it be some wrong done by the owner subsequent to the making of the contract which caused the fall. Liability of the builder does not rest upon a guaranty of the specifications, but upon his failure to perform his contract to complete and deliver the structure."

The decision in the *Lonergan* case, which is reviewed in the preceding paragraph, was cited with approval in *McDaniel* v. *City of Beaumont,* 92 S.W. 2d 552, Tex., 1936. The principal point at issue was the claim by the contractor that the material called for by the specifications for plaster did not produce the results required, that the construction superintendent and the architect knew the specified material was not suitable in the local climate, and that the material was not recommended for use in that area by the manufacturer. The court decided against the contractor and stated that, under the law in Texas, the owner did not warrant the sufficiency of the plaster specifications. Whereas the owner has the right to offer to bidders any kind of plans and specifications, each bidder must decide for himself whether or not he can erect and deliver the structure shown in the plans and produced according to the specifications. "The contractor is obligated to do the work according to the plans and specifications which have been made a part of the contract. By agreeing to construct a building according to plans and specifications furnished, he is deemed to represent that he understands them and impliedly warrants that he can erect the building according to them, and he cannot excuse himself for non-performance on the ground that they are defective."

Although the established rule in Texas was also approved in *City of Houston* v. *L. J. Fuller, Inc.,* 311 S.W. 2d 284, Tex., 1958, the court found in this case that the city was liable for extra work when it became necessary to do the work over after breaks occurred in a new sewer line, because the specified and installed shells were found by the city inspectors to be insufficient for the purpose intended.

In *Bentley* v. *State,* 41 N.W. 338, Wis., 1899, a statute authorized a commission to employ an architect to prepare plans and specifications for the construction of an addition to the State Capitol, and to superintend the work as the representative of the state. The specifications were made a part of the contract upon the award, and when the work was well advanced, a part of the structure fell. The contractor restored the affected portion of the walls and sued the state for the cost of the restoration. The court decided in favor of the contractor and ruled that the state stood as guarantor for the sufficiency of the specifications furnished by the architect. The decision commented: "According to such facts, the state undertook to furnish suitable plans and specifications, and required the plaintiffs to conform thereto, and assumed control and supervision of the execution thereof, and thereby took the risk of their efficiency. What was thus done, or omitted to be done, by the architect, must be deemed to have been done or omitted by the state. Moreover, we must hold, notwithstanding the English case cited—*Thorn* v. *London,* L.R. 1 App. Cas. D.C. 120—that the language of the contract is such as to fairly imply an undertaking on the part of the state that such architect had sufficient learning, experience, skill and judgment to properly perform the work thus required of him, and that such plans, drawings and specifications were suitable and efficient for the purpose designed. There seems to be no lack of able adjudications in support of such conclusions. *Clark* v. *Pope,* 70 Ill. 128; *Daegling* v. *Gilmore,* 49 Ill. 248; *Schwartz* v. *Saunders,* 46 Ill. 18; *Seymour* v. *Long Dock Co.,* 20 N.J. Eq. 396; *Sinnott* v. *Mullin,* 82 Pa. 333; *Smith* v. *Railroad Co.,* 36 N.H. 459; *Railroad Co.* v. *Van Dusen,* 29 Mich. 431; *Burke* v. *Dunbar,* 128 Mass. 499; *Bridge Co.* v. *Hamilton,* 110 U.S. 108."

The great weight of judicial authority is that the act of the owner in furnishing the plans and specifications amounts to a warranty of their fitness and sufficiency and that the contractor is entitled to recover for damages in relying on the warranty in preparing his bid or in performing the work in reliance upon the implied warranty. The general rules of law are stated in *United States* v. *Spearin,* 248 U.S. 132, 1918, as follows: "Where one agrees to do, for a fixed sum, a thing possible to be performed, he will not be excused or become entitled to additional compensation, because unforeseen difficulties are encountered. *Day* v. *United States,* 245 U.S. 159; *Phoenix Bridge Co.* v. *United States,* 211 U.S. 188. Thus, one who undertakes to erect a structure upon a particular site, assumes ordinarily the risk of subsidence of the soil. *Simpson* v. *United States,* 172 U.S. 372; *Dermott* v. *Jones,* 2 Wall. 1. But if the contractor is bound to build according to plans and specifications, prepared by the owner, the contractor will not be responsible for the consequences of defects in the plans and specifications. *MacKnight Flintic Stone Co.* v. *The Mayor,* 160 N.Y. 72, 54 N.E. 661; *Filbert* v. *City of Philadelphia,* 181 Pa. 530, 37 At. 545; *Bentley* v. *State,* 73 Wis. 416, 41 N.W. 338. See *Sundstrom* v. *State of New York,* 213 N.Y. 68, 106 N.E. 924. This responsibility of the owner is not overcome by the usual clauses requiring builders to visit the site to check the plans and to inform themselves of the requirements of the work, as shown by *Christie* v. *United States,* 237

U.S. 234; *Hollerbach* v. *United States,* 233 U.S. 165; and *United States* v. *Utah, etc. Stage Co.,* 199 U.S. 414, where it was held that the contractor should be relieved if he was misled by erroneous statements in the specifications."

In *McCree & Company* v. *State of Minnesota,* 91 N.W. 2d 713, Minn., 1958, the court stated the applicable rule and exception in *Friederick* v. *County of Redwood,* 190 N.W. 801, Minn., 1922, as follows: "Where a contractor makes an absolute and unqualified contract to construct a building or perform a given undertaking it is the general, and perhaps universal, rule that he assumes the risks attending the performance of the contract, and must repair and make good any injury or defect which occurs or develops before the completed work has been delivered to the other party. But where he makes a contract to perform a given undertaking in accordance with prescribed plans and specifications, this rule does not apply. Under such a contract he is not permitted to vary from the prescribed plans and specifications even if he deems them improper and insufficient. . . . Where the contract specifies what he is to do and the manner and method of doing it, and he does the work specified, his engagement is fulfilled and remains liable only for defects resulting from improper workmanship or other fault on his part, unless there be a provision in the contract imposing some other or further obligation."

An unusual illustration of the rule of the risk obligation of the contractor is found in *Garofano Constr. Co.* v. *State of New York,* 206 Misc. 760, N.Y., 1954, affirmed 4 N.Y. 2d 748, where the contractor sued the state for damages that were sustained during the progress of the work. A fire was caused by a truck, which crashed into the timber support of a bridge under construction. The falsework to support the poured concrete was specified to be held in place by steel beams to afford a span without obstructing traffic. Because of restrictions, due to the Korean War, then in effect, it was necessary to revise the plan and use wooden beams with a center support. In the fire the wooden falsework was destroyed, and the bridge structure was damaged. The court, in deciding against the contractor, held that the design prepared by the contractor was not the responsibility of the state, and the contractor was under contractual obligation to deliver a completed and undamaged bridge.

5.4 Meaning of approximate quantities

"Approximate quantities" has been defined by the court to mean not absolutely equal, but contemplates "only a negligible deviation." In the absence of fraud or misrepresentation, the contractor cannot claim that he was misled as to the quantity of a particular item. In *Dunbar & Sullivan Dredging Co.* v. *State of New York,* 259 App. Div. 440, N.Y.,

1940, affirmed 291 N.Y. 652, the decision stated: "A contractor who submits a gross bid for the entire performance of a given work assumes the risk as to the nature and quantity of the work to be performed, even though approximate estimates of the quantities which are materially wrong have been prepared by the public authorities for the guidance of the bidders. *Lentilhon* v. *City of New York,* 102 App. Div. 548. A bidder is not entitled to rely on the plans alone, but is required to resort to the contract and the specifications and he is under the duty to examine the work and make investigations for himself. *Barash* v. *Board of Education,* 226 App. Div. 249, 250, affirmed 255 N.Y. 587."

Where a contract was made for a waterworks basin, the drawings showed "approximate original grade"; the general average excavation was about 9 feet with a fill of about 9 feet, and there was enough spoil from the excavation to make the specified backfill. The contractor located the bottom of the basin to be about 3 feet instead of 9 feet, for which the contractor claimed that he had incurred extra expense in moving the dirt from the nearest available location to make the backfill. The court found in *Pitt Const. Co.* v. *City of Alliance, Ohio,* 12 F. 2d 28, Ohio, 1926, that the drawing plainly showed the distance of 9 feet, and the words "about" and "approximately" were not applicable to the substantial misrepresentation and was a discrepancy for which the city was responsible, not the contractor. The contract expressly required the bidders to examine the site of the work, to make all necessary investigations as to the character of the soil and the difficulties involved in the work. The court rejected as unsound the city's defense that these provisions for examination of the site by the contractor caused him to assume the risk of discrepancy. The "contractor was entitled to accept and to formulate his bid in reliance upon the representation of fact by the other party. In substance and effect we cannot distinguish," the court said, "that case from this. See also *U.S.* v. *Smith,* 256 U.S. 11; *Faber* v. *New York,* 222 N.Y. 255." The opinion by the court also pointed out: "The suit is not brought to recover anything earned under the contract or for extra work of the character contemplated by the contract; it is brought to recover damages for the misrepresentation by which the contract was induced—or, to express the same substance in another form, to recover damages for not furnishing the agreed site. *Bates* v. *Rogers,* 274 F. 659; *Grace Co.* v. *Chesapeake Co.,* 281 F. 904."

Where the court held that a public commission had caused a substantial misrepresentation to be made as to subsurface conditions, it approved an award of damages to the contractor in *Pennsylvania Turnpike Commission* v. *Smith,* 39 At. 2d 139, Pa., 1944. The plans and specifications indicated that the material to be excavated would consist principally of loose earth and approximately 50,000 cubic yards of limestone, which was visible at one end of the section. Each bidder was required by the "Information to Bidders" to visit the site to ascertain the existing conditions and was also informed that the quantities of work and materials to be furnished were approximate, for which the commission did not assume responsibility. Subsurface conditions information was indicated as being based upon usual procedures but with the possibility that actual conditions may differ. The plans showed a factor of fill plus shrink, which indicated that the character of the material to be excavated was mostly loose earth. The engineers for the commission made subsurface investigations and found it to be predominantly rock, which should have been shown on the plans. The contractor first discovered the limestone rock when he began to excavate, whereupon he filed a claim and stated that he had relied upon the subsurface information on the plans because there was insufficient time allowed in which to make his own subsurface investigation. The court pointed out that "declarations in government specifications that no guaranty as to the accuracy of description is intended, and the admonition to bidders that they must decide as to the character of the material to be dredged and make their bids accordingly, does not prevent the contractor from relying upon the government specifications and maps as to the character of the material to be encountered. In *O'Neill Construction Co. Inc.* v. *Philadelphia,* 6 At. 2d 525, this court reviewed *Hollerbach* v. *United States,* 233 U.S. 165; *Christie* v. *United States,* 237 U.S. 234; and *United States* v. *Atlantic Dredging Co.,* 253 U.S. 1, as to their self-exoneration clauses, and states in 6 At. 2d 528: 'There were self-exonerating clauses of substantially similar phraseology involved in *Passaic Valley Sewerage Comrs.* v. *Holbrook, Cabot & Rollins Corpn.,* 6 F. 2d 721; *Pitt Const. Co.* v. *City of Alliance, Ohio,* 12 F. 2d 28; *Jackson* v. *State,* 210 App. Div. 115, N.Y., affirmed 241 N.Y. 563; *Ganley Bros. Inc.* v. *Butler Bros. Bldg. Co.,* 212 N.W. 602. In all of these cases it was held that such provisions afforded no shield against conscious misrepresentation or anything but mere inaccuracies and innocent mistakes.' " The decision added that "In no event is this information to be considered as a part of the contract. Since the plans were, by reference, a part of the contract, this statement was equivalent to notice that the information furnished by the borings was not to be considered as a part of the plans. In effect, therefore, the borings were as much obliterated from the plans as if physically erased, and their appearance thereon was relegated to the status of private memorandum for the city's own purpose."

5.5 Additions to scope of contract

Extra work and additional work: When used in connection with public construction contracts, the term "extra work" means the performance of work and the furnishing of required labor and materials outside and entirely independent of and not necessary to complete the contract or something done or furnished in excess of the requirements of the contract, not contemplated by the parties, and not controlled by the contract. (*Kansas City Bridge Co.* v. *State,* 250 N.W. 343, So. Dak., 1933; *Blair* v. *U.S. et al.,* 66 F. Supp. 405, Ala., 1946.)

"Additional work" is that which results from a change or alteration in the plans concerning work that has to be done under the contract.

A distinction exists between "extra work" and "additional work." "Extra work" usually arises outside of and entirely independent of the contract and not required in its performance, whereas "additional work" is that which is necessarily required in the performance of the contract, not unintentionally omitted from the contract, and not reasonably implied and necessary to the completion of the work. (*15-A Words and Phrases 46; Shields* v. *City of New York,* 84 App. Div. 502, N.Y., 1903; *City of Baton Rouge, La.* v. *Robinson,* 127 F. 2d 693, 1942.) The necessity for "additional work" usually results from a change or alteration in plans concerning work that has to be done under a contract and might arise from conditions that could not be anticipated and that were not open to observation and could not be discovered until the specified work under the contract was actually undertaken.

In *Coryell* v. *Dubois Borough,* 75 At. 25, Pa., 1909, the contract for the construction of a dam classified the work in different items, estimated the approximate quantities, and also stated that additional work would be paid for at the contract price for each item so affected, whereas "extra work," for which no price was fixed in the contract, has to be work not covered by the plans and specifications. According to the estimates, rock bottom would be at a depth of about seven feet, but when rock bottom was not found as expected, the contractor was ordered by the engineer in charge to continue excavating, which resulted in double the amount of the original estimate. The contractor claimed this extra expense should be compensated on the basis of reasonable value and not on the bid price for this classification of work. The court ruled that the extra excavation was, according to the contract provisions, additional work performed at the bid price for that classification. "When a contractor relies upon a bill of estimates furnished by the municipality, the law implies no agreement on the part of the latter that the estimates are correct. The mere fact that the amounts are estimates puts the contractor upon inquiry. If he is misled by any mis-statement of fact in connection with the estimates or by deceit of any kind calculated to induce him to accept the estimates as veritable, he would have ground for equitable relief, but nothing of the kind is here alleged."

The determination of a claim by a contractor for compensation for extra work depends upon the contract; the meaning and intention of the contract provisions, in the light of the definition of the term "extra work," are usually questions of fact. Too, most public construction contracts require compliance by the contractor in accordance with express formalities, as, for example, a written order by the engineer in charge or a written notice by the contractor to the responsible public official before the contractor undertakes to perform the alleged "extra work." An omission of such a formality by the contractor may well defeat his claim for compensation, although, in certain instances, the waiver of a written order by the engineer in charge can obligate the government agency, notwithstanding such waiver.

In a leading case, *United States* v. *Spearin,* 248 U.S. 132, 1918, the court stated the basic rules that apply to a claim for extra work, as follows: "Where one agrees to do, for a fixed sum, a thing possible to be performed, he will not be excused or become entitled to additional compensation, because unforeseen difficulties are encountered. *Day* v. *United States,* 245 U.S. 159; *Phoenix Bridge Co.* v. *United States,* 211 U.S. 188. Thus one who undertakes to erect a structure upon a particular site, assumes ordinarily the risk of subsidence of the soil. *Simpson* v. *United States,* 172 U.S. 372; *Dermott* v. *Jones,* 2 Wall. 1. But if the contractor is bound to build according to plans and specifications prepared by the owner, the contractor will not be responsible for the consequences of defects in the plans and specifications. *MacKnight & Flintic Stone Co.* v. *Mayor of New York,* 160 N.Y. 72; *Filbert* v. *Philadelphia,* 181 Pa. 530; *Bentley* v. *State,* 73 Wis. 416; *Sundstrom* v. *State of New York,* 213 N.Y. 68. This responsibility of the owner is not overcome by the usual clauses requiring builders to visit the site, to check the plans, and inform themselves of the requirements of the work, as is shown by *Christie* v. *United States,* 237 U.S. 234; *Hollerbach* v. *United States,* 233 U.S. 165; and *United States* v. *Utah, etc. Stage Co.,* 199 U.S. 414, 424, where it was held that the contractor should be relieved, if he was misled by erroneous statements in the specifications."

Contractor's obligations and work guarantee: After a contractor had completed a project satisfactorily, but before final acceptance of the contract, an unusual two-week rainfall had so softened the "upper lifts" of an embankment that it was necessary to rework and recompact them to bring them up to the specifications. After the contractor refused to do the work without additional compensation, it was agreed that he would proceed, each party would keep records of cost, and, after completion, it would be decided whether he would be compensated for the extra work. The public authority eventually denied the claim, and the contractor sued to recover for his work. The court held that the contractor was entitled to compensation for the extra work upon the ground that when a contract specifies the manner and method of performance of the work, the contractor is liable only for defects resulting from improper workmanship, although an express provision in the contract to the contrary would impose that liability on the contractor. "When the principal object of the contract is to obtain a result . . . the risk of accomplishing such purpose or result is on the builder" (*Glass* v. *Wiesner,* 238 P. 2d 712, Kan., 1951). Where, however, the contract provides for the performance of a given undertaking in accordance with a prescribed plan and specification, this rule does not apply "because the contractor is not permitted to vary from the prescribed plan and specifications, even if he deems them improper and insufficient; and therefore cannot be held to guarantee that work performed as required by him will be free from defects, or withstand the action of the elements, or accomplish the purpose intended. Where the contract specifies what he is to do and the manner and method of doing it, and he does the work specified in the manner specified, his engagement is fulfilled and he remains liable only for defects resulting from improper workmanship or other fault on his part, unless there be a provision in the contract imposing some other or further obligation. *Friederick* v. *Redwood County,* 190 N.W. 801; *Schliess* v. *City of Grand Rapids,* 90 N.W. 700" (*Kansas Turnpike Authority* v. *Abramson,* 272 F. 2d 711, Kan., 1960).

Variations in plans and extra work: Salt Lake City v. *Smith,* 104 F. 457, Utah, 1900 is a leading case on the subject of compensation for extra work that is created by a variation in the plans. The lawsuit involved a contract for furnishing materials and doing the necessary work, except that work which required excavations, to construct a covered water conduit approximately six miles long. The line of the conduit for the last mile was changed materially by the city engineer after the contractor had started his work, from comparatively level ground to a course of deep ravines and through hills, which required expensive tunnels, concrete lining for the tunnels, laying heavy iron pipes inside the tunnels, and constructing large culverts, for which the contractor asked more than $97,000 for the extra work. The city claimed that the line of the last mile had not been surveyed or located when the bids and the contract were made and that the contractor was fully informed on the subject. The city claimed also that the city engineer had determined the price for the extra work as permitted by the contract and that the contractors had been paid on that basis. There was conflict of testimony relative to the contractor's notice of the absence of a survey or location of the last mile of the conduit, but there were no stakes on the ground to show the line. The court, in commenting on the conduit as shown in the plans and specifications and for which the contract was made, said:

"The stipulation that the contractor shall do such extra work in connection with that described in the agreement as the city engineer and the board of public works may direct is as effectually limited by this fact to such extra work of proportionally small amounts as was necessary to the construction of the contemplated conduit as it would have been if this restriction had been written in the agreement in so many words.... The stipulation that 'the city shall have the right to make any alterations that may hereafter be determined upon as necessary or desirable,' and the contractors shall be paid for increased quantities at the contract prices, is subject to a like limitation. That provision, not unusual in agreements with cities and other corporations, is limited in its meaning and effect, by reason, and by the object of the contract, to such modifications of the contemplated work as do not radically change its nature and cost.... Since they did not contemplate or intend to contract concerning it (i.e. the different nature and cost of the work in the last mile of the project) when they made their agreement, it was new and different work from that covered by their contract and the plaintiff (the contractors) were entitled to recover its reasonable value."

The decision pointed out that the theory applicable to this case was that the city had no right to require the contractors to perform large quantities of work "radically different in its character, nature and cost' from that originally contemplated by the parties when they made their contract," and that, if it had required such work, the contractors were entitled to recover its reasonable value.

"A contract," the court stated, "must be read and interpreted in the light of reason and of the subject contemplated by the parties. The stipulation common to many corporation contracts, that contractors may be required to perform extra work at the price named in the agreement or fixed by an engineer, is limited by the subject matter of the contract to such proportionally small amounts of extra work as may become necessary to the completion of the undertaking contemplated by the parties when the contract was made, and work which does not fall within this limitation is new and different from that covered by the agreement, and the contractor may recover the reasonable value thereof notwithstanding the contract. The customary provisions in such contracts that the corporation or its engineer may make any necessary or desirable alterations in the work, and that the contractor shall receive the contract price or a price fixed by the engineer for the work or materials required by the alteration, is limited in the same way, by the intention of the parties when the contract is made, to such modifications of the work described in the contract as do not radically change its nature and cost. Material quantities of work required by such alteration, that are substantially variant in character and cost from that contemplated by the parties when they made their agreement, constitute new and different work, not governed by the agreement, for which the contractors may recover its reasonable value.... The stipulation in such contracts that all questions, differences, or controversies which may arise between the corporation and the contractor under or in reference to the agreement and the specifications, or the performance or non-performance of the work to which they relate, shall be referred to the engineer, and his decision thereof shall be final and conclusive upon both parties, does not give the engineer jurisdiction to determine that work which is not done under the contract or specifications, and which is not governed by them, was performed under the agreement and is controlled by it, and his decision to that effect is not conclusive upon the parties. Neither an engineer nor a judge who has no jurisdiction of a question can confer jurisdiction of it upon himself by erroneously deciding that he has it."

The principles stated in the *Salt Lake City* case discussed in the previous paragraph are also supported by the following decisions, which are also cited frequently on this subject: *Henderson Bridge Co.* v. *McGrath,* 134 U.S. 260, Ind., 1890; *Wood* v. *Ft. Wayne,* 119 U.S. 312, Ind., 1886; *Wolff* v. *McGavock,* 29 Wis. 290, 1865; and *Fuccy* v. *Coal & Coke Rwy. Co.,* 83 S.W. 301, W. Va., 1914.

A provision in a public construction contract that was based on unit item prices and that permitted the engineer in charge to "increase or diminish the quantities to any extent," was held by the court to mean "within ordinary and reasonable limits and not as authorizing the engineer to eliminate more than half the work." The court in *Drainage Dist. No. 1* v. *Rude,* 21 F. 2d 257, Neb., 1927, said: "If municipal corporations may enter into contracts calling for definite construction and are then permitted to reduce or increase the amount of construction to any extent they see fit, all the safeguards thrown around the making of such contracts become futile; the contract can only be let at an excessive figure, because no contractor can tell how many men or how much equipment or material to furnish, or whether he will be called upon to do the most or the least profitable work involved, and the door is completely open to manipulation and fraud."

Preliminary requirements of a valid claim and justification: Public construction contracts usually provide that extra work is authorized when ordered in a particular manner and when agreed upon before the work is performed. The courts have frequently been requested to declare the meaning of such stipulations. Yet, the preliminary requirements to support a valid claim by a contractor for extra work have been ruled judicially in most states to be reasonable in their nature and are given effect. (*Huntington* v. *Force,* 53 N.E. 443, Ind., 1899; *Watterson* v. *Mayor, etc.,* 61 S.W. 782, Penn., 1901; *Dolman* v. *Board of Comrs. etc.,* 226 P. 240, Kan., 1924; *O'Leary* v. *Board of Port Comrs. etc.,* 91 So. 139, La., 1922; *Orpheum Theater* v. *Kansas City Co.,* 239 S.W. 841, Mo., 1922; *Hoskins* v. *Powder L. and I. Co.,* 176 P. 124, Oregon, 1918; *City of Salisbury* v. *Lynch-McDonald Const. Co.,* 261 S.W. 356, Mo., 1924; *Kinney* v. *Mass. B. & Ins. Co.,* 175 N.Y.S. 398, 1919; *Vaughan Const. Co.* v. *Virginian Ry. Co.,* 103 S.E. 293, W.Va., 1920.) Although the trial of a claim may disclose that the work was really extra work of substantial cost, that cost cannot be recovered if the work order was not issued in the proper manner by the person authorized to issue it. (*Plumley* v. *U.S.,* 226 U.S. 545, 1913.)

The justification for the express stipulations in a construction contract concerning extra work, the manner of ordering it done, and the right of the contractor to payment for the work were discussed in *Wilson* v. *Salt Lake City,* 174 P. 847, Ut., 1918, as follows: "Unjust and exorbitant demands are so often made for extra labor by those undertaking the work of constructing public improvements that the text-writers have frequently taken occasion to comment concerning the claims of contractors for extra labor performed. We quote: 'Municipal corporations have so frequently been defrauded by exorbitant claims for extra work under contracts for public improvements that it has become usual to insert in contracts a provision that the contractor shall not be entitled to compensation for extra work unless it has been ordered in a particular manner.' *19 Ruling Case Law 1077, section 362.*" Judicial recognition has also been given to the right of the contractor to recover damages for any loss he suffers by reason of delay due to the refusal of the engineer or architect in charge of the work to issue the order. (*Baltimore* v. *Clark,* 128 Md. 291, 1916.)

When a contractor found it necessary to dig holes in rock at a site where the estimate indicated shale and he had to construct temporary roads to establish a camp and to provide access for another source of supply because of the variation in the soil conditions, he made a claim for compensation for extra work not included in the contract. The specifications provided for extra work when necessary or desirable, at an agreed price or on a force account basis, to be arranged for before the work was done. The contractor failed to make the necessary arrangements, and as there was no order by the engineer in charge to do the extra work, the court rejected the contractor's claim. (*Kansas City Bridge Co.* v. *State,* 250 N.W. 343, So. Dak., 1933.)

Waiver of requirement for written order: In a claim for payment for extra work on a public construction contract, it appeared by the provisions of the contract that any additions or changes in the work would be authorized in writing, and at a price to be agreed upon beforehand. The board of commissioners in charge of the project retained an architect with full authority to supervise the work. He ordered extra work to be done, but without written directive or agreement beforehand as to price. The contractor made the change as ordered, and upon completion of the building, it was accepted as changed, by the board of commissioners. The court held in favor of the contractor, and the municipality was held to be liable for the extra work on the ground of waiver of the written order. In construing the contract, the court recognized the inferences to be derived naturally from the language of the written document, as well as the intent and the object of the parties in entering into the contract and the result intended to be accomplished. (*Gibson County* v. *Mothwell Iron & Steel Co.,* 24 N.E. 115, Ind., 1890.)

In another case, the owner's representative in charge of the work ordered several changes that increased the cost considerably. The contractor told the owner's representative that he expected to be paid for complying with the order. The owner resisted the contractor's claim for extra work on the ground that such a claim was not supported by the issuance of a written order before the work was started, as called for by the contract. The court ruled that the owner's representative knew when he ordered the extra work that it would not be done without compensation. It was also held that the requirement of a written order was waived by the owner because he assented to and permitted the extra work to be done without objection. (*McLeod* v. *Genius,* 47 N.W. 473, Neb., 1890.)

In *Abells* v. *City of Syracuse,* 7 App. Div. 501, N.Y., 1896, the contract provided for extra work upon a written order at the price already established in the contract or at actual reasonable cost to the contractors, as determined by the representative of the municipality, plus 15 per cent of the cost. It was necessary to construct a vertical wall because abutting property owners resisted any encroachment, whereupon the city engineer with the knowledge and consent of the Superintendent of Public Works, verbally ordered the contractor to proceed with the wall construction. The Common Council ratified the order, accepted the estimate, and ordered the amount to be paid; however, it later attempted, by resolution, to omit the vertical wall from the final estimate because of the omission of a written order. During the trial the city claimed the work of the vertical wall was not permissible as extra work, but was a project for which competitive bids should have been obtained, as required by its charter regulating public construction. The court held that the city had no right to encourage the contractor to do the work, then reap the benefits of the improvement and try to deny payment to the contractor because the municipality did not have the power to proceed as alleged. The provision in the contract was held to have been waived by responsible representatives of the city.

In a contract for the construction of a water works, it was provided that no claim for extra work would be allowed unless it was performed as the result of a written order by the engineer and the trustees of the water works. The plan was altered by the trustees, and the work to accomplish the alteration was authorized by the trustees. This authorization was held by the court to be the equivalent of a written order by the trustees and the engineer. (*Wood* v. *Fort Wayne,* 119 U.S. 312, 1886.)

The waiver of the written order for extra work was involved in a project for a municipal breakwater. The contract required any changes in the work to be performed as the result of a written order. When it was discovered during the work that the quantity of stone had to be increased above that stated in the original contract, the Common Council called upon the Harbor Commission for advice. When the decision was made to proceed with the work, the Commissioner of Public Works made a determination to pay for the extra work, but a taxpayer sought to restrain the city from making the payment. There were no written orders for the extra work; however, after reviewing the action by the city officials and boards, the court held that the requirement for a written order was an administrative provision and not required by law. As such, the city could and did waive the requirement, which was its clear intention according to the discussions and action by its representatives. (*Park* v. *Great Lakes Dredge & Dock Co. et al.,* 192 N.W. 1012, Wis., 1923.)

"Where the owner requested the contractor, without a written order, to furnish a large number of items of work and materials, and the same were furnished by the contractor and accepted by the owner, and part of them were paid for, a provision in a contract that the contractor would make no charge for extra work unless ordered in writing by the owner or its engineer, was waived" (*Douglass & Varnum* v. *Village of Morrisville,* 95 At. 810, Vt., 1915).

Whether or not the omission of a written order for extra work constitutes a waiver of the writing is a question of fact for a jury to determine. Other evidence may be considered to show that, although the work was being performed without a written order, the representatives of the governmental agency disregarded the requirement intentionally. (*Emslie* v. *Livingston,* 51 App. Div. 628, N.Y., 1910.)

A subcontractor on a project for the construction of a lighthouse tower was refused a written agreement as to the value of extra work by the authorized inspector and agent of the contractor on the site, although required by the contract. The subcontractor was successful in his contention because the court held that the contractor had waived the requirement. "The waiver may be express or it may be implied from facts and circumstances, as where the party relying upon the condition has accepted or appropriated the property or fruits of the labor of the other party." The court opinion also pointed out that a party or his authorized agent who neglects or prevents the compliance with such a requirement cannot take advantage of his own wrongful act. (*Moran* v. *Schmidt,* 67 N.W. 323, Mich., 1896.)

A comprehensive and clarifying summary of the principles of the important subject of a waiver of a written order in public construction work is contained in the case of *Campbell Bldg. Co.* v. *State Road Commission,* 70 P. 2d 857, Utah, 1937. The definition of "extra work" in the specifications included two classes: (1) new and unforeseen items not covered by the unit price bid; and (2) additional work of the same class and character covered by the unit bids, but only where the amount is in excess of 20 per cent of the total amount of work listed in the classification. No written orders were issued and no agreement in advance concerning prices was made, but the contractor claimed he was ordered verbally by the engineer in charge to proceed with the items of work; he also alleged that the state had waived a written order when work orders were signed after the items of work were done, rather than before. The court denied the contractor's claim and stated the following rules:

"The general rule is that a provision in a contract that all extra work, in order to be paid for, shall be ordered by the architect or engineer in writing, may be waived by the owner. *Zarthar* v. *Saliba,* 185 N.E. 367, Mass. Whether there has been a waiver is a question of fact to be determined as other facts, dependent on the circumstances of the particular case. 9 C.J. 846; *Massachusetts Bonding & Ins. Co.* v. *Lentz,* 9 P. 2d 408, Ariz.; *Lord Const. Co.* v. *United States,* 28 F. 2d 340.

"An additional question injects itself into the case on account of the state being a party to the contract, and that is whether it is within the competency of the engineer or his assistants to waive for the state any provisions in the written contract. Assuming, without deciding, that the State Road Commission may with the consent of the other party change or modify the written contract or waive any of its provisions inserted for the benefit of the state, may such be done by words or conduct of the state road engineer or any other agent or employee of the State Road Commission? There is no evidence that any of the claimed extra work was authorized by formal action of the State Road Commission or that it was approved as extra work by the Commission after being done. The evidence merely tends to show that it was approved by the engineer or one of his assistants and the work was done by the contractor without any formal protest or request that an order in writing be given it. In each instance the work was measured and paid for by the state on the theory that it was within the terms of the contract and not extra work, and the amount of work done was paid on the basis of unit bid prices.

"There are three questions involved: (1) Whether the work was extra work as defined by the contract; (2) Whether the engineer actually waived or intended to waive for the state the requirement that the order should be in writing; and (3) Whether it is competent for the engineer to waive, even if the facts show that he intended so to waive the contract requirement with regard to written work orders.

"We think the engineer had not authority to waive on behalf of the state the requirements in the written contract. He undoubtedly had no authority to enter into a new or different contract, and it would follow that he had no authority to waive the provisions in this one. The contract specified what his duties and powers were and this was well known to the contractor. It is generally held that an architect or engineer in charge of construction work does not have authority to waive a provision requiring written extra work orders. *Wiley* v. *Hart,* 132 P. 1015, Wash.; *Massachusetts Bonding & Ins. Co.* v. *Lentz,* 9 P. 2d 408, Ariz. The rule is not more liberal with respect to the powers of an engineer acting for the state in the absence of express powers conferred either by the contract or by the statute. The statute empowers the State Road Commission to bind the state by written contracts and it is only on such written contracts that it may be sued. The state cannot be held for the acts of its engineer beyond the powers conferred by law or the written contract. *Clark County Const. Co.* v. *State Highway Comm.,* 58 S.W. 2d 388, Ky.; *California Highway Comm.* v. *Riley,* 218 P. 579,

Cal.; 29 C.J. 610. Any person doing business with the state by way of contract or otherwise must take notice of the limitations on the authority of the officers or agents of the state, since they may act only within the scope of their lawful powers. *Clark County Const. Co.* v. *State Highway Comm.,* 58 S.W. 2d 388, Ky.; *California Highway Comm.* v. *Riley,* 218 P. 579, Cal. The state road engineer cannot waive a provision in the contract that extra work, before it can be paid for, must have been authorized and the prices fixed by a work order in writing. *Kansas City Bridge Co.* v. *State,* 250 N.W. 343, So. Dak.; *Ambaum* v. *State,* 141 P. 314, Wash.; *Crane Const. Co.* v. *Commonwealth,* 195 N.E. 110, Mass.

"With respect to the fact as to whether a waiver was intended, the evidence must be of a clear and satisfactory character and clearly show a distinct agreement that the work be deemed extra work and a definite agreement with the owner to pay extra for such extra work. *Lord Const. Co.* v. *United States,* 28 F. 2d 340; *Bjerkeseth* v. *Lysnes,* 22 P. 2d 660. This is a fact question to be determined by the court or the jury as the case may be. *Hunt* v. *Treka Terrazzo & Mosaic Co.,* 11 P. 2d 521, Okla."

Verbal promise not binding to government agency: The verbal promise by a representative of the government agency or the engineer in charge to pay for extra work does not entitle the contractor to payment when his contract provides for a written order for extra work. Unless there is a valid waiver of the requirement of a written order, the oral promise or direction is not binding on the state or the municipality. (*Contra Costa Constr. Co.* v. *Daly City,* 192 P. 178, Cal., 1920; *Stuart* v. *Cambridge,* 125 Mass. 102, 1878; *O'Brien* v. *Fowler,* 11 At. 174, Md., 1870.) Contractors should not assume that there is a waiver of the requirement of a written order merely because the governmental representative has knowledge that he is doing the extra work, or if that representative acquiesces in the extra work by the contractor. The proof of such a waiver of the requirement of a written order is disclosed by the circumstances in each case, and the basic reasoning in favor of the requirement is the importance of recognizing the written contract. Modification of the contract by parole or oral evidence is not a simple procedure.

Voluntary performance of work: When a contractor performs services and furnishes material not required under the terms of the contract and not ordered as the result of a valid direction from the engineer in charge, the contractor does not have the right to be paid for the services and material as extra work. (*Anderson* v. *State of New York,* 103 Misc. 388, N.Y., 1918.)

The provisions of a highway contract that was governed by a statute required that payment for contingencies would be made only in accordance with the original contract or a supplemental contract. The engineer in charge ordered extra work, but no supplemental contract was made for it. The court denied the claim of the contractor for the extra work, because the conduct of the engineer in charge, or the contractor, or both of them cannot bind the state to an obligation for extra work when the express requirement to create such an obligation is by a supplemental contract. The contractor had the right to refuse to obey the order of the engineer in charge; if he does obey, his act is voluntary and the performance of the work does not obligate the state. (*Stanton* v. *State of New York,* 103 Misc. 221, N.Y., 1918, affirmed 187 App. Div. 963.)

Compliance under protest: When a contractor on public construction work believes that a valid order by the engineer in charge is for extra work, his right to compensation for his services and materials is dependent upon a prompt protest and the assertion of a claim for payment. In a leading case on the subject of compliance by the contractor, but under his protest, the decision in *Borough Const. Co.* v. *City of New York,* 200 N.Y. 149, N.Y., 1910, held that a proper court action lies against the city for breach of the contract. The lawsuit was to recover for extra work performed during construction of a sewer; the engineer in charge ordered the contractor to lay a portion of the sewer above the city datum line in expensive cement and to prepare an elevator for use by city officials when inspecting the work. The engineer's order also required the contractor to illuminate the entire length of the sewer. The contractor protested the order as being outside the contract, but he did as he was ordered and then sued the city for damages for a breach of the contract. It is a settled rule of law that such an action can be maintained by the contractor, and the court pointed out "that within certain limits a contractor who is ordered by the proper representatives of the municipality to furnish materials or to do work as covered by his contract which the former thinks are not called for by such contract, may under protest do as directed and subsequently recover damages because he has been so required even though it should turn out that the contractor was right and that the official had no right to call on him to furnish such materials and do such labor. Decisions of this court have so conclusively established the principle that under such circumstances the contractor may treat the conduct of the municipality acting through its representative as a breach of contract and recover damages, that it is only necessary to summarize these without argument."

Engineer's order and opportunity for collusion: There is room for abuse of the principle that damages may be recovered for a breach of contract when a contractor has been unlawfully directed to furnish materials and do work not included in the contract. Collusion between a contractor and the engineer in charge is possible in situations where the order for the work was never contemplated by the contract documents. On this point, the court said in the *Borough Construction Co.* case that:

"The underlying justice of the principle is that where a municipal representative having authority to speak for it and supposed to be familiar with such matters in apparent good faith and with a show of reason requires a contractor to do certain things as covered by his contract, the contractor although protesting against the requirement ought not to be compelled to refuse obedience and incur the hazard of becoming a defaulter on his contract even though it shall subsequently turn out that he was right and the municipal representative wrong in the dispute. The theory involves the idea that the requirement of the municipal representative finds some reasonable basis in the contract and that the question whether his demand is proper or improper is one which may be the subject of some doubt and debate and in respect of which the contractor might prove to be mistaken if he should refuse to do what was required of him, and there is no justification for applying it when the municipality's representative requires something which is so palpably and manifestly beyond the provisions of the contract that the contractor would not be confronted by any of the legal perils of an erroneous decision if he should refuse to obey.

"These considerations seem to suggest the general rule that where the municipal representative, without collusion and against the contractor's opposition, requires the latter to do something as covered by his contract, and the question whether the thing required is embraced within the contract is fairly debatable and its determination surrounded by doubt, the contractor may comply with the demand under protest and subsequently recover damages even if it turns out that he was right and that the thing was not covered by his contract, and, on the other hand, if the thing required is clearly beyond the limits of the contract the contractor may not even under protest do it and subsequently recover damages. While this rule is only a general one and may not be determinative of every conceivable case, it seems to furnish a test by which to decide phases of the question which will ordinarily present themselves, and it may both be illustrated by and applied to the facts established in this case."

Application to public work: The courts have recognized the injustice that sometimes attends the limitations against recovery for extra work to the extent that, in government work, a contractor may proceed as directed, under protest, and recover the cost. (*Gearty* v. *City of New York,* 171 N.Y. 61, N.Y., 1902.)

In *Todd Dry Dock Eng. & Repair Corp.* v. *City of New York,* 54 F. 2d 490, N.Y., 1931, it was also decided that there is a basis for damages on a breach of contract, and the opinion of the court indicated: "This doctrine is well settled and we have ourselves followed it. *American Pipe Co.* v. *Westchester County,* 292 F. 941. It is confined to public corporations, cities, towns, and the like, and is rather a part of local and municipal, than of general or commercial law. *Detroit* v. *Osborne,* 135 U.S. 492; *American Suiety Co.* v. *Billingham Natl. Bank,* 254 F. 54; *Boston* v. *McGovern,* 292 F. 705, 712. We accept it as controlling, regardless of whether we could ourselves have reached the same result unaided."

The opinion in the *Todd Dry Dock* case pointed out a distinction from the *Borough Construction Co.* case, which was discussed in a previous paragraph, because the demand by the representative of the municipality in the *Borough* case was "flagrantly beyond the specifications," and the installation of the elevator and of the lighting or illuminating of the sewer for the accommodation of the city officials to inspect the work had nothing to do with the sewer contract. "The contractor incurred no legal risk in refusing, however much he might suffer indirectly from the outraged sensibilities of those in charge." Whereas, in the *Todd* case, the disputes concerned the subject matter, and the only question was whether or not the orders from the city to the contractor were within the contract provisions.

When a contractor was ordered by the city's representative to take up and re-lay a section of expensive block pavement, which was installed over a period of sixty days, and the order was given only a few days after the city had issued a certificate of satisfactory performance of the affected part of the work, the contractor protested the order, but laid the pavement again as required. In a suit for breach of contract against the city for an arbitrary exercise of power, the contractor did not seek extra compensation under the contract, but sued for damages for breach of contract, and thus did not waive the claim because he obeyed the engineer's order and made no demand at that time for extra compensation. The court held that when a contractor is ordered by an engineer on public construction to take up a pavement, which was done in this case, the contractor may stop work, claim that he has complied with the contract specifications and sue to recover for labor and materials furnished under the contract plus anticipated profits, or he may replace the pavement as directed and sue to recover on the ground that he was compelled unlawfully to do the work a second time. (*Gearty* v. *City of New York,* 171 N.Y. 61, N.Y., 1902.)

Choice of available procedures: When a government agency and a contractor disagree on an order by the engineer, which the contractor claims is for extra work and not work required by the contract, which of two available rights or remedies does the contractor who complies with the order utilize? Does the contractor waive the right to treat a breach of a contract as a discharge of the contract, or does he retain the right to sue for damages caused by the breach? The weight of authority holds that the contractor may waive his right to consider the contract discharged at that time because the order by the engineer caused a breach of the contract; however, even though he waives that right, he does not waive his right to recover damages caused by the breach of the contract. Therefore, a contractor may comply with the engineer's order, even though the former protests that the order breaches the contract because it relates to work not contemplated by the contract, and by complying, the contractor does not waive his right to recover damages caused by the breach. The same questions arise when there is a dispute between the contracting parties in which the state or municipality feels that the contractor has breached the contract for lack of good performance or for an unauthorized change in the work. The governmental agency could terminate the contract or it could let the contractor go on with the work and then offer to pay for the work that was done, but subject to payment of damages to the state or municipality because the contractor did not do the work as agreed. The waiver of the right to terminate a contract by either of the contracting parties is a question of fact and is based on knowledge of the circumstances in each case. (*Purington Paving Brick Co.* v. *Metropolitan Paving Co.,* 4 F. 2d 676, Mo., 1925; *Sperry & Hutchinson Co.* v. *O'Neill-Adams Co.,* 185 F. 231, N.Y., 1911; *Bennecke* v. *Insurance Co.,* 105 U.S. 355, 1881; *Dunn* v. *Steubing,* 120 N.Y. 232, 1890; *Frankfurt-Barnett Co.* v. *William Prym Co.,* 237 F. 21, N.Y., 1916; *Blair* v. *United States,* 147 F. 2d 840, Ark., 1945.) The rule as to a choice of procedure that is available to both the contractor and the governmental agency, as applicable, is stated in the *Blair* decision as follows: "Where there has been a breach, the party not at fault may elect to continue performance and accept payment from the other party. If he does so, he manifestly waives his right to treat the contract as repudiated, but he does not thereby waive his right to sue for damages sustained because of the breach."

chapter 6

How to read a public construction contract

The formation of public construction contracts consists of an offer and an acceptance, and they are governed by the same principles as contracts between private parties. The offer consists of a proposal by the contractor to perform certain work shown by the drawings and specifications; the acceptance of the offer is the award of a contract by a responsible representative of the governmental body, in accordance with the statutes that govern the procedure.

In practically all jurisdictions, the public construction contract must be in writing, and the form must be guided by the related statutes in the particular locality. Representative public officials are empowered to bind the governmental unit, provided the laws governing the action are complied with; if the formation of a contract does not conform to the established requirements, the result is usually a void document. The contractor must, for his own protection, insure that the required procedure has been followed completely, particularly as to the advertising of the notice to bidders, the award to the lowest responsible bidder, and any other steps prescribed by statute or rule.

6.1 Validity of a public construction contract

There is a general presumption that the awarding officer has complied with the necessary preliminaries including the governing statutes. Within the scope of that presumption, the execution of the contract by the responsible official is, of course, binding upon the governmental unit which he represents; however, the contractor assumes a risk as to the validity of the document. Therefore, it is practical to say that a prospective contractor on public construction work should be informed of the requirements and should assure himself of compliance.

The case of *Atlanta Construction Co.* v. *State of New York,* 103 Misc. 233, 235, N.Y., 1918 discloses significant rulings on the formation of state contracts. The claim against the state arose by reason of a note at the bottom of the plans, which were part of the contract documents. The note stated that the source of supply of item 40 for 7,800 cubic yards of stone was "one-half mile north of station 88 plus 00 and 133 plus 00 on road to be improved." The stone was not available at the place stated, and the longer haul to procure the material cost the contractor almost $4,000 extra. The court held that the note on the plans became a part of the contract. "We are fully aware," said the court, "that these contracts are so drawn as to excuse the State from almost any statement or representation which it may make in reference to matters similar to the one here involved, but we believe that there should be a limit placed on this method of leading a contractor astray to his damage and loss. There is no reason why the English language, when used by officials of the State, should not have the same meaning and significance which is attached to it when used in contracts between individuals. There was no ambiguity in any way about this statement. It was positive, direct and peremptory." The opinion continued that "The State is called upon in contracting with its citizens, to set a standard which for fairness, justice, equity, honesty and plain frank statement of its purpose, without subterfuge or circumlocution, shall be beyond all criticism as being in any way possible of deception. We find as a fact that the language of the contract actually misled the contractor. The contract was perfectly susceptible of the construction the contractor gave it. It was deceived by this act of the State to its loss. Justice requires that the State should make it good."

When an individual enters into a government contract, he is considered to know the limit of authority of the official who is acting as the government representative. (*McMahon* v. *State of New York,* 178 Misc. 865, N.Y., 1942.)

Although the State has received the benefit of work under the contract, the unauthorized acts of the public official do not stop the state from declaring the contract to be invalid. (*People* v. *Santa Clara Lumber Co.,* 213 N.Y. 61, 1914; *Williams Oil Co.* v. *State of New York,* 198 Misc. 907, N.Y., 1950.) If the public officer discovers his error in exercising unlawful authority, he may immediately stop further performance by the contractor and refuse to carry on the project any further. (*People ex rel. Ottman* v. *Comr. of Highways,* 27 Barb. 94, N.Y., 1858, affirmed 30 N.Y. 470.)

When the legislature places limitations or restrictions upon the expenditure of public funds, the state cannot be bound except as provided by the law relating to the use and availability of that money. No state official may waive those limitations or restrictions. In *Belmar Contr. Co.* v. *State of New York,* 233 N.Y. 189, 1922, the court said: "The state has chosen to enact something similar to the statute of frauds for its own protection. Those dealing with it do so knowing this fact and at their own risk. If there is no contract, there is no liability. However inequitable the conduct of the state may be, it has said that it shall only be responsible upon one condition and consequently the claimant must show that the condition has been complied with. Nor may this rule be evaded any more than the provisions of the statute of frauds may be evaded on any theory that while the contract itself is unenforcible the contract to make this contract is valid."

6.2 Government and power to make contracts

In the United States, the powers of sovereignty between the federal government and each of the states are divided. The preamble of the Federal Constitution provides that the national government was established by the people of the United States and was granted enumerated powers either expressly or by necessary implication. However, each state is sovereign as to its own constitutional powers and is independent as to the exercise of such powers.

Although the United States as "a body politic and corporate, capable of attaining the objects for which it was created," is limited in its powers, it does have the power as the general right of sovereignty and capacity, within constitutional powers, to enter into contracts, although not expressly directed or authorized to do so by any legislative act. (*Van Brocklin* v. *Anderson,* 117 U.S. 151, 1886.)

When the United States enters into a contract with its citizens, the federal government is controlled by the same laws that govern the citizen. The obligations that would be implied against the citizens under the same circumstances will be implied against the United States. (*United States* v. *Bostwick,* 94 U.S. 65, 1877.) A government contract is to be interpreted and the intention of the contracting parties is to be ascertained as in contracts between individuals to give it effect in accordance with the terms of the document. (*Hollerbach* v. *United States,* 233 U.S. 165, 1914.)

In *Carr* v. *State,* 26 N.E. 778, Ind., 1891, the court stated, concerning a state contract, that "In entering into the contract, it (the state) laid aside its attributes as a sovereign, and bound itself substantially as one of its citizens does when he enters into a contract." The decision also pointed out: "Its contracts are interpreted as the contracts of individuals are, and the law which measures individual rights and responsibilities measures, with few exceptions, those of a state whenever it enters into an ordinary business contract. *Gray* v. *State* (1880) 72 Ind. 567; *State ex rel. Carpenter* v. *Ralston* (1914) 182 Ind. 150; *Grogan et al.* v. *City of San Francisco* (1861) 18 Cal. 590; *People* v. *Stephens et al.* (1878) 71 N.Y. 527; *Schunnemunk Construction Co.* v. *State of New York* (1918) 116 Misc., N.Y. 770; *Coster* v. *Mayor, etc.* (1871) 43 N.Y. 399; *Hartman* v. *Greenhow,* (1880) 102 U.S. 672."

When a state binds itself to a contract in the same manner as an individual, it eliminates the power to cancel or modify its own contract. (*Hall* v. *Wisconsin,* 103 U.S. 5, 1880.)

In *Davenport* v. *Buffington,* 97 F. 234, 1899, the court said: "Nations, states and municipalities have and exercise two classes of powers,—one governmental, by which they rule their people; the other proprietary or business, by which they carry on their business affairs as legal personalities. The same fundamental principles of justice, of law, and of equity govern them in the exercise of their powers of the latter class which control the acts of private individuals."

It is important for those who enter into such (government) contracts to recognize the following:

"The state is only a corporate name for all the citizens within certain territorial limits. The whole people acting as a public corporation have a right to enter into contracts and make purchases. In doing so, however, they must act through some agency. They may choose to act through the legislature, which is the highest representative authority through which the people can act. And where an agreement is entered into by a state through an act of its general assembly, its terms are to be found in the provisions of the act to which it owes its creation. Its intent can in no other way be ascertained; and whatever of the substance of the contract is therein expressed enters into the obligation, which is mutually binding. The legislature may, however, there being no constitutional restriction, delegate its authority to contract to officers, commissions, boards or committees. And the action of the officers or members of a board in the matter of awarding a contract is usually considered to be the action of the state; their determination is its determination. The law generally requires public officers who are charged with letting contracts for public work to accept the lowest bid therefor, and to make the contract accordingly. According to some decisions when such bidder has fully complied on his part with the requirements of the law, he may by writ of mandamus, compel the officers to make the contract for him. The better doctrine, is that the duties of officers intrusted with the letting of contracts for works of public improvement to the lowest responsible bidder are not duties of a strictly ministerial nature, but involve the exercise of such a degree of official discretion as to place them beyond control of the courts by mandamus" (25 Ruling Case Law 392).

In *People* v. *Stephens et al.,* 71 N.Y. 549, 1878, the court said: "The state in all its contracts and dealings with individuals must be adjudged and abide by the rules which govern in determining the rights of private citizens contracting and dealing with each other. There is not one law for the sovereign, and another for the subject. But, when the sovereign engages in business enterprises and contracts with individuals, whenever the contract, in any form, comes before the courts, the rights and obligations of the contracting parties must be adjusted upon the same principle as if both contracting parties were private persons. Both stand upon equality before the law, and the sovereign is merged in the dealer, contractor and suitor." This decision in New York State was cited with approval in support of the judicial ruling in *Chapman* v. *State,* 38 P. 457, Cal., 1894. The power and capacity of a state to enter into a contract were confirmed in *People ex rel. Graves* v. *Sohmer,* 207 N.Y. 460, 1913.

Counties, cities, towns, and villages are created by the respective state legislatures, and the power to engage in contracts for public improvements is granted by the legislature or is implied from the grant of these powers. (*Heilbrun* v. *Cuthbert,* 23 S.E. 206, Ga., 1895; *Wicks* v. *Lake City,* 208 P. 538, Ut., 1922.)

The general rule that a municipality has only the powers that have been conferred upon it by the state, was commented upon in *Lakota Oil & Gas Co.,* 116 P. 2d 761, Wyo., 1941, as follows: "Some courts, however, are more liberal than others in considering as to what powers are implied. In *Memphis* v. *Memphis Water Co.,* 5 Heisk, Tenn., 495, it was held that the power is inherent in a city to supply itself with water. In Indiana it is held that municipalities have inherent power to provide light and water for public purposes. *Underwood* v. *Fairbanks, Morse & Co.,* 185 N.E. 118, Ind.; *Crawfordville* v. *Braden,* 28 N.E. 849, Ind. . . . Most of the authorities however, do not seem to concede any such inherent powers. . . . *Van Eaton* v. *Town of Sidney,* 231 N.W. 475, Iowa; *Hyatt* v. *Williams,* 84 P. 41, Cal.; *Whiting* v. *Mayor,* 172 N.E. 338, Mass.; *McRae* v. *Concord,* 6 N.E. 2d 366, Mass.; *South Texas Public Service Co.* v. *Jahn,* 7 S.W. 2d 942, Tex."

Generally, the constitutional provisions of a state or the legislative intent in the related statutes confer upon municipalities the power to make contracts for public works for the health, safety, and welfare of the inhabitants.

6.3 Government's power to discontinue a construction contract

As a general rule, while a contract is executory or being performed, a governmental unit has the right to direct that the work be stopped; then it can be subjected to the payment of damages for stopping performance of the contract at that point. When the public agency directs that the work be interrupted and stopped, and the effect of the order is a renunciation or revocation of the contract, the contractor should comply with the directive or order, particularly because the continued performance of the work could be considered to be for the purpose of increasing the contractor's claim for damages. (*Morgan-Garner Elect. Co.* v. *Beelick Knob Coal Co.,* 112 S.E. 587, 590, West Va., 1922; *Gibbons* v. *Bente,* 53 N.W. 756, Minn., 1892; *Hinckley* v. *Pittsburgh Bessemer Steel Co.,* 121 U.S. 264, 1887; *Heaver* v. *Lanahan,* 22 At. 263, Md., 1891; *Comstock* v. *Droney Lumber Co.,* 71 S.E. 255, West Va., 1911.)

A leading case that is referred to with approval in a number of states on the subject of the right of a governmental body to suspend or discontinue a public construction contract is *Lord et al.* v. *Thomas,* 64 N.Y. 107, 1876. The facts in that lawsuit were that commissioners were appointed, pursuant to a statute, to erect a reformatory building, and they entered into a contract with Aldrich for the brick and stone work of the buildings. Aldrich assigned the contract to Lord with the approval of the commissioners, and Lord undertook to perform the work under the contract. Several years later, after foundation walls had been laid, and a portion of the front wall and other work had been done, a new statute was passed, pursuant to which the commissioners were suspended and Thomas was appointed one of the superintending builders. When he took charge of the public institution, he changed the plans and specifications to complete a certain portion of the work, and then proceeded to advertise for proposals as required by the statute relating to his appointment. The proposal was for the erection and completion of the portion of the work for which the new plans and specifications had been prepared. The lawsuit was commenced to restrain Thomas from entering into a new contract under the later statute. The contract held by Lord was inapplicable to the new plans and specifications, and the action by Thomas under the new statute was, in effect, a refusal to proceed under the original plans and specifications, and was also a breach of the contract held by Lord. The court ruled that the intention of the later statute was to provide for an award to the lowest responsible bidder after advertising for proposals and for the completion of the construction at a cost not exceeding the appropriation. The court refused to restrain Thomas from proceeding with the work because he was acting as an agent of the state pursuant to authority conferred by the statute under which he was appointed. In the opinion of the court, which dismissed the action started by Lord, it was pointed out that:

"The State cannot be compelled to proceed with the erection of a public building, or the prosecution of a public work at the instance of a contractor with whom the State has entered into a contract for the erection of a building or the performance of work. The State stands, in this respect, in the same position as an individual, and may at any time abandon an enterprise which it has undertaken and refuse to allow the contractor to proceed, or it may assume the control and do the work embraced in the contract, by its own immediate servants and agents, or enter into a new contract for its performance by other persons, without reference to the contract previously made, and although there has been no default on the part of the contractor. The State in the case supposed would violate the contract, but the obligation of the contract would not be impaired by the refusal of the State to perform it. The original party would have a just claim against the State for any damages sustained by him due to the breach of the contract.... A law of the State suspending or discontinuing a public work or providing for its performance by different agencies from those theretofore employed is not, therefore, subject to any constitutional objection because the change would involve a breach of contract with a contractor with whom it had entered into a contract for doing it."

A contract between the State of Arkansas and a contractor for the construction of a capitol building was involved in the case of *Caldwell* v. *Donaghey,* 156 S.W. 839, Ark., 1913. The building operations progressed until they were interrupted because the legislative body failed to appropriate money for payments under the contract. The contractor boarded up the openings of the uncompleted building, as directed by the commission. The next legislature passed an act discharging the contractor, the architect, and the commissioners, and created a new commission to adjust any controversy with the contractor. Another statute was passed by the same legislature to carry the work forward, appropriate money to pay former contractors, and pay for the new work authorized by the law. The new commission broke the locks on the uncompleted parts of the building, took possession of the premises, and tore out some of the work that was done under the original contract. The contractor claimed he had the right to retain possession until the building was completed according to his contract, and also, that the legislative acts were unconstitutional as they impaired the obligation of the State's contract. The court held that if the new statutes violated the contract, they did not impair the obligation of the contract, and the legislature has the same right as an individual has to violate the contract. The opinion referred to the case of *Lord et al.* v. *Thomas,* 64 N.Y. 107, which is discussed in the preceding paragraph, and stated that the ruling by the New York courts also applies in the *Caldwell* case, to the effect that the State cannot be compelled to proceed with the work of construction, and may at any time abandon the project and refuse to allow the contractor to do the work. The contractor in such a case would have a just claim for damages because of the breach of the contract.

Also, in *McMaster* v. *State of New York,* 108, N.Y. 542, 1888, the State refused to proceed with the construction of five buildings for which a contract had been made pursuant to statutory authority. The court ruled that the State was bound by its contracts, just as an individual would have been bound. It might violate them, but could not repudiate or destroy the obligation of them. In awarding damages to the contractor, the court said: "The contractors were clearly bound to commence performance on their part and to continue to final performance, and it would be quite unreasonable to hold that while they were at all times bound to be ready and able to perform, the state was at liberty at any time for an indefinite period to arrest and abandon performance which might again be resumed at pleasure."

In *Flynn Const. Co. et al.* v. *Leininger,* 257 P. 374, Okla., 1927, the State Highway Commission awarded a contract to Flynn to construct a bridge. The commission had advertised for bids; the plaintiff was the lowest responsible bidder. He proceeded with the work pursuant to the award and completed all the substructure and a portion of the superstructure. The contractor received progress payments, but upon the commission's refusal to pay the balance and its attempt to nullify the contract, the contractor started the action to enforce the payment. The proposal form required that each bidder must state his own date of completion, instead of accepting a date set by the commission. The commission claimed that the contract in this lawsuit, and all similar contracts that were made pursuant to proposal forms prepared by the commission and not showing the required date of completion, were void, notwithstanding the good faith of the parties and the complete absence of any fraudulent practice in the awards. The point of the commission's claim was that the bids, being for different periods of time for completion of the work, were not based or estimated upon the same undertaking and, therefore, were not competitive according to the related statute. Although the court stated its approval of a uniform bidding procedure whereby all bidders would have the same stipulated completion date, it held that this case was an exception to the general rule in competitive public bidding that requires the same completion date for all bidders in order to permit competition on the same basis of contract performance. The opinion emphasized that consideration must be given in this case both to the status of the contract and the advanced stage of the work, and pointed out that: "To nullify and set aside such contract at this advanced stage of completion would be a manifest injustice to plaintiffs and an inevitable loss to the state and inconvenience to the public." The contractor was authorized to complete the contract, and it was held that he was entitled to be paid for the work.

Another illustration of the application of the rule of law relative to the abandonment of a public construction project by the governmental agency is in the case of *Hays* v. *Port of Seattle, et al.,* 251 U.S. 233, 1919. The contractor agreed with the State of Washington to excavate certain waterways in Seattle Harbor for a stated price, in accordance with an 1893 statute. After the contractor had started the work, the public official in charge decided to change the form of a bulkhead which was authorized in the contract. The work had to be suspended until new plans and specifications were prepared, but it was never resumed after the suspension, because each party insisted that it was the other's duty to furnish the new plans and specifications. After long delay, a state statute was passed in 1911 to abandon the project and to vest the title to the lands in the city. The contractor sued the governmental agency to restrain the enforcement of the later statute on the ground that it impaired the obligation of the existing contract. The court ruled that the new statute did not impair the obligation of the contract. The decision stated: "Upon the first constitutional point (the impairment of the contract in violation of section 10 of Article 1 of the Federal Constitution), it is important to note the distinction between a statute that has the effect of violating or repudiating a contract previously made by the State and one that impairs its obligation. Had the legislature of Washington, pending performance or after complete performance by complainant (the contractor), passed an act to alter materially the scope of his contract, to diminish his compensation, or to defeat his lien upon the filled lands, there would no doubt have been an attempted impairment of the obligation. The legislation in question had no such purpose or effect. It simply, after seventeen years of delay without substantial performance of the contract, provided that the project should be abandoned and title to the public lands turned over to the municipality. Supposing the contract had been abandoned by complainant himself or terminated by his long delay, its obligation remained as before, and formed the measure of his right to recover from the State for the damages sustained."

A judicial guide to the procedure a contractor should follow when a public construction contract has been abandoned is found in *Atlantic Bitulithic Co.* v. *Town of Edgewood,* 137 S.E. 223, West Va., 1927. The contract was for furnishing labor and materials to pave certain streets, but the municipality refused to permit the work to be done. The contractor obtained an injunction against interference with his operations by public officials and completed as much work as the money available to the municipality would pay for. He then sued for an additional sum as damages resulting from the delay due to the attempted repudiation of the contract. The court denied the contractor's claim for the additional damages and pointed out that when a contract is repudiated before any work is done under its provisions, the contractor may elect to keep the contract alive for the benefit of both parties and must be ready and able to perform the work, and at the end of the time for complete performance, he may sue under the contract; or he may decide that the repudiation put an end to the contract, depriving him of the opportunity of performance, and sue for the profits that he would have made if the contract had been completed.

A public body or agency can annul a construction contract but cannot destroy the obligation of the contract. (*Danolds* v. *State,* 89 N.Y. 36, 1882.)

An illustration of an impairment of the obligation of a public contract is found in *Northern P.R. Co.* v. *Minnesota,* 208 U.S. 583, 1908. The municipality had formally agreed to maintain a certain viaduct, but municipal legislation was later passed, and the railroad with which the agreement had been made could be forced to maintain the viaduct at its own expense and in accordance with municipal plans. The court held that this legislation added new duties and obligations on the railroad, which was more than a mere repudiation of the municipality's agreement, and the legislative intent was to impair the obligation of the contract. Such a statute is unconstitutional because it creates antagonistic power that affects the contract rights.

When a municipality made a contract giving an exclusive right to furnish water and then constructed its own water system, the repudiation of the contract impaired the obligation of the contract. (*Vicksburg Waterworks Co.* v. *Vicksburg,* 185 U.S. 65, 1902; *Walla Walla* v. *Walla Walla Water Co.,* 172 U.S. 1, 1898.)

6.4 Definition of public works

A judicial definition of "public works" is stated in *Carter* v. *City and County of Denver,* 160 P. 2d 991, Colo., 1945 as follows: " 'Public works,' in connection with contracts respecting such works, is a work in which the state is interested, or every species and character of work done for the public, and for which the taxpaying citizens are liable, or work by or for the state and by or for a municipal corporation and contractors therewith" (35 Words and Phrases 151).

Although the *Century Dictionary* definition of "public works" includes, with judicial approval, "All fixed works constructed for public use, as railways, docks, canals, waterworks, roads, etc." (*Ellis* v. *Grand Rapids,* 82 N.W. 244, Mich., 1900; *Penn Iron Co.* v. *William R. Trigg Co.,* 56 S.E. 329, Va., 1907), there are variations in the meaning of "public works" as provided for in statutes. In Massachusetts, generally, towns do not own the title or the fee to the bed of town highways, and the easement of public travel belongs to the general public. In *McHugh* v. *City of Boston,* 53 N.E. 905, Mass., 1899, the question in the lawsuit was the right of a subcontractor to recover the cost of labor on a public highway from the municipality under a statute that provided that such a right existed for "a person to whom a debt is due for labor performed in constructing any building, sewer, drain, waterworks or other public works owned by a city. . . ." The court ruled that the public way was not owned by the municipality, that the language of the particular statute did not include the public highway, and that the contractor had no right of action against the municipality.

A pumping station for a system of waterworks of a municipality is considered a "public works" project within the meaning of statutory language of "any bridge, street, road, sidewalk, park, public ground, ferry boat or public works" of a city. (*Winters* v. *Duluth,* 84 N.W. 788, Minn., 1901.)

Where the authority of a local board which was created to supervise all public work of a city and to pay for such work out of revenue or special taxes, was challenged, the court decided that the disposal of garbage "whether done by machinery, or hauling and dumping in the Mississippi River, is the performance of public work" (*State* v. *Butler,* 77 S.W. 560, Mo., 1903).

In *Schneck* v. *City of Jeffersonville,* 52 N.E. 212, Ind., 1898, the city raised funds to be used "in aid of public improvements or public works" for constructing a county courthouse and jail when the county seat was transferred to the affected city. The court said that "the clause 'public improvements or public works' cannot be so extended or construed as to authorize the city to render aid, by donation in money or bonds, in locating therein the seat of justice and constructing the necessary county buildings."

6.5 Words and their meaning in public construction contracts

The meaning of the language in a contract is of prime importance in the successful operation and administration of the public contractor's business. The basic rule that must be observed to avoid disastrous results is: *Read the contract before you sign it, and then don't sign until after you have read it.*

It has been stated frequently in the law of contracts that the context of the document discloses the meaning of the terms. *Webster's Unabridged Dictionary* defines "context" as "the weaving together of words in language." This principle of law declares that the legal interpretation of a contract is governed, not by a single phrase or paragraph standing by itself, but by the whole of a document, that is, by its general composition.

When contracting parties are in court because of a dispute or disagreement over the meaning and intent of the contract, the decision of the court usually provides an interpretation of the writing between the parties, whenever the contract provisions are ambiguous. There is no real basis for the general belief by uninformed persons that the court will make a contract for the interested parties. The judge will construe the written contents in a fair and reasonable manner but will not reconstruct the contract provisions or rewrite them to meet situations that developed after the document was formalized.

Public construction contracts are composed of several parts, and the specifications or the form of the agreement usually names the parts that are included in the contract documents, although the parts may not all be assembled in a single binder or cover. For example, the form of agreement used by the New York State Department of Public Works consists of comparatively few provisions, all of which are of major importance; however one paragraph, in particular, identifies the different parts of the contract documents as being inclusive of the published notice to bidders, the information for bidders, the specifications with amendments, the plans and drawings, the form of the agreement, and the surety bonds for completion and for labor and materials. The full obligations of the parties to the contract are included in all of the named parts, and the contractor as well as the governmental representatives are bound by the provisions of all of the parts as evidence of the intention of the parties.

Clear language is necessary so that interpretation or construction of the document will not be required. Disputes will arise, however, as to the meaning of certain words and phrases, which one of the parties claims have another meaning or intention.

Government construction contracts are construed and interpreted in the same way as those "between individuals with a view to ascertaining the intention of the parties and to give it effect accordingly, if that can be done consistently with the terms" (*Hollerbach* v. *U.S.*, 233 U.S. 165, 1914).

6.6 The intention or "meeting of the minds" in the written contract

It is reasonable to assume that the parties in a contract accepted that relationship with honest motives and fair purpose. The practical approach, then, is the initiation and the development of the understanding between the parties; it is essential then, to achieve a firm and positive contract, to write clearly and objectively the terms to be observed and the obligations to be performed by the respective parties. A simple axiom to guide the contract formulation is: *Say what you mean.* Verbal promises and undertakings, and any mental reservations made prior to the execution of the formal contract do not, as a rule, survive after the contract has been written and signed. The probative value of verbal promises depends, too, upon the statutes of frauds that are in effect in most jurisdictions, and which identify as a matter of law, the promises and agreements that must be reduced to writing in order to be binding upon the parties involved.

The subject of the construction of contracts has been reviewed and discussed by the courts in all jurisdictions. The following comment from *Salt Lake City* v. *Smith,* 104 F. 457, 1900, is relevant because of its definitive expressions:

"The purpose of a written contract is to evidence the terms on which the minds of the parties to it met when they made it, and the ascertainment of those terms, and the sense in which the parties to the agreement used them when they agreed to them, is the great desideratum and the true end of all contractual interpretation. The express terms of an agreement may not be abrogated, nullified, or modified by parol testimony; but, when their construction or extent is in question, the meaning of the terms upon which the minds of the parties met when they settled them and their intention in using them must be ascertained, and when ascertained they must prevail in the interpretation of the agreement, however broad or narrow the words in which they are expressed. In the discovery of this meaning, the intention, the situation of the parties, the facts and circumstances which surrounded and necessarily influenced them when they made their contract, the reasonableness of the respective claims under it, and above all, the subject-matter of the agreement and the purpose of its execution, are always conducive to and often as essential and controlling in the true interpretation of the contract as the mere words of its various stipulations.

"These are rules for the construction of contracts which commend themselves to the reason and are established by repeated decisions of the courts, and they must not be permitted to escape attention in the consideration of the contract which this case presents."

The following is a pertinent judicial admonition to read the contract before signing: "When written contracts fail to express the agreement upon which the minds of the parties actually met, by reason of fraud or mutual mistake, equity affords a remedy by way of reformation. Courts of law must enforce written contracts according to the language used by the contracting parties, giving, however, to such language a rational interpretation and one which will, so far as possible, effectuate their mutual intention, and not defeat the object and purpose sought to be accomplished. The intention of the parties as expressed by, and not divorced from their language, is what a court must seek to discover in construing a contract. It is elementary, of course, that, where language is plain and its meaning clear, there is no room for construction. In cases of doubt, however, as to the meaning of the language of contracts, preliminary negotiations, subject-matter, and surrounding circumstances should be considered, not to vary the terms or change the language of the contract, but to enable the court to determine in what sense words of doubtful meaning were used" (*Drainage Dist. No. 1* v. *Rude,* 21 F. 2d 257, Neb., 1927).

The intention of the written contract is challenged, at times, by one of the parties because of the choice of the words in the document, which may hold different meanings to different persons. One judicial guide to an answer to such a challenge is found in *Eighth Ave. Coach Corp.* v. *City of New York,* 286 N.Y. 84, 1941, as follows: "Words considered in isolation may have many and diverse meanings. In a written document the word obtains its meaning from the sentence, the sentence from the paragraph, and the latter from the whole document, all based upon the situation and circumstances existing at its creation."

When a document reaches a court for the purpose of contract interpretation or construction, it is customary for the court to "put itself in the place of the parties when their minds met upon the terms of the agreement, and then from a consideration of the writing itself, its purpose and the circumstances which conditioned its making, and making an endeavor to ascertain what they intended to agree to do, upon what sense or meaning of the terms they used, their minds actually met" (*U.S.F. & G. Co.* v. *Board of Comrs.*, 145 F. 144, 1906).

In *Hawkins* v. *United States,* 96 U.S. 607, 1877, the court said: "Verbal agreements between parties to a written contract, made before or at the time of the execution of the contract, are in general inadmissible to vary its terms or to affect its construction, the rule being that all such verbal agreements are to be considered as merged in the written instrument. But oral agreements subsequently made, on a new and valuable consideration and before the breach of the contract, in cases not falling within the Statute of Frauds, stand upon a different footing; as such agreements may, if not within the Statute of Frauds, have the effect to enlarge the time of performance or may vary any other of the terms, or may waive and discharge it altogether."

Judicial comment on the rules of contract construction with particular reference to the intention of the parties is found in *Harnett Co. Inc.* v. *Thruway Authority,* 3 Misc. 2d 257, N.Y., 1956. The contractor was awarded contracts to construct restaurants at three different locations. At two of the locations, the preparatory work of filling and grading of the site had already been completed by other contractors. At the third site, undergrowth and trees had to be removed. The specifications required the removal of existing trees where indicated on the plans or where the trees would interfere with the work, even though these trees were not indicated on the plans. The Authority paid the contractor to remove muck from the site; in order to perform this extra work under a supplemental agreement it was necessary to remove the trees, but the Authority rejected the contractor's claim for the tree removal. The contractor's claim for damages was made for the cost of additional work in removing more than 700 trees and about three acres of undergrowth. He also contended that the specifications in the contract did not assume that he would remove the trees, but that this work was an extra required by the Authority as necessary to make the grade of the site conform to that shown on the plans. The site plans showed the new grade to be the existing grade. The court held in favor of the contractor and ruled:

"The intention of the parties must be found in the language used to express such intention. (*Hartigan* v. *Casualty Co. of America,* 227 N.Y. 175.) In construing the contract and contract papers before us, the same rules of construction are applicable as between individuals—*Jackson* v. *State,* 210 App. Div. 115, affirmed 241 N.Y. 563; *People ex rel. Graves* v. *Sohmer,* 207 N.Y. 450, motion for reargument denied 208

N.Y. 581, and cases cited therein; and since the contract was drawn by the State of New York, if the language used is capable of more than one construction, the court must resolve all doubts against the person who uses the language, and most beneficially to the promisee. (*Gillet* v. *Bank of America,* 160 N.Y. 549, 555.) If the court finds as a matter of law that the contract is ambiguous, evidence of the intention and acts of the parties plays no part in the decision of the case. Plain and unambiguous words and undisputed facts leave no question of construction except for the court. The conduct of the parties may fix a meaning to words of doubtful import. It may not change the terms of the contract. (*Brainard* v. *New York Central R.R. Co.,* 242 N.Y. 125; *Restatement of Contracts, sec. 235, subd. e.*)

"It is a well-settled rule of law that the practical construction which the parties give to a contract is of great weight in determining what the agreement really is. (*Stewart* v. *Barber,* 182 Misc. 91; *Woolsey* v. *Funke,* 121 N.Y. 87; *Carthage Tissue Paper Mills* v. *Village of Carthage,* 200 N.Y. 1.) In that connection it must be noted that the basic contract does not concern merely the Batavia, N.Y. contract site, (the one involved in this claim) but deals also with the Manchester, N.Y. and East Syracuse, N.Y. contract sites. In connection with the last two enumerated sites no question which had to be litigated arose under the contract with reference to the grading and preparation of the sites, all of the necessary preparatory work at such sites having been done by others than claimant and before the latter started its contract work, denoting a construction of the contract by the parties thereto with reference to the amount of preparation work required and by whom it was to be done in keeping with claimant's contention with reference to the Batavia, N.Y. contract site. To be noted also is the fact that claimant was given a supplemental agreement on the Batavia, N.Y. contract site which we have hereinbefore discussed; and that in order that the work provided for by the supplemental contract could be done, it was necessary to remove the aforementioned trees and undergrowth and unless said removal is contemplated by the provisions of the basic contract, said item, too, should have been allowed as an extra and should have been included in a supplemental agreement. . . .

"The Authority stresses the point that if claimant's president had visited the Batavia, N.Y. contract site before making and placing its bid, he would have seen the site as it actually was and would have acted accordingly and on the basis thereof. A complete answer to that is that the grade and elevation of the site as set forth in the drawings having been found to be representations as hereinbefore concluded, upon this record the Authority cannot be heard to say it is not responsible. (*Faber*

v. *City of New York,* 222 N.Y. 255.) Nor, under the circumstances herein, is the Authority relieved from liability for its failure to furnish claimant a site capable of receiving the work to be performed by claimant under its contract. (*A. W. Banko, Inc.* v. *State of New York,* 186 Misc. 491, 496.)"

In *Lyman-Richey Sand & Gravel Co.* v. *State,* 243 N.W. 891, Neb., 1932, the court stated: "It may be accepted as a universal rule that a contract upon a printed form prepared by and regularly used by the state for such highway contracts and containing a printed condition of doubtful meaning, such disputed provision should be construed more strictly against the party who prepares and furnishes such form of contract than against the other party."

In a case where the specifications for floating equipment required each bidder to state a price for constructing and delivering scows, the contractor who received the award added 2½ per cent for sales tax when he submitted his bill; the state refused to pay it. The contract did not mention sales tax, and the question before the court was whether or not the buyer is liable to the seller for sales tax although the contract made no provision for it. The court held that the particular sales tax was imposed on the retailer and not the consumer, and the retailer had the option of collecting or of not collecting it from the consumer. The bid was for a stated price without mention of the sales tax, and it was decided by the court that the intention was to pay the price stated in the bid. (*Pacific Coast Engr. Co.* v. *State,* 244 P. 2d 21, Cal., 1952.)

A written instrument containing uncertain or ambiguous provisions will be held most strongly against the party who drew it and most favorably to the other party in order to avoid placing one party at the mercy of the other one. (*Wilson & English Const. Co.* v. *N.Y.C.R.R. Co.,* 240 App. Div. 479, N.Y., 1934; *Bintz* v. *City of Hornell,* 268 App. Div. 742, N.Y., 1942, affirmed 295 N.Y. 628; *M. Barash* v. *State of New York,* 2 Misc. 2d 680, N.Y., 1956.)

The meaning of clear and unambiguous language in a contract is not subject to extraneous facts but is to be determined by such language. This rule holds true even though the result is harsh and unreasonable, provided the meaning of the words used in the writing discloses a clear intention of the parties. (*Matter of Delaware Co. Elec. Corp.* v. *City of New York,* 257 App. Div. 526, N.Y., 1951, affirmed 304 N.Y. 196.)

Any variance in the meaning of words is also subject to the rule that special clauses or specific clauses will prevail over general clauses. This rule is intended to aid in determining the intention of the parties when they executed the document and it points out the requirement of considering the whole contract. (*English* v. *Shelby,* 172 S.W. 817, Ark, 1915.)

6.7 Choice of words and usage and custom

Generally, proof is not admissible to show a custom or usage that is not consistent with the language in a contract and that contradicts that language either expressly or by necessary implication. An exception to this general rule exists, however, when the meaning of the language in a contract is not ascertainable unless outside evidence of usage and custom is furnished to explain the meaning and intention of the parties. The proof may be either written or verbal evidence on the theory that the parties knew of the existence of the custom and contracted in reference to the accepted usage of the particular language in a contract. (*Robinson* v. *United States,* 80 U.S. 363, 1871.) This rule was followed in *Hostetter* v. *Park,* 137 U.S. 30, 1890, in which the court said: "It is well settled that parties who contract on a subject matter concerning which known usages prevail, incorporate such usages by implication into their agreements, if nothing is said to the contrary."

When a contract called for marble slabs two inches thick, the material when polished and shipped was not of the thickness specified in the contract. The producer of the material showed that the marble pieces when sawed were of the specified thickness before polishing, and that, in the trade, a specified thickness means the thickness of the slabs as sawed, before polishing. The court in the case of *Evans* v. *Western Brass Mfg. Co.,* 24 S.W. 175, Mo., 1893 held in favor of the producer of the marble and stated that "The general rule undoubtedly is that parol evidence cannot be admitted to contradict, add to or vary a written contract, and it is the duty of the court to construe the writing. . . . But it is equally well settled that proof of usage is often admitted to interpret the meaning of the language used, for under many circumstances the parties may be supposed to contract with reference to a usage or custom, as they are presumed to use words in their ordinary signification." Whereas usage cannot be permitted to introduce an inconsistency with the contract provisions, the court commented: "But it does not follow that evidence of usage can only be received where the words of the contract are ambiguous. Such evidence is often received to show that words are used in a sense different from their ordinary meaning." This statement was based "on the theory that the parties knew of the usage or custom, and contracted in reference to it, and in such cases the evidence does not add to or contradict the language used, but simply interprets and explains its meaning."

In an action on a contract to sell and deliver tungsten powder "free from copper, tin and all other impurities," it was shown that the material contained tin. The seller introduced evidence to show that by custom or usage of the trade these words meant that the amount of tin was not sufficient to impair the quality of the finished product. The decision pointed out that general custom in a trade or business established for long enough time to become known to those in the trade or business is binding upon contracting parties, unless the contract stipulates that its terms and provisions are without regard to usage and custom. (*Electric Reduction Co.* v. *Colonial Steel Co.,* 120 At. 116, Pa., 1923.)

The case of *Frye* v. *State of New York,* 192 Misc. 264, N.Y., 1948, involved the meaning of the specifications, which stated: "New conductor piping shall be of galvanized extra strong mild steel pipe with standard weight cast iron fittings." There was no provision whether the fittings should be "screwed" or "caulked," but the State ordered the installation of screwed fittings instead of standard weight cast iron caulked fittings, for which the contractor claimed damages for extra labor and material costs. The defense of the State was that the fittings must be the screwed type because caulked fittings on steel pipes do not permit sufficient room in the bells to caulk the joints. The State also contended that it is a custom or usage of the trade to use screwed fittings to meet the specifications as written in this contract. The court said:

"Certainty is an essential element in all contracts, otherwise it is impossible to determine as to what the parties have agreed. The fact that extrinsic evidence must be resorted to to remove a latent ambiguity or obscurity will not render the contract invalid (*McIntosh* v. *Miner,* 53 App. Div. 240; *Fish* v. *Hubbard,* 21 Wend. 651; *Lent* v. *Hodgman,* 15 Barb. 274; *Newhall* v. *Appleton,* 114 N.Y. 140; *Hart* v. *Thompson,* 10 App. Div. 183; *Barnard Bakeshops* v. *Dirig,* 173 Misc. 862), nor will the use of an indefinite phrase, if it can be given a reasonable and just limitation, when taken in connection with the subject matter of the contract. So language, which to some extent may be ambiguous, can often be made certain by reference to the intent of the parties, the purpose sought to be effected, and the means employed to that end. (*Heisel* v. *Volkmann,* 55 App. Div. 607.)

"Every legal contract is to be interpreted in accordance with the intention of the parties making it, and custom or usage (with the limitations hereinafter noted), when it is reasonable, uniform, well settled, not in opposition to fixed rules of law, not in contradiction of the express terms of the contract, is deemed to form a part of the contract and is deemed to enter into the intention of the parties. Parties are held to contract in reference to the law of the State in which they reside, for all men, being bound to know the law, are presumed beyond dispute, to contract in reference to it. And so they are presumed to contract in reference to the usage of the particular place or trade in or as to which they enter into agreement, when it is so far established and so far known to the parties that it must be supposed that their contract was made in reference to it. Evidence of usage is received when a written contract is under consideration to explain expressions used in a particular sense, by particular persons, as to particular subjects, to give effect to language in a contract as it was understood by those who made use of it. (*Walls* v. *Bailey,* 49 N.Y. 464; *Rickerson* v. *Hartford Fire Ins. Co.,* 149 N.Y. 307; *Fox Film Corp.* v. *Springer,* 273 N.Y. 434; *Gravenhorst* v. *Zimmerman,* 236 N.Y. 22.) Usage or custom may also render sufficiently certain a phrase otherwise indefinite in its meaning (*McIntosh* v. *Miner, supra*), and may explain and make definite a writing otherwise vague and of doubtful meaning. (*Stulsaft* v. *Mercer Tube & Mfg. Co.,* 288 N.Y. 255.)

"A party, however, cannot be bound by usage unless he either knows or has reason to know of its existence and nature. Accordingly, one who seeks either to define language or to annex a term to a contract by means of usage must show either that the other party is actually aware of the usage, or that the existence of the usage in the business to which the transaction relates is so notorious that a person of ordinary prudence in the exercise of reasonable care would be aware of it. If so notorious, actual knowledge of it is immaterial.

"It is a question of fact whether a party has reason to know that the other party intends his words or other acts to be governed by a usage. The burden of establishing that such is the case is on the party so asserting. Though the question is one of fact, the existence of usage, like other facts, may be so well known that a court will take judicial cognizance of it. (*Restatement of Law of Contracts, sec. 247, comment b; Walls* v. *Bailey, supra; Rickerson* v. *Hartford Fire Ins. Co. supra.*) Usage is a matter of fact, not of opinion and it is to be established or negatived in all of its essentials, as well as to knowledge as to any other, by the same character and weight of evidence as are necessary to maintain other allegations of fact. It may be established by presumptive, as well as by direct evidence. Nor, on the other hand, is it exempt from the difficulty that a presumption may not prevail as against direct evidence to the contrary of it. The jury may presume, from all the circumstances of the case, that knowledge or notice existed. (*Walls* v. *Bailey, supra; Rickerson* v. *Hartford Fire Ins. Co., supra.*)

"Having in mind the preceding discussion, the admission herein of evidence of trade usage, which evidence was offered by the State, was proper under the circumstances, and we have considered and weighed it in arriving at our decision herein. We are of the opinion, however, that the attempt of the State to establish the custom or trade usage hereinbefore referred to has fallen far short of the mark required by the cases. No custom or trade usage was shown herein with the strength or clarity which the rules relating to custom or trade usage require. (*Gravenhorst* v. *Zimmerman, supra; Walls* v. *Bailey, supra; O'Donohue* v. *Leggett,* 134 N.Y. 40; *Fairchild Engine & Airplane Corp.* v. *Cox,* 50 N.Y.S. 2d 643; *Wise & Co. Inc.,* v. *Wecoline Products, Inc.* 286 N.Y. 365.)

"As to the contention of the State that standard weight cast iron fittings under the specifications herein must be screwed fittings because to use standard weight cast iron caulked fittings on steel pipes leaves insufficient room in the bells to caulk the joints, there is uncontradicted evidence in the case to the effect that two such 'caulked' joints were made and installed by claimant's employees during the course of the work under the instant contract, and apparently were inspected, passed and accepted by the State's inspector stationed on the job. Furthermore, there is proof in the record that claimants and their employees at other times and on other contracts had made and installed caulked fittings under circumstances similar to those of the instant case. We are of the opinion that the State has not established its contention hereinbefore mentioned by a fair preponderance of the credible evidence."

Where a contract called for the installation of pipe with a stipulated pressure resistance and also designates the required material to be "extra heavy" pipe, the contractor is not entitled to recover the cost of providing the material that was expressly specified; there is no need in such a case to explain the custom of the trade when such language is used. In *Armstrong* v. *State of New York,* 111 Misc. 297, N.Y., 1920, the State Architect ordered the contractor to remove the pipe that did not meet the specifications and to install the stipulated "extra heavy" pipe. The contractor offered evidence that, in the usage and custom of the trade, the words "extra heavy" pipe do not mean anything. This evidence was held by the court to be an attempt to change the express terms of a contract. "It has been repeatedly held" said the court "that the usage in relation to matters embraced in a contract when it is reasonable, uniform, well settled, not in opposition to fixed rules of law and not in contradiction of the express terms of the contract, and when it is so far established and known to the parties that it may be supposed the contract was made in reference there, is deemed to form a part of it; and evidence is always permissible to explain the meaning usage has given to words or terms as used in a particular trade or business. But it is not competent to prove a custom or usage inconsistent with the express terms of a contract. Custom or usage is only resorted to, to explain the intent of the parties to a contract, when it cannot be ascertained without extrinsic evidence; never to contravene express stipulations. Where the terms of a contract are expressed in clear language, custom or usage of the trade cannot be resorted to or permitted to change the terms of contract. *Collender* v. *Dinsmore,* 55 N.Y. 200; *Newhall* v. *Appleton,* 114 id. 140; *McIntosh* v. *Pendleton,* 75 App. Div. 621; *Howell* v. *Dimock,* 15 id. 102."

In *Lenart Constructors, Inc.* v. *State of New York,* 6 Misc. 2d 473, N.Y., 1957, the contract required the contractor to obtain permission from the State to subcontract for any portion of the work, and the provisions expressly stated that no consent would be given for more than 50 per cent of the total original contract value. In a dispute after the State refused to grant permission for certain subcontracting, the court ruled that such a restrictive provision is valid in a contract, and, the opinion added, there was no implied obligation on the State to consent to delegation of performance, notwithstanding the custom of the industry to require subcontracting. "No such requirement may be implied as a provision of this contract on the evidence presented. The restrictions on delegation of performance were clear and unambiguous and voluntarily entered into by the claimant."

6.8 The meaning of the term "working days"

The meaning of the term "working days" was discussed in *Mumm Contr. Co.* v. *Village of Kenmore,* 104 Misc. 268, N.Y., 1918. The contract called for the construction of a sewer to be started on a certain date and completed on or before the expiration of one hundred working days. The contractor claimed that it meant those days when the work could be done properly, weather conditions considered. The Village officers stated that the term meant calendar days, Sundays and holidays excepted. One hundred calendar days, not counting Sundays and holidays, would carry the time to February 13, 1917; the contract was actually completed on April 29, 1917.

The court said: "No determination is found in the courts of this state as to the meaning of the term working days, but generally speaking it means the days occupied in employment as distinguished from holidays and Sundays. This, however, is not conclusive, because in some lines of business it is disregarded. Night shifts are common in many occupations, so that the meaning of the term must be the *time* employed within twenty-four hours, whether day or night. In maritime law it has two meanings. While at sea all days are alike regardless of holidays or Sundays, because in the very nature of the business seamen are on duty and subject to call at all times. But when in port, holidays and Sundays are in many cases respected because the work of loading and unloading depends upon the usages of the men ashore, (*Hagerman* v. *Norton,* 105 F. 996) but weather conditions are usually not considered. In *Houghe* v. *Woodruff,* 19 F. 136, it was held that working days allowed for unloading did not include rainy days, in the unloading of a cargo of salt. It is seen, therefore, that no general rule can be adopted, and that consideration should be given to the circumstances surrounding a given case. Following these suggestions I am of the opinion that in the present case the term working days means days when the work of excavation for and the construction of a sewer could be reasonably expected to proceed, weather conditions in this severe winter climate being considered."

Under the provisions of the Labor Law of the State of New York (section 220) "eight hours shall constitute a legal day's work for all classes of employees in this state except those engaged in farm and domestic service." Of equal importance and also in the same section 220, is the declaration that each contract "to which the state or a public benefit corporation or a municipal corporation or a commission appointed pursuant to law is a party" shall stipulate that no workman be permitted or required to work more than eight hours in any one calendar day or more than five days in any one week. Extraordinary emergency as defined in section 220 does permit the commissioner of labor to grant a dispensation for additional hours, but at premium wages. The usual dispensation authorizes work up to ten hours per day for six days in any one week.

Section 2143 of the Penal Law of the State of New York provides that all labor on Sunday is prohibited excepting the "works of necessity and charity. In works of necessity or charity is included whatever is needful during the day for the good order, health or comfort of the community."

It does seem reasonable because of the provisions of the statutes above mentioned, that in the absence of a definition of the term in the contract itself, a "working day" means the period of eight hours in any one calendar day with a limit of five days in any one week, exclusive of Sundays and subject to weather conditions. The employment of extra shifts, the granting of a dispensation by the Industrial Commissioner, and work on Sundays, are not usually considered and determined before the contract is entered into, provided, of course, the language of the contract otherwise covers the period of work.

The term "calendar day" is defined by section 19 of the General Construction Law of New York State as follows: "A calendar day includes the time from midnight to midnight. Sunday or any day of the week specifically mentioned means a calendar day."

6.9 Validity of decisions by engineer in charge

Government construction contracts usually authorize the public contracting officer to make decisions of questions of fact, and, in each instance, the language of the contract is the guide as to whether a particular decision is one of fact or one of legal construction. The following pertinent comment is found in 137 American Law Reports 537: "There is no question that parties to a contract are competent to make a stipulation of this kind and its provisions when made are binding upon them. But the competency of the parties to so stipulate, as the courts have many times pointed out, is limited to the decision of questions of fact arising under the contract, such as the quantity and quality of materials delivered, whether the work performed meets contract requirements, causes of delay in the performance of the work, etc. These are questions of fact, the correct solution of which is usually largely dependent on the professional knowledge and skill of the engineer. They are questions which the parties to a contract may properly submit to the determination of the contracting officer or head of the department, and lawfully agree to be bound by his decision." A disputed question of law – the proper construction of the contract – is outside the jurisdiction of the contracting officer or head of the department and is a matter for the courts to determine as to the law of the contract.

Where a contract for road construction provided: "The decision of the State Highway Engineer upon any question connected with the execution of this agreement . . . shall be final and conclusive," a dispute arose to the meaning of the term "aggregate," as one of the materials to be used. Would it permit the use of stone aggregate or some other kind, as slag? The term "aggregate" was the general term used in the contract, while the standard specifications, which were a part of the contract by reference, contained the term "slag" in connection with "construction methods." The engineer ordered slag, which was more expensive, and the contractor sued for the difference in cost to him. The court ruled that the contract provisions authorized the engineer to make the decision, and in the absence of bad faith, fraud or gross mistake, his decision is final and conclusive. The court said: "There are many decisions to the effect that where parties to a building or construction contract designate a person who is authorized to determine questions relating to its execution and stipulate that his decision shall be binding and conclusive, both parties are bound by his determination of *those matters which he is authorized by the contract to determine,* except in case of such gross mistake as would necessarily imply bad faith, or a failure to exercise an honest judgment." The court pointed out that "decisions by the United States Supreme Court and the courts of 26 states are listed as subscribing to the doctrine, and only the courts of Indiana are mentioned as holding to the contrary," and concluded that the rule is "of almost universal acceptation" (*State Highway Dept.* v. *MacDougald Constr. Co.,* 6 S.E. 2d 570, Ga., 1939). The fact that the designated umpire is an officer or employee of one of the contracting parties does not invalidate the decision, as the parties to the contract were aware of the relationship when they expressly designated the umpire. (*Edwards* v. *Hartshorn,* 82 P. 520, Kan., 1905; *Norcross* v. *Wyman,* 72 N.E. 347, Mass., 1904.) The parties, in an effort to avoid litigation, anticipated that disputes might arise and made the engineer or architect the exclusive arbitrator of their differences. (*Empson Packing Co.* v. *Clawson,* 95 P. 346, Colo., 1908; *Boettler* v. *Tendick,* 11 S.W. 497, Tex., 1889; *Marsch* v. *Southern New England R. Corp.,* 120 N.E. 120, Mass., 1918.)

Where the language of the requirements of the contract is plain and unmistakable, the stipulation as to a decision by the engineer or architect is not applicable because a decision upon that basis could result in a new contract, which would not be proper or binding. A stipulation of this nature is subject to strict construction.

There is a general rule that the final certificate of the engineer is binding and conclusive upon the contractor, unless there is a showing of fraud, bad faith, or mistake on the face of the certificate. In the case of *Uvalde Contracting Co.* v. *City of New York,* 160 App. Div. 284, N.Y., 1914, it was held that the rule does not apply in every case, particularly where the engineer has erroneously interpreted the contract. In the *Uvalde* case, the engineer ordered the contractor to overlap the application of pavement asphalt, and when the final certificate was prepared by the engineer, he included the actual area of the work but eliminated the substantial amount of overlapping. The decision quoted the distinguishing rule from the case of *Burke* v. *Mayor,* 7 App. Div. 128, N.Y., 1896, as follows: "But we fail to find any case which holds that where the engineer has, upon an erroneous construction of the contract and of the rights of the parties thereunder, deliberately excluded from his final certificate work which has been done by the contractor, although he may have acted with an honest purpose, the contractor is precluded from showing that he has done such work, and that it is included in the terms of his contract. . . . It is undoubtedly true that where there is any dispute in regard to the work or its character, the certificate is final if honestly given. But where a contract calls for the performance of work which the contractor has done, and the engineer, upon an erroneous construction of the contract, has excluded it from his final certificate, it is clear that the contractor has a right to recover, notwithstanding the provisions of the contract in regard to the final certificate."

6.10 Third party beneficiaries

Contractor's obligation: Public construction contracts in many jurisdictions provide that the contractor must be responsible for all damage resulting from his work. The provision usually also includes the contractor's express obligation to indemnify the State or the municipality, as the case may be, against all lawsuits and damages and to hold it harmless from all related costs due to the contractor's operations. When the contractor's responsibility is clearly stated in such language, he can be called upon to pay damages for injuries for which the State or the municipality would not be bound to pay. His liability for such damages is fixed, in appropriate cases, whether the damage was caused by his negligence or without his negligence. The contractor's responsibility is based upon the rights of "third party beneficiaries."

When the contractor enters into a public contract that contains the provisions described in the preceding paragraph, he consents to and accepts the intention to pay such damage. The contractor who assumes that responsibility is answerable to an individual, for example, to an abutting property owner, for damages resulting from the work under the contract, even though the contractor would not have been liable otherwise. The contract is made for the benefit of the public. The inhabitant of the State or the municipality is, in effect, a party to the public contract containing the provisions referred to above, although not individually named in the public document. Such provisions constitute a direct promise by the contractor to pay, as in the example, the abutting property owner for damages sustained as the result of the contractor's operations.

Principle of contractor's obligation: Although most of the United States of America have established the rights of third party beneficiaries and variations in the application of the rule relating to these rights have been disclosed in other States, an exhaustive analysis of the decisions of the courts was made by the Committee on Contracts of the American Law Institute. The result of that study is found in section 145 of volume 1 of the *Restatement of the Law of Contracts:*

"Beneficiaries under Promises to the United States, a State, or a Municipality.

"A promisor bound to the United States or to a State or municipality by contract to do an act or render a service to some or all of the members of the public, is subject to no duty under the contract to such members to give compensation for the injurious consequences of performing or attempting to perform it, or of failing to do so, unless,

(a) an intention is manifested in the contract, as interpreted in the light of the circumstances surrounding its formation, that the promisor shall compensate members of the public for such injurious consequences, or

(b) the promisor's contract is with a municipality to render services the nonperformance of which would subject the municipality to a duty to pay damages to those injured thereby."

As an illustration of the clause, the Committee on Contracts stated on page 174 of volume 1 of the *Restatement of the Law of Contracts,* as follows:

"A, a municipality, enters into a contract with B, by which B promises to build a subway and to pay damages directly to any person who may be injured by the work of construction. Because of the work done in the construction of the subway, C's house is injured by the settling of the land on which it stands. D suffers personal injuries from the blasting of rock during the construction. B is under a contractual duty to C and D."

The availability of the "third party beneficiary" rule is premised upon the basic principle that a contractor is liable for injuries caused by his negligence in the performance of the work. Some States and political subdivisions are immune from liability for negligence when their own employees are doing the work of construction, but this immunity does not extend to private contractors whose negligence causes damages to third persons. In this respect, the contractor has a responsibility because of a wrongful act and not because he has a contract with the public body. (43 Am. Jur. 825.) In those States where the "third party beneficiary" rule has been established, the rule is based on the principle of public policy essential to public welfare. (*Jenkins* v. *C. & O. Ry. Co.,* 57 S.E. 48, West Va., 1907.)

Judicial pro and con: A municipal contract to build a new bridge across the Passaic River required the contractor to provide and maintain a temporary footbridge. He was to assume all risks pertaining to its construction and use, and to provide watchmen and lighting for the temporary facility. A pedestrian was injured in a manner that implied failure to light the footbridge properly. The lawsuit (*Styles* v. *F. R. Long,* 57 At. 488, N.J., 1904) was based upon the obligations of the contractor as stated in the contract and the court held that the construction work was for the benefit of the public. It was stated judicially, however, that the contractor was not responsible to every citizen of the municipality, as there were no contractual relationships between the contractor and the third parties. The court pointed out: "Whether, in any particular case, a right of action in favor of third persons exists, is a question of difficulty, upon which the cases are hard to reconcile upon general principles. It is, I think, safe to say that there is no right of action where the duty of another person to exercise care intervenes between the neglect of the defendant and the injury of the plaintiff." In this case, the court, said the county officials, were required to maintain public bridges, and New Jersey law authorizes an action against them for neglect to perform a duty.

In New York State, the case of *Cook* v. *Dean et al.,* 11 App. Div. 123, 1896, was based on comparable facts, but the decision that was affirmed by the Court of Appeals, 160 N.Y. 660, was so opposite to the ruling in the case of *Styles* v. *F. R. Long,* as to be considered antagonistic to the New Jersey decision. In the New York case, the contractor was also required to erect a temporary bridge and to keep it in good repair until the permanent bridge was constructed. The plaintiff's husband was killed when the temporary bridge fell. The court said: "The bridge was a public thoroughfare, kept and maintained for the use of the general public, and was used daily by a large number of persons. The obligation to repair was, therefore, for the benefit of the general public, who had an interest therein. The covenant contained in the contract was absolute, and imposed upon the defendants an absolute duty to keep the structure in repair, and safe for the purpose for which it was used, so far as proper repairs thereon would make it safe. Under those circumstances defendants (the contractor) became liable for neglect of duty in connection with the obligation assumed, and such neglect created a right of action in favor of a person who suffered injury therefrom as a consequence thereof."

The opinion in the *Cook* v. *Dean* case declared that the obligation of the muncipiality to maintain the public way is not the basis of responsibility in this lawsuit. When the contractor assumed the duty of maintaining the temporary structure and of keeping it safe for public use, he accepted the obligation of the public officials in that regard. Thus the contractor became bound to perform the obligation and to pay damage to a person who was injured by the contractor's fault or negligence.

Legal right to claim promise for own benefit: A comprehensive summary of the application of the "third party beneficiary" rule is contained in *Pennsylvania Cement Co.* v. *Bradley Contracting Co.,* 7 F. 2d 822, 1925. The Bradley Company made a contract with the City of New York to construct the Lexington Avenue subway line. The contract required the contractor to pay any damage to foundations or walls of adjacent buildings incurred during the construction. The liability thus assumed was stated to be "absolute and is not dependent upon any question of negligence," and called upon the contractor to be "solely responsible for all physical injuries to persons or property occurring on account of and during the performance of the work hereunder, and shall indemnify and save harmless the city from liability" for all claims of damages. When the work was being performed in front of property owned by one Bolger, the structure sustained $3,000 damage. The Bolger claim based on the contract provisions was disallowed. Upon appeal from that disallowance to the United States Circuit Court of Appeals, Second Circuit, the question was Bolger's rights under the contract as an abutting owner, thus making him one of the persons for whom the damage provisions were written in the contract.

The court discussed the matter as follows:

"In England and in this country in some of the States, it is held, subject to some exceptions that where two parties enter into a contract, in which one of them promises to do something for the benefit of a third person, the only persons who can sue upon the contract are the parties who made it. *Price* v. *Eaton,* 4 B and Ad. 433; *Atwood* v. *Burpee,* 77 Conn. 42, 58 At. 237; *Sampson* v. *Commonwealth,* 202 Mass. 326, 88 N.E. 911; *First M. E. Church* v. *Isenberg,* 246 Pa. 221, 92 At. 141; *Edwards* v. *Thoman,* 187 Mich. 361, 153 N.W. 806. But the English rule to its full extent does not prevail in the United States, and the prevailing rule in this country allows a third person to sue on such a contract, subject, however, to qualifications. In *Hendrick* v. *Lindsey,* 93 U.S. 143, 149, 23 L. ed. 855, 1876, the Supreme Court discussing the proposition that a third person not a party to a contract made for his benefit could not sue upon it, said: 'This would be true, if the promise were under seal, requiring an action of debt or covenant; but the right of a party to maintain assumpsit (a lawsuit to enforce a promise, like the payment of a debt) on a promise not under seal, made to another for his benefit, although much controverted, is now the prevailing rule in this country.'

"But the weight of authority in the United States is that an action cannot be maintained by one not a party, simply because he will be incidentally benefited by performance. The contract must have been entered into for his benefit, and he must have some legal or equitable interest in its performance. *Constable* v. *National Steamship Co.,* 154 U.S. 73; *Davis* v. *Patrick,* 122 U.S. 138; *St. Louis Second National Bank* v. *Grand Lodge,* 98 U.S. 123. In *Pennsylvania Steel Co.* v. *New York City Ry. Co.,* 198 F. 721, this court, referring to the right of one not a party to the contract to sue thereon, said: 'It is not enough that the contract may act to his benefit. It must appear that the parties intend to recognize him as the primary party in interest and as privy (having legal relationship) to the promise.'

"The leading case on this subject in the State of New York is the well-known case of *Lawrence* v. *Fox,* 20 N.Y. 268. That case was decided in 1859. The facts were that A loaned $300 to B, stating to him at the time that he owed that amount to X. Thereupon it was agreed between A and B that the latter was to pay the sum loaned to X. This was not done, and X brought an action against B. It was held that the action could be maintained. In *Vrooman* v. *Turner,* 69 N.Y. 280, the court explained the principle upon which *Lawrence* v. *Fox* was decided and said: 'In either case there must be a legal right founded upon some obligation of the promisee, in the third party, to adopt and claim the promise as made for his benefit.' "

The Bradley Company in the *Pennsylvania Cement Co.* case expressly contracted with the City that the plans and specifications would not involve "any danger to the foundations, walls or other parts of adjacent buildings" unless the work was done negligently by the contractor. The contractor also agreed that he would make good, at his own expense, any damage "to any such foundations, walls, or other parts of adjacent buildings," and that he would be solely responsible for all physical injuries to persons or property occurring as the result of and during the performance of the work. Further, in the event of any question arising concerning the fulfillment of the contract, the determination of the engineer in charge would be conclusive.

Thus it follows: "As his contract imposed duties which he owed to the city, which included duties to the owners of abutting property, it was not fulfilled until he had performed whatever he owed under the contract." The United States Circuit Court of Appeals approved the right of Bolger to sue under the contract, although he was not a party to it, whether or not the damages resulted from negligence on the part of the contractor. Bolger's claim was denied in this action, however, because of the omission of the contract requirement that the engineer in charge determine that the damage was caused by the work of subway construction.

In 1940 the Supreme Court of Minnesota considered the case of *LaMourea* v. *Rhude et al.,* 295 N.W. 304, which was for damages caused by blasting operations in constructing a sewer. The contract provided for excavation in solid rock with heavy charges of explosives; the contractor agreed to be liable for any damages to public or private property or for injuries to persons during the operations. The claim of the property owner was approved by the court as being a recognition of section 145 of the *Restatement of the Law of Contracts* which is quoted in this chapter of this volume.

In commenting on variations in the application of the rules, the opinion stated: "The courts of Michigan and Arizona have expressly adopted the rules of section 145. *Bator* v. *Ford Motor Co.,* 269 Mich. 648; *Cole* v. *Arizona Edison Co. Inc.,* 53 Ariz. 141. And two lower New York courts have indicated approval, *Wilson* v. *Oliver Costich Co. Inc.,* 231 App. Div. 346; *Creedor* v. *Automatic Voting Machine Corp.,* 243 App. Div. 339, following a line of earlier cases in that State which allowed recovery in situations covered by that section and analogous to the instant case. *Little et al.* v. *Banks,* 85 N.Y. 258; *Smyth* v. *City of New York,* 203 N.Y. 106; *Rigney* v. *N.Y.C. Ry. Co.,* 217 N.Y. 31; *Schnaier* v. *Bradley Constr. Co.,* 181 App. Div. 538."

The opinion also pointed out: "The contrary doctrine, now altogether if not quite outmoded, was put in the main upon factors of consideration and privity (contract relationship). *Jefferson* v. *Asch,* 53 Minn. 446; *Kramer* v. *Gardner,* 104 Minn. 370; *Clark* v. *Clark,* 164 Minn. 201."

Rights controlled by express intention: Cases involving public construction contracts are controlled by the special rule that declares that the rights of a third party beneficiary are to be upheld when the contracts include definite agreement by the contractor to protect the inhabitants of that state or municipality. The liability of a contractor to a third party, regardless of the former's negligence, rests upon the express intention of the contracting parties that the contractor will be responsible for damage whether or not he is at fault.

This intention of the contracting parties has been considered from time to time by the New York courts; the different rulings point out the distinctions created in expressing the contractual obligation. One of the judicial precedents that has become a leading affirmative authority on the subject, both in New York and other jurisdictions, is *Coley* v. *Cohen,* 289 N.Y. 356, 1942. Special provisions in the contract included: "All blasting necessary on this contract shall be done with the express provision that the contractor shall be and is hereunder responsible for any and all damages and claims arising from such blasting or by accidental explosions and for the defense of all actions arising from such causes." It was held by the Court of Appeals that "The term 'accidental explosion' in the clause under consideration indicates an intent to cover every situation as a result of which damage is done. As used in the contract that expression is intended to include unavoidable accidental explosions for which the contractor ordinarily would not be liable." On the other hand, a contract provision whereby the contractor agreed to "protect" persons and property and to take all "necessary precautions," did not obligate the contractor under the third party benefit rule. (*Weinbaum* v. *Algonquin Gas Transmission Co. et al.,* 20 Misc. 2d 276, N.Y., 1954, affirmed 285 App. Div. 818.)

The New York State Public Works specifications were amended in 1955 to read as follows (new matter is italicized):

"Damage. All damage, direct or indirect, of whatever nature resulting from the performance of the work or resulting to the work during its progress from whatever cause, including omissions and supervisory acts of the State, shall be borne and sustained by the Contractor, and all work shall be solely at his risk until it has been finally inspected and accepted by the State. The Contractor, however, shall not be responsible for damages resulting from faulty designs as shown by the plans and specifications nor the damages resulting from wilful acts of Department officials or employees and *nothing in this paragraph or in this contract shall create or give to third parties any claim or right of action against the Contractor or the State beyond such as may legally exist irrespective of this paragraph or contract.*

"The Contractor shall indemnify and save harmless the State from suits, actions, damages and costs of every name and description resulting from the work under this contract during its prosecution and until the acceptance thereof, and the State may retain such moneys from the amount due the contractor as may be necessary to satisfy any claim for damages recovered against the State. The contractor's obligations under this paragraph shall not be deemed waived by the failure of the State to retain the whole or any part of such moneys due the Contractor, nor shall such obligation be deemed limited or discharged by the enumeration or procurement of any insurance for liability for damages imposed by law upon the contractor, sub-contractor or the State."

Identical provisions are found in the State Architect's Standard Specifications and relate to public buildings projects in New York State.

The amendment to the "Damage" provisions quoted above was an issue in the case of *Costa* v. *Callanan,* 15 Misc. 2d 198, N.Y., 1958. An action was instituted by one who claimed to be a third party beneficiary and sought to recover damages from a contractor under a public construction contract. The contractor's defense to the action was that the amendment to the "Damage" provision, included in his contract, negated any intention of vesting the plaintiff with direct protection as a "third party beneficiary" and thereby entitling him to payment for damages without proof of negligence on the part of the contractor. The court ruled against the claimant and concluded that the provision, as amended, was an effective bar to an action for damages without proof of negligence.

Public policy, property owners, and contractors: Public officials, aware of the "third party beneficiary" rule, have the important function of establishing public policy on this subject for their particular unit of government. When public construction work damages private structures, whether by careful or by careless operations of the contractor, the owners will naturally look to the public officials for prompt redress. When damage is the result of a contractor's carelessness, the property owner has the fundamental right of action to recover damages from the contractor on the theory of legal liability. However, when every known precaution is taken by the contractor, and property is damaged without his fault, the right of recovery could be governed by the "third party beneficiary" rule, provided the language of the contract discloses that intention.

Where the work of construction, under a contract that clearly obligated the contractor to the "third party beneficiary" rule, was carried on with care, abutting property owners claimed that the comparatively light explosive charges caused cracks to appear in the walls of their houses. The contractor believed that the services of specialists in blasting operations, together with the installation of every known precautionary device, would avoid difficulty in his obligation to the property owners. He secured legal advice that the claimants would not be required to prove any negligent operation in his work in order to recover damages, and he suddenly realized the extent of his financial responsibility on the mere proof by the claimants that he had blasted at the site of the work adjacent to the property claimed to have been damaged.

The contractor's protest to the public officials with whom he made the contract disclosed that he carried liability insurance and contractual liability insurance coverage to protect him for legal liability, including the indemnification of the State, as a compliance with related provisions in the contract. It seemed, however, that there was no available liability insurance to protect him against damages under the "third party beneficiary" rule, principally because there is no basis of experience or of estimating the limits of such a liability insurance policy.

Unless a contractor is financially able and willing to assume the risk of paying such damages from his own resources, it is probable that he will include a "cushion" in his bid price to protect himself against that risk. If he is called upon to pay such damages, the undisclosed "cushion" will serve its purpose; on the other hand, that sum would be a windfall to him if the project could be completed without any claim or payment under the "third party beneficiary" rule. Such a "cushion" would necessarily raise the cost of the work to the State or municipality without producing a commensurate accomplishment of the project.

These observations by the author are to point out the basic considerations that responsible public officials must weigh when they undertake the serious function of establishing public policy for the prudent expenditure of public funds when third parties are involved by virtue of applicable law.

6.11 Negligence and liability for damages

Because of their sovereignty, most states are immune from suits for negligence and for damages to persons or property. Related provisions in state constitutions or statutes must be examined to determine if this immunity is applicable to the state and its municipalities. In New York, for example, the immunity of the state is waived by the following provision in section 8 of the Court of Claims Act:

"The state hereby waives its immunity from liability and action and hereby assumes liability and consents to have the same determined in accordance with the same rules of law as applied to actions in the supreme court against individuals or corporations, provided the claimant complies with the limitations of this article. Nothing herein contained shall be construed to affect, alter or repeal any provision of the workmen's compensation act."

The contractor engaged in public work is liable for damages caused by his own negligence, and frequently the contract includes a provision by which the contractor agrees to indemnify and to save a governmental unit harmless from all claims arising by reason of the work done by the contractor. The general liability of the contractor is not based upon the contract with the public agency, but it is dependent upon an act or omission that is a breach of his duty in conducting his operations under the contract.

The public construction contract usually stipulates that the contractor will observe reasonable rules of safety to protect both his employees and the public. Usually, there are specifications that require the contractor to erect and maintain barriers, install lights, engage the services of uniformed patrolmen and watchmen, and to take safety precautions.

In an action to recover damages for personal injury, it appeared that a contractor engaged a private trucker to haul sand, gravel, and rock boulders from a river bank to the site of the job. The trucks were loaded by employees of the contractor, using equipment owned by the contractor. A truck, which was not identified, hit a rock about the size of a large orange lying in the street in front of a store and hurled it 40 feet into a store door where it struck the plaintiff and caused her serious personal injury. (*Matsumoto* v. *Arizona Sand & R.* Co., 295 P.

2d 850, Ariz., 1956.) The court said that it can be inferred from the evidence that the rocks scattered along the road could be caused only by overloading the trucks. A sheriff's deputy had warned that overloading the trucks would create a dangerous condition on the public way. The decision quoted from section 414 of *Restatement of the Law of Torts:* "One who entrusts work to an independent contractor, but who retains the control of any part of the work, is subject to liability for bodily harm to others, for whose safety the employer owes a duty to exercise reasonable care, which is caused by his failure to exercise his control with reasonable care."

The court held that this contractor owed a "duty to the public generally not to knowingly overload the trucks hauling its material from the river bank to the road construction work to the extent that rocks falling therefrom might create a dangerous or hazardous condition upon the highway. This is true regardless of the requests or demands of the truck drivers to overload said trucks. The defendant (contractor) having the duty to load the trucks would be responsible for a negligent performance of that duty.

"Under the rule laid down in the case of *Palsgraf* v. *Long Island R. Co.,* 248 N.Y. 339 which we adopted in *Tucker* v. *Collar,* 79 Ariz. 141, we hold that if defendant (contractor) in loading the trucks in question, acting as a reasonable person, was able to foresee or should have foreseen an unreasonable risk of harm to plaintiff or one in plaintiff's situation as a result of rocks falling from the trucks onto said highway, caused by overloading the same, defendant, under such circumstances, would be negligent and liable for damages to the person injured as a proximate result of such negligence. *West* v. *Cruz,* 251 P. 2d 311."

Liability insurance in highway projects: The immunity from liability for negligence and damages found in the principle of the sovereignty of states and which, in many jurisdictions, applies also to municipalities does not apply to a contractor on a public construction project who has committed a tort that has caused damages to persons or property. Recently, many governmental bodies have called upon their contractors to carry appropriate liability insurance with reasonable limits. The requirements of liability insurance must be comprehensive, adequate for the purpose, but reasonable in cost. The general contractor must also consider his obligations under a contractual liability clause in his contract to indemnify and hold harmless the government unit, together with the obligations of his subcontractor with respect to liability for negligence in doing the work under a subcontract. Of prime importance in the administration of public construction activities, the program of liability insurance coverages for the benefit of the governmental agency must include the responsibility of the public body in the scope of the liability insurance protection, because of the operations of its contractor and his subcontractor.

After years of experience with various types of liability insurance policies, the New York State Department of Public Works has developed a well-defined and comprehensive program of coverages. The Public Works Specifications provide as follows:

> INSURANCE. The Contractor shall procure and maintain at his own expense and without expense to the State, until final acceptance by the State of the work covered by the contract, insurance for liability for damages imposed by law, of the kinds and in the amounts hereinafter provided, in insurance companies authorized to do such business in the State covering all operations under the contract whether performed by him or by sub-contractors. Before commencing the work the Contractor shall furnish to the Superintendent a certificate or certificates of insurance in form satisfactory to the Superintendent showing that he has complied with this paragraph, which certificate or certificates shall provide that the policies shall not be changed or cancelled until thirty days written notice has been given to the Superintendent. The kinds and amounts of insurance are as follows:
>
> (a) Workmen's Compensation Insurance. A policy covering the obligations of the Contractor in accordance with the provisions of Chapter 41 of the Laws of 1914, as amended, known as the Workmen's Compensation Law, covering all operations under the contract, whether performed by him or by his sub-contractors. The contract shall be void and of no effect unless the person or corporation making or executing same shall secure compensation coverage for the benefit of, and keep insured during the life of said contract, such employees in compliance with the provisions of the Workmen's Compensation Law. (State Finance Law, Section 142);
>
> (b) Liability and Property Damage Insurance. Unless otherwise specifically required by special specifications, each policy with limits of not less than:

Highways including Federal Aid Projects
and other State contracts

Bodily Injury Liability		Property Damage Liability	
Each Person	Each accident	Each accident	Aggregate
$100,000	$300,000	$100,000	$300,000

Projects Affecting Railroad Rights-of-way,
Including Structures and Facilities Thereon

Bodily Injury Liability		Property Damage Liability	
Each Person	Each accident	Each accident	Aggregate
$150,000	$300,000	$150,000	$300,000

for all damages arising during the policy period, shall be furnished in the types specified, viz.:

(1) Contractor's liability insurance issued to and covering the liability for damages imposed by law upon the Contractor with respect to all work performed by him under the agreement;

(2) Contractor's liability insurance issued to and covering the liability for damages imposed by law upon each sub-contractor with respect to all work performed by said sub-contractor under the agreement;

(3) Contractor's protective liability insurance issued to and covering the liability for damages imposed by law upon the Contractor with respect to all work under the agreement performed for the Contractor by sub-contractors;

(4) Protective liability insurance issued to and covering the liability for damages imposed by law upon The People of the State of New York and the Superintendent of Public Works and all employees of the Superintendent of Public Works both officially and personally, with respect to all operations under the agreement by the Contractor or by his subcontractors, including omissions and supervisory acts of the State;

(5) Completed Operations' liability insurance issued to and covering the liability for damages imposed by law upon the Contractor and each sub-contractor arising between the date of final cessation of the work and the date of final acceptance thereof, out of that part of the work performed by each;

(6) Owners', Landlords' and Tenants' liability insurance issued to and covering the liability for damages imposed by law upon The People of the State of New York and the Superintendent of Public Works and all employees of the Superintendent of Public Works both officially and personally, with respect to temporarily opening to vehicular traffic any portion of the State highway under the agreement, until the construction or reconstruction pursuant to the agreement has been accepted by the State.

(7) In connection with contracts involving operations on the right-of-way of a railroad, including structures and facilities thereon, Railroad Protective liability insurance in the name of the railroad is required. Unless otherwise provided by special specifications, such insurance shall afford coverage equivalent to the following:

(a) The policy shall cover the liability imposed upon the insured (railroad) by law for damages, including damages for care and loss of services, because of bodily injury, sickness or disease, including death at any time resulting therefrom, sustained by any person or persons, and damages because of injury to or destruction of property, including the loss of use thereof, caused by accident and arising out of any acts or omissions of the following in the performance of any operations under the contract for the designated projects:

(I) The Contractor and his sub-contractor;

(II) Flagmen, watchmen and other protective employees of the insured, except those specified in (III) and (IV) below, specifically loaned or assigned by the insured to the work performed by the contractor or his subcontractor, provided the cost of services of such employees is specifically to be borne by the contractor or sub-contractor or by a governmental authority;

(III) Supervisory employees of the insured while performing services at the job site with respect to the operations of the contractor or his subcontractors;

(IV) Employees of the insured while operating, attached to or engaged on, at the job site, work trains or other railroad equipment exclusively assigned to the contractor or his subcontractor by the insured.

provided, however, that a proximate cause of the accident is any act or omission of any individual included in (I), (II), (III) or (IV) above.

(b) The policy shall also cover the liability imposed upon the insured by law for damages, including damages for care and loss of services, because of bodily injury, sickness or disease, including death at any time resulting therefrom, caused by accident and sustained at the job site by any employee of the contractor or his sub-contractors.

(c) The terms "contractor" and "sub-contractor" as used in (a) and (b) above do not include the named insured.

(d) The coverage shall be subject to the following exclusions:

(I) Any obligation for which the insured or any carrier as his insurer may be held liable under any workmen's or unemployment compensation, disability benefits or similar laws. The Federal Employer's Liability Act, U. S. Code (1946) title 45, Sections 51-60 shall not for the purpose of this insurance be deemed to be any similar law.

(II) Any liability assumed by the insured under any contract or agreement other than contracts of carriage as a common carrier. Contracts of carriage as a common carrier as used herein shall not include those between the insured and the contractor or his sub-contractors.

(III) Any liability arising out of accidents occurring after operations have been completed or abandoned at the place of occurrence thereof, other than accidents caused by the existence of tools, uninstalled equipment and abandoned or unused materials, but operations shall not be deemed incomplete because improperly or defectively performed or because further operations may be required pursuant to a service or maintenance contract. This exclusion does not apply until the State notifies the insured of acceptance of the project.

(IV) Injury to or destruction of property owned by the insured.
Each policy shall provide that it shall not be changed or cancelled until thirty days' written notice has been given to the railroad and to the Superintendent of Public Works.

With the collaboration of the National Bureau of Casualty Underwriters and the Mutual Insurance Rating Bureau, the "Certificate of Insurance" shown here has been formally adopted by the insurors. This form has eliminated the necessity of delivering the policies to the state, because the liability insurance policy forms are standardized. This is desirable because filing space for retaining many policies is at a premium in most government office buildings.

Liability and fire insurance for public buildings work: The same rules of liability that relate to highway work apply to contractors and government units when the contract is for public building construction. Administrative policy must determine the advisability of having liability insurance when a project is located within the perimeter of institutional grounds or other public lands where visitors are allowed to enter but where there is no regular flow of pedestrian or vehicular traffic. When public building construction begins next to a public sidewalk or street, then the specifications should require the liability insurance policies that are listed and defined in (1), (2), (3), (4), and (5) of the previously quoted "Public Works Specifications" of the New York State Department of Public Works. Although the law requires less care for trespassers than for the protection of people entitled and invited to proceed and to pass on or along the site of the work, many prudent contractors follow a regular business policy of carrying liability insurance on all jobs, regardless of the requirements or the omission of requirements in the specifications.

Fire insurance coverage during the construction of public buildings is a subject that requires the same thoughtful consideration as liability insurance. The extent of the risk of loss by fire while a contractor is performing his contract is determined, as a rule, by the provisions of the contract documents relating to the obligation of the contractor to deliver a completed structure. Many contracts impose the risk of a fire loss upon the contractor until the completed structure has been accepted by the public body. During the period of construction, progress payments are usually made on the basis of the value of the work performed, but in most instances the progress payments are not construed as the acceptance of any part of the work. In case of fire damage to the building during construction, the contractor will, as a rule, be obligated by his contract to restore the damaged work and the installations that have been affected by the fire at his own cost. The advisability of adequate fire insurance with appropriate coverages is obvious, unless the

CERTIFICATE OF INSURANCE FOR CONSTRUCTION AND RECONSTRUCTION OF STATE HIGHWAY PROJECTS

To New York State Department of Public Works:

The subscribing insurance company certifies that insurance of the kinds and types and for limits of liability covering the work herein designated, has been procured by and furnished on behalf of the insured contractor named in Item 1.

1. *Name of Insured* ....

   Address of Insured ....

2. *Location and Description of Work* ....

....

....

3. *Kinds and Types of Insurance*

Insurance	*Policy Number*	*Effective Date*	*Expiration Date*	*Coverage*	*Limits of Liability*		
					Each Person	*Each Accident*	*Aggregate*
(a) Contractor's Liability				B. I.			xxxxxxx
				P. D.	xxxxxxx		
(b) Contractor's Protective Liability				B. I.			xxxxxxx
				P. D.	xxxxxxx		
(c) Contractor's Protective Liability furnished by general contractor in name of The People of the State of New York and/or the Superintendent of Public Works.				B. I.			xxxxxxx
				P. D.	xxxxxxx		
(d) Completed Operations Liability				B. I.			
				P. D.	xxxxxxx		
(e) Owners', Landlords', and Tenants' Liability insurance furnished by general contractor in name of The People of the State of New York and/or the Superintendent of Public Works.				B. I.			xxxxxxx
				P. D.	xxxxxxx		xxxxxxx
(f) Workmen's Compensation				xxxx	xxxxxxx	xxxxxxx	xxxxxxx
(g) Disability Benefits				xxxx	xxxxxxx	xxxxxxx	xxxxxxx
(h)				B. I.			
				P. D.	xxxxxxx		

Such insurance as is herein certified applies to all operations of said insured in connection with the work herein described at the locations stated, and is written in accordance with the company's regular policies and endorsements, subject to the company's applicable manuals of rules and rates in effect, except

....

....

The subscribing company agrees that no policy referred to herein shall be changed or canceled until thirty days' written notice has been given to the New York State Department of Public Works.

This Certificate is furnished in accordance with and for the purpose of the specifications of the New York State Department of Public Works for the construction and reconstruction of State highways, covering the operations herein described.

....
(Name of Company)

Dated.................... By ....
(Authorized Representative)

Certificate of insurance for construction and reconstruction of state highway projects

contractor is willing to assume the risk of a fire loss and pay for restoring the building from his own resources. In New York State, the "Standard Specifications of the State Architect" provide for the following comprehensive, yet realistic, coverage, which has the approval of the representatives of most of the fire insurance companies that conduct business in the state:

"Each contractor for the work of construction and for heating, sanitary, electric, refrigeration, elevators and all other installations shall maintain until the date of the final certificate, fire insurance on the Builder's Risk Completed Value Form, with extended coverage, on the value of all work and materials included in his contract. Whenever applicable, the Contractors' Interest Completed Value Form may be used. The extended coverage endorsement may include a loss deductible clause of $50.00.

"Insurance will not be required on work included under the subdivision of excavation work, work buried in the ground except underground wiring, work within an existing building, nor on fire escapes including enclosures that are being erected on an existing building. However, insurance will be required on all tunnel construction and work therein.

"Should any building or buildings included in this contract be occupied by the State prior to the issuance of final certificate, the contractor shall so notify the State Architect, and fire insurance on the occupied building or buildings only may be cancelled as of date of occupancy as determined by the State Architect. Fire insurance on the remainder of the buildings or other work of the contract shall be maintained in force by the contractor until final acceptance or occupancy by the State.

"All policies shall be issued by fire insurance companies authorized to conduct such business under the laws of the State of New York, shall be written for the benefit of the State of New York and for the contractor as their interests may appear, and shall run until the date of the final certificate. Policies expiring on a fixed date before final acceptance must be renewed and refiled before such date. The policy shall be filed in the office of the State Architect before the contractor applies for any payment on the work which is to be insured pursuant to this paragraph."

6.12 Municipal zoning regulations and state construction projects

Municipalities exercise police powers of the state, which are delegated to the municipalities for the welfare, safety, and health of the public. Under the police power, cities and villages are empowered to enact reasonable ordinances and regulations to preserve health, eliminate nuisances, regulate the use and storing of dangerous articles, and similar purposes. "The police power is not impaired by the 14th Amendment of the Constitution of the United States," but property is held by a person subject to the exercise of police power and he is entitled to use that property reasonably, without injury to others. A zoning ordinance is a valid exercise of the police power. (*Barbier* v. *Connolly,* 113 U.S. 27, 1887.)

The immunity of governmental agencies from local zoning regulations applies, as a rule, when the use of the property is for a governmental function. A variation in such immunity is found in some states when the property is to be used for an activity that is proprietary or of a corporate character. In Ohio and Georgia, it has been held that local zoning ordinances do not apply to governmental projects for which the governmental agency has the right of exercising the power of eminent domain or condemnation. The case of *State ex rel. Helsel* v. *Board of County Commissioners,* 79 N.E. 2d 688, Ohio, 1947, involved land which had been acquired by eminent domain for an airport. The court held that this was a use "of public property for public purposes" and the adjoining property owners were not entitled to an order to set aside the acquisition of the land because the village zoning ordinance prohibited the use of the property for an airport. It was decided that to give the municipality the right to invoke its zoning ordinance would permit the municipality to restrict the right of eminent domain. A similar objection was raised in *State* v. *Allen,* 107 N.E. 2d 345, Ohio, 1952 as to the consent of municipalities and other political subdivisions through which the Ohio Turnpike was to pass. The court decided that a zoning restriction "cannot be construed as applying to the State or any of its agencies vested with the right of eminent domain in the use of the lots for public purposes." The erection of a fire station by the City of Savannah in an area zoned for residential purposes was approved by the court in the case of *Mayor, etc. of Savannah* v. *Collins,* 84 S.E. 2d 454, Ga., 1954. The court said: "To rule otherwise would offend that provision of our constitution which declares the right of eminent domain shall not be abridged."

The case of *Water Works Board of Birmingham* v. *Stephens,* 78 So. 2d 267, Ala., 1955 supported the applicability of a city zoning ordinance to the erection of a multimillion-gallon water tank in a residential zone. The court decided that the sale of water by the city is a business or a function that is proprietary in nature, and the city must observe its own zoning regulations.

The construction of a water tower by a village in an area zoned for residential purposes was stopped in the case of *Baltis* v. *Westchester,* 121 N.E. 2d 495, Ill., 1954. Also, in *Taber* v. *Benton Harbor,* 274 N.W.

324, Mich., 1937, the zoning ordinance for a residential area was applicable and forbade the construction of a steel water tower and tank. The function of operating and maintaining a waterworks system was held to be of a proprietary or business character and the municipality, in such cases, is bound by its own zoning rules.

The construction of a sewage disposal plant in a residential zone was forbidden by the court, as a proprietary function of the county, in *Jefferson County* v. *City of Birmingham,* 55 So. 2d 196, Ala., 1951. The court said that the operation of a sewage disposal plant is a governmental function with relation to questions of tort liability and not where zoning questions are involved. "Cases in other jurisdictions hold that where zoning is involved, the operation of a sewage disposal plant is proprietary and not governmental. *O'Brien* v. *Town of Greenburgh,* 239 App. Div., N.Y. 555."

Exemption from a zoning regulation may be granted expressly by the statute itself. In New York, for example, section 6 of the State Public Buildings Law provides, in part, that: "No municipality of the state shall have power to modify or change plans or specifications for the erection, reconstruction, alteration or improvement of state buildings, or the construction, plumbing, heating, lighting or other mechanical branch of work necessary to complete the work in question, nor to require that any person, firm or corporation employed on any such work shall perform said work in any other or different manner than that provided by said contract and specifications, nor to obtain any other or additional authority or permit from such municipality, department or person as a condition of doing such work, nor shall any condition whatever be imposed by any such municipality in relation to the work under the supervision of the superintendent of public works, but such work shall be under the sole control of such superintendent in accordance with the drawings, plans, specifications and contracts in relation thereto; and the doing of any work for the state by any person, firm or corporation in accordance with the terms of such contract, plans or specifications shall not subject said person, firm or corporation to any liability or penalty, civil or criminal, other than as may be stated in such contract and specifications or incidental to the proper enforcement thereof. . . ."

The statutory power of the New Jersey Highway Authority to construct and maintain service areas on the Garden State Parkway with relation to local zoning requirements, was discussed in *Town of Bloomfield* v. *New Jersey Highway Authority,* 113 At. 2d 658, N.J., 1955. The court pointed out that the Highway Authority was expressly authorized, by the statute that created it to construct and operate the project as an essential governmental function of the state, and the legislature has the power to grant immunity from local zoning and building restrictions. The history and the terms of the statute do not disclose even a suggestion to subject the Authority to such restrictions.

In *C. J. Kubach Co.* v. *McGuire,* 248 P. 676, Cal., 1926, a regulation in the city charter restricting the height of buildings to 150 feet, was held to be inapplicable to the construction of a city hall that exceeded that limitation. A garage to store village police cars and snow-fighting equipment and to provide facilities for highway employees could not be excluded from an area that was zoned for private dwellings. (*Nehrbas* v. *Incorporated Village of Lloyd Harbor,* 140 N.E. 2d 241, N.Y., 1957.) A county zone regulation was held to be inapplicable to the construction of a branch office and treatment center of a state hospital. (*Davidson County* v. *Harmon,* 292 S.W. 2d 777, Tenn., 1956.) The construction of a county jail was immune from a city zone ordinance in the case of *Green County* v. *Monroe,* 87 N.W. 2d 827, Wis., 1958. A state office building with space for private tenants was held to be excepted from a city zoning ordinance which excluded such buildings. (*Charleston* v. *Southeastern Constr. Co.,* 64 S.E. 2d 676, West Va., 1950.)

In the *Charleston* case, the city sued the contractor and the State Building Commission for an injunction to restrain them from constructing a state office building on land purchased by the state for the purpose. In 1939, an ordinance was adopted by the city, after the State Building Commission was created by the State Legislature. The state statute provided that buildings to be constructed by the Commission must be "subject to such consent and approval of the City of Charleston in any case as may be necessary." The Commission submitted plans and specifications for the new building to the City Building Inspector and made application for a permit, which was refused because the height of the proposed building exceeded the limit specified in the local ordinance. The court held that the building to be constructed by the Building Commission is a public building and is not subject to the zoning ordinance. Although the opinion pointed out that the local ordinance was inconsistent with the state statute and is, therefore, null and void as to the subject of the lawsuit, the consent clause was not considered a "valid express grant of the State." The zoning ordinance was passed after the state statute creating the Building Commission had been adopted, and the zoning regulation cannot make the Legislature subservient to every future action of the City, and, too, such police power of the state cannot be granted or delegated to the city. Further, the court commented on the provisions of the zoning regulation in question, and as it provided that "no *person* shall erect or construct any building . . .," the use of the word "person" includes private corporations, and "is not applicable in this jurisdiction to State agencies which are an arm of the State, or even to public corporations, which are not State agencies."

chapter 7

Obligations of the contractor and the surety

The public's interest is best served when its construction contracts provide for general direction and supervision by the government engineer and also require the completion of the entire project in a workmanlike manner to the satisfaction of the responsible public representative. This obligation of performance by the contractor anticipates that the engineer in charge will act in good faith.

7.1 General principles of performance

Some public construction contracts give the government engineer in charge the right to decide if the work has been accomplished satisfactorily, without his being compelled to state the reasons for a decision of unsatisfactory performance by the contractor. Other contracts provide for a decision of performance that is based upon reasonable satisfaction, such as meeting certain standards or operational tests. Either procedure is a matter of contract obligation, and, when clearly expressed, either is binding upon the parties to the contract.

Generally, a contractor is entitled to the entire contract price upon complete performance of the work. If he performs only a part of the work and then leaves or abandons it without the consent or the fault of the government unit or its representative, the contractor is not entitled to payment for the part of the contract that he performed.

"The law requires exact performance as a condition precedent to recovery. It is no answer that the material supplied is equivalent or superior to that provided in the contract. The defendant (owner) has the absolute right to the thing specified, and need not pay for something which the claimant (contractor), his experts, or the court may conclude is equally valuable and useful." (*Stanton* v. *State of New York,* 103 Misc. 221, N.Y., 1918, affirmed 187 App. Div. 963; *Ward* v. *Kilpatrick,* 85 N.Y. 413, 1881.)

7.2 Substantial performance

The exception to the rule of exact performance is found in the principle of "substantial performance." This is a recent principle, peculiar to building construction contracts generally, and has been evolved by the courts to mitigate the severity of the rule of exact performance in appropriate cases. Substantial performance is limited in its application and is well defined. A contractor may recover for the performance of a contract, not only when he is able to show exact performance, but also, in proper instances, when substantial performance will suffice. The rule of substantial performance is applicable where a builder has intended to comply with the contract in good faith and has substantially complied with it, although there may be a slight defect or defects caused by inadvertence or unintentional omissions. He may recover the contract price, less the damage on account of such defects. (*Woodward* v. *Fuller,* 80 N.Y. 312, 1880.)

The principle of substantial performance is based on justice and is applicable when the failure to perform completely is slight and trivial and when the contractor has tried to comply with the terms of the contract. Literal compliance in every detail is not required, and the contractor is entitled to payment, under the rule, if he has substantially performed the contract, less the reasonable cost of remedying trivial defects and omissions. There is no exact answer to the question of substantial performance, but consideration is given to the contractor's intentions, the amount of work performed, and the benefit received by the owner. (*Restatement of Contracts, section 275.*)

The basis of the rule is that it would be unfair that the owner should "receive and keep a part of that which he contracted for, and pay nothing for it because he did not receive the whole." The performance "must be of a substantial part of the contract, and the acceptance must be under such circumstances as to show that the party accepting knew or ought to have known, that the contract was not being fully performed." (*Hennessy* v. *Preston,* 106 N.E. 570, Mass., 1914.)

"Where the rule of substantial performance prevails it is essential that the (builder's) default should not have been willful; and the de-

A.I.A. DOCUMENT NO. A-311
(Formerly Form 107) 1958 Edition

PERFORMANCE BOND

This document approved and issued by The American Institute of Architects
1735 New York Avenue, N. W., Washington 6, D. C.

KNOW ALL MEN BY THESE PRESENTS:

That ....,
(Here insert the name and address or legal title of the Contractor)
....

as Principal, hereinafter called Contractor, and ....

....,
(Here insert the legal title of Surety)
....

as Surety, hereinafter called Surety, are held and firmly bound unto ....

....,
(Here insert the name and address or legal title of the Owner)
as Obligee, hereinafter called Owner, in the amount of ....

.... Dollars ($..............................),
for the payment whereof Contractor and Surety bind themselves, their heirs, executors, administrators, successors and assigns, jointly and severally, firmly by these presents.

WHEREAS, Contractor has by written agreement dated ....

entered into a contract with Owner for ....

....

....

in accordance with drawings and specifications prepared by ....

....
(Here insert full name and title)
which contract is by reference made a part hereof, and is hereinafter referred to as the Contract.

A.I.A. DOCUMENT NO. A-311
(Formerly Form 107) 1958 Edition

Performance bond (face)

fects must not be so serious as to deprive the property of its value for the intended use nor so pervade the whole work that a deduction in damages will not be fair compensation." (*Williston—"Contracts,"* section 805.)

The discussion of substantial performance in *Jacob & Youngs, Inc.* v.

NOW, THEREFORE, THE CONDITION OF THIS OBLIGATION is such that, if Contractor shall promptly and faithfully perform said contract, then this obligation shall be null and void; otherwise it shall remain in full force and effect.

The Surety hereby waives notice of any alteration or extension of time made by the Owner.

Whenever Contractor shall be, and declared by Owner to be in default under the Contract, the Owner having performed Owner's obligations thereunder, the Surety may promptly remedy the default, or shall promptly

1) Complete the Contract in accordance with its terms and conditions, or

2) Obtain a bid or bids for submission to Owner for completing the Contract in accordance with its terms and conditions, and upon determination by Owner and Surety of the lowest responsible bidder, arrange for a contract between such bidder and Owner, and make available as work progresses (even though there should be a default or a succession of defaults under the contract or contracts of completion arranged under this paragraph) sufficient funds to pay the cost of completion less the balance of the contract price; but not exceeding, including other costs and damages for which the Surety may be liable hereunder, the amount set forth in the first paragraph hereof. The term "balance of the contract price," as used in this paragraph, shall mean the total amount payable by Owner to Contractor under the Contract and any amendments thereto, less the amount properly paid by Owner to Contractor.

Any suit under this bond must be instituted before the expiration of two (2) years from the date on which final payment under the contract falls due.

No right of action shall accrue on this bond to or for the use of any person or corporation other than the Owner named herein or the heirs, executors, administrators or successors of Owner.

Signed and sealed this .... day of .... A.D. 195......

IN THE PRESENCE OF:

.... (Principal) (Seal)

.... (Title)

.... (Surety) (Seal)

.... (Title)

Performance bond (reverse)

Kent, 129 N.E. 889, N.Y., 1921 provides a clear basis for the rule and is frequently referred to in many states. The corporation built a country residence for the defendant, Kent, at a cost of more than $77,000. About one year after completion, the owner complained of defective performance. Some of the pipe was not "standard pipe" or "Reading" manufacture; it was made in other factories, and the architect, as a result of the owner's complaint, ordered the contractor to do the work over. This meant the demolition of substantial parts of the structure in order to substitute other pipe for the kind that was installed. The lawsuit was begun by the contractor to collect the balance of $3483.40 on his contract. There was no proof of fraud in omitting the prescribed pipe; it was the result of oversight and lack of attention by the plumbing subcontractor. The architect had not noticed the discrepancy when he inspected the work. The court held that "It is never said that one who makes a contract fills the measure of his duty by less than full performance. They (the courts) do say, however, that an omission both trivial and innocent will sometimes be atoned for by allowance of the resulting damage, and will not always be the breach of a condition to be followed by a forfeiture. *Spence* v. *Ham,* 163 N.Y. 220; . . . *Bowen* v. *Kimball,* 203 Mass. 364." The court allowed the owner the difference in value, which would be either nominal or nothing. "The measure of allowance is not the cost of the reconstruction. 'There may be omissions of that which could not afterwards be supplied exactly as called for by the contract without taking down the building to its foundations, and at the same time the omission may not affect the value of the building for use or otherwise, except so slightly as to be hardly appreciable.' *Handy* v. *Bliss* 204 Mass. 513; *Foeller* v. *Heintz,* 137 Wis. 169; *Oberlies* v. *Bullinger,* 132 N.Y. 598. The rule that gives a remedy in cases of substantial performance with compensation for defects of trivial or inappreciable importance, has been developed by the courts as an instrument of justice. The measure of the allowance must be shaped to the same end."

The rule of substantial performance was involved when the foundations of a public school building were improper and poorly constructed because the contractor had not complied with the specifications. The proportions of cement and gravel were not correct, the walls were not built of the required density of material, the rafters were not properly braced, the cornices were open so that light showed through, and the outside walls were not in plumb. In the lawsuit *Kasbo Const. Co.* v. *Minto School Dist.,* 184 N.W. 1029, No. Dak., 1921, it was shown that the difference in cost between the building as erected and as it was specified was about $4500, although the construction could not be remedied for that amount; the witness stated that the building would have to be rebuilt in order to meet the specifications. The court action was started by the contractor to recover the balance of the contract price; the school district counterclaimed for $9,000. The decision granted the counterclaim and pointed out that the measure of damages for failing strictly to construct the building, is, when there has been substantial performance, the difference between the value of the building when constructed and what its value would have been if constructed in compliance with the specifications and with reasonably sound material and skilled workmanship. (*Walter* v. *Huggins,* 148 S.W. 148, Mo., 1912; *White* v. *McLaren,* 24 N.E. 91, Mass., 1899; *Hartford Mill Co.* v. *Hartford Tobacco Warehouse Co.,* 121 S.W. 477, Ky., 1909; *Small* v. *Lee Brothers,* 61 S.E. 831, Ga., 1908.)

Generally, when defects in construction can be remedied without taking down and rebuilding a substantial portion of the building, the owner is entitled to deduct from the contract price, the expense he incurs to correct the condition. (*Walsh Constr. Co.* v. *Cleveland,* 271 F. 701, 1920; *School Directors* v. *Robinson,* 65 Ill. App. 298, 1896; *Gustor* v. *Clark,* 157 N.W. 49, Mich., 1916.)

When a contractor intentionally, without the consent of the owner, substituted brick for stone, the contractor is liable to the owner for the cost of replacing the brick with stone, although the value of the completed building is slightly reduced by the used of brick instead of stone. (*Pence* v. *Dennie,* 182 P. 980, Cal., 1919.)

When a contract for construction of a building expressly required the contractor to prove that he had paid for all labor and materials, his failure to submit the proof, without a valid excuse, is a failure to comply with a prerequisite, and he is not entitled to the benefit of the rule of substantial performance. Because the contract contained such an absolute and unconditional provision, the owner was not obligated to demand the proof from the contractor. (*Witherell* v. *Lasky,* 286 App. Div. 533, N.Y., 1955.)

No formula to determine substantial performance exists; in each case it is a question of degree based upon the related factors. For example, in a contract for which an engineer prepared detail drawings of a structure, he refused to make modifications or to supervise the work as agreed. The court held that he had substantially performed the contract for which he should be paid the agreed price, less the expenses incurred by the other contracting party for making the changes. When an owner reaps the benefit of the work under a contract because of the valuable services of the contractor, the owner is expected to pay the contract price less any damages he suffered by failure of complete performance, provided substantial performance leaves only those defects or defaults that are unintentional and trivial. (*Antonoff* v. *Basso,* 78 N.Y. 2d 604, 1956.)

A building contractor, sued for breach of contract, apparently had deviated from the specification as follows: (1) the exterior walls had no wood sheathing and consisted of construction paper covered with chicken wire with an application of stucco; (2) one floor furnace with a capacity of

35,000 Btu. was installed, instead of two furnaces of 30,000 Btu. capacity each; and (3) sheet rock with taped joints was used for the interior wall instead of gypsum lath and plaster. The defense in this case, *Shell* v. *Schmidt,* 330 P. 2d 817, Cal., 1958, was that the contractor had made substantial performance. The court held that the contractor should pay as damages the cost of making the work conform to the contract, rather than the difference between the value of the building as planned and its value as built, and quoted from *Thomas Haverty Co.* v. *Jones,* 197 P. 105, Cal., 1921: "The rule of substantial performance has now been greatly relaxed, and it is settled especially in the case of building contracts, where the owner has taken possession of the building and is enjoying the fruits of the contractor's work in the performance of the contract, that if there has been a substantial performance thereof in good faith, when the failure to make full performance can be compensated in damages, to be deducted from the price or allowed as a counterclaim, and the omissions and deviations were not willful or fraudulent, and do not substantially affect the usefulness of the building for the purposes for which it was intended, the contractor may, in an action upon the contract recover the amount unpaid of his contract price, less the amount allowed as damages for the failure in strict performance."

It is a well-established rule that the owner is entitled to receive what he has contracted for or its equivalent. If the defects or deviations are such that can be corrected without destroying a substantial part of the benefits to the owner resulting from the contractor's work, the "equivalent" is the cost of having the work done to conform with the contract. If this correction makes it necessary to undo a substantial part of what the contractor has done, however, and his deviation was in good faith or the owner has taken possession, the owner is entitled to recover the difference in value between the result he contracted for and the product as accomplished by the contractor. (*Am. Jur., Vol. 9, sec. 152.*) In some states, though, the allowable damages are the cost of remedying the breach of contract or the difference in value as described above, whichever is lower. (*Texas Pacific Coal & Oil Co.* v. *Stuard,* 7 S.W. 2d 878, Tex., 1928; *Bigham* v. *Wabash-Pittsburgh Term. Ry.,* 72 At. 318, Pa., 1909.)

In *Cardell* v. *City of Perry,* 207 N.W. 775, Iowa, 1926, the lawsuit was over a street improvement assessment to which objections had been raised because the contractor failed to comply with the plans and specifications in the matter of longitudinal expansion joints. The testimony was that the joints installed were composed of a much lower proportion of asphalt than specified and did not constitute a substantial performance. The decision of the court was that the "wrong done (to the taxpayers) represents a very minor, rather than a major, proportion of the whole," and held the assessment to be valid. The court commented: "There is no evidence of actual fraud in the instant matter, nor is there any indication of fraudulent purpose. Nevertheless, property owners are entitled to that for which they are called upon to pay. Public improvement contracts and the manner of their fulfillment demand the closest scrutiny. The substitution of materials, slipshod inspection of methods, and numerous other deviations from standard which so frequently occur, call for constant vigilance."

The equitable rights of a contractor and the responsibility of the public officials were discussed in connection with substantial performance of a street improvement in *City of Earlington* v. *Powell,* 10 S.W. 2d 1060, Ky., 1928. The curbs and gutters were not constructed as specified, the City Council refused to accept the work as not being substantial performance, and it called upon the contractor to correct the work. The court ruled that the statute governing the project empowered the council to decide when a contract has been performed or breached, but did not authorize a rejection of "work of substantial value for mere defects or insufficiences not amounting to a total failure to perform the contract. It plainly contemplates that the council may require specific defects to be remedied, and insufficiencies supplied, or it may modify the cost to conform to the facts, or it may do both, if the exigencies demand it. If the statute should be construed to authorize one party to a construction contract to reject the work entirely and make its action conclusive, it would place arbitrary power in the hands of that party." The court said that the finished work was of such magnitude and value that it could not be wholly rejected, and there was substantial performance "to impose upon the city the duty of proceeding under the statute to provide a fund for the payment of the contractor according to the statute and conforming to the standard contemplated by the contract."

In *Transfer Realty Co.* v. *City of Superior,* 147 N.W. 1051, Wis., 1914, it is stated that: "Substantial performance of a contract . . . is substantial justice, where no damage is suffered by failure to literally perform." In this case the court commented on a point made by counsel concerning the power of the public official to compromise this lawsuit, as follows: "We need not pass upon the question whether compromise, except as a basis for substantial performance, is within the power of the board, because the adjustment was not a compromise of a disputed claim in the ordinary sense of that term. It was more in the nature of acceptance of the work on the terms of substantial performance. Such adjustments are equitable and just and are held binding by the court where full benefits have been received under them."

Where there may have been "slight deviations from the contract in respect to the size and quality of gravel employed," the court approved an assessment for a public street improvement. In *Hendry* v. *City of Salem,* 129 P. 531, Oregon, 1931, it was held that the "evidence tends to show that as a net result the property holders have as good and substantial a street as they would have had in the event of a strict and literal compliance with the contract. There was a substantial compliance with the plans and specifications. *Barkley* v. *Oregon City,* 33 P. 978."

7.3 Anticipated default in completion

The right of a state or municipality to cancel a contract because its representative anticipated a delay in completion was discussed in *Wakefield Construction Co.* v. *City of New York,* 157 App. Div. 35, N.Y., 1913, affirmed 213 N.Y. 633. The contract was for complete construction of a sewer within 600 working days, not counting Sundays and holidays. The contract designated the chief engineer of the city arbiter of all disputes; he had the right to certify to the mayor that in his opinion the contractor was not completing the work in accordance with the contract, and the mayor had the power to cancel the contract. The work was commenced in November 1909. In May, 1910, the contractor received notice that his contract was canceled. The court ruled that although the notice of cancellation was unequivocal as to its direction to the contractor to discontinue the work, it did not specify which of the grounds in the contract was the cause for cancellation. It also appeared to the court that the certification by the chief engineer did not state that in his opinion the performance of the work had been unnecessarily delayed; instead, he said that it then seemed unlikely that the contractor would be able to proceed with due diligence in the future. The jury was asked to decide whether the performance of the contract was unreasonably or unnecessarily delayed, and whether the chief engineer had exercised honest judgment. The reports to the chief engineer were made by his subordinates; in the absence of his own investigation, the court stated that his certificate "would not be conclusive because it would not represent his honest judgment based upon existing facts." The jury found that the chief engineer's certificate was an arbitrary act. Winter weather conditions and delays in staking out the work contributed largely to the loss of time in progressing the job. The contractor was awarded damages by the court, and the verdict was affirmed by the higher courts.

Public construction contracts and related statutes usually provide for cancellation of a contract when the governmental official has correct information with which to determine that the contractor is not proceeding in accord with the contract. In order to arrive at such a determination, a written complaint should be sent to the contractor with a list itemizing the areas of faulty progress. A conference should be scheduled within a reasonable time after the delivery of the complaint, so that the contractor will have an opportunity to explain or defend his position and the status of the work. For example, in the *Wakefield* case, discussed previously, the details brought out during the trial showed clearly that the weather conditions during the period when the contractor was charged with dilatory practice prohibited the trench work for the laying of pipes and concrete. Only one-fifth of the prescribed time for completion had expired at the time of the cancellation, and severe weather conditions had prevailed most of the time. The opinion commented that the evidence failed to show that the contractor had been delinquent up to the time of the cancellation, but showed only that the contractor was not ready at the beginning of the favorable season to continue the work with sufficient speed. The contract provisions only authorized definite action for unreasonable delay, not for anticipated delay. It is probable, that a conference with the contractor before the determination had been made in the *Wakefield* case, would have avoided a costly and time-consuming lawsuit that took a period of several years. Reasonable action by public officials is the guide to fewer problems in contract relationship, and the conference method is worth the time and attention it requires, as distinguished from arbitrary action that drags the contending parties through involved court trials.

7.4 Use of surety bonds

The modern way of obtaining security for public construction contracts is a surety bond or bonds issued by a qualified surety corporation. Surety bonds guarantee that the contractor will perform his work according to the contract and will pay for labor and material. If the contractor defaults in performance, the surety guarantees to pay the cost of completing the work. If the workmen or material supplies are not paid, the surety bond is a guarantee that their valid claims will be paid even though, generally, liens cannot be filed against property owned by the public. Unless a related statute expressly authorizes the filing of a mechanic's lien on public property, there can be no effective filing of a notice of such a lien. Courts construe the general mechanic's lien statutes to exclude the property of the sovereign or a municipality, unless it is expressly mentioned in the law. The reasoning for such a construction is found in public policy that is opposed to the sale of public property to satisfy liens, and especially when such property is not available for sale under an execution following a judgment against

LABOR AND MATERIAL PAYMENT BOND

This document approved and issued by The American Institute of Architects 1735 New York Avenue, N. W., Washington 6, D. C.

Note: This bond is issued simultaneously with another bond in favor of the owner conditioned for the full and faithful performance of the contract.

KNOW ALL MEN BY THESE PRESENTS:

That ....,
(Here insert the name and address or legal title of the Contractor)

....

as Principal, hereinafter called Principal, and ....

....
(Here insert the legal title of Surety)

....

as Surety, hereinafter called Surety, are held and firmly bound unto ....

....,
(Here insert the name and address or legal title of the Owner)

as Obligee, hereinafter called Owner, for the use and benefit of claimants as hereinbelow defined, in the amount of .... Dollars ($..............................),
(Here insert a sum equal to at least one-half of the contract price)
for the payment whereof Principal and Surety bind themselves, their heirs, executors, administrators, successors and assigns, jointly and severally, firmly by these presents.

WHEREAS, Principal has by written agreement dated ....

entered into a contract with Owner for ....

....

....

in accordance with drawings and specifications prepared by ....

....
(Here insert full name and title)
which contract is by reference made a part hereof, and is hereinafter referred to as the Contract.

Labor and material payment bond (face)

the state or the municipality. (*Bates* v. *Santa Barbara County,* 27 P. 438, Cal., 1891; *Rathbun* v. *State,* 97 P. 335, Idaho, 1908; *Secrest* v. *Delaware County,* 100 Ind. 59, 1885; *Barrett Mfg. Co.* v. *New Orleans,* 63 So. 505, La., 1913; *Friedman* v. *Hampden County,* 90 N.E. 851, Mass., 1910; *Bell* v. *New York,* 11 N.E. 495, N.Y., 1887; *General Bonding &*

NOW, THEREFORE, THE CONDITION OF THIS OBLIGATION is such that if the Principal shall promptly make payment to all claimants as hereinafter defined, for all labor and material used or reasonably required for use in the performance of the Contract, then this obligation shall be void; otherwise it shall remain in full force and effect, subject, however, to the following conditions:

1. A claimant is defined as one having a direct contract with the Principal or with a subcontractor of the Principal for labor, material, or both, used or reasonably required for use in the performance of the contract, labor and material being construed to include that part of water, gas, power, light, heat, oil, gasoline, telephone service or rental of equipment directly applicable to the Contract.

2. The above named Principal and Surety hereby jointly and severally agree with the Owner that every claimant as herein defined, who has not been paid in full before the expiration of a period of ninety (90) days after the date on which the last of such claimant's work or labor was done or performed, or materials were furnished by such claimant, may sue on this bond for the use of such claimant, prosecute the suit to final judgment for such sum or sums as may be justly due claimant, and have execution thereon. The Owner shall not be liable for the payment of any costs or expenses of any such suit.

3. No suit or action shall be commenced hereunder by any claimant.

a) Unless claimant, other than one having a direct contract with the Principal, shall have given written notice to any two of the following: The Principal, the Owner, or the Surety above named, within ninety (90) days after such claimant did or performed the last of the work or labor, or furnished the last of the materials for which said claim is made, stating with substantial accuracy the amount claimed and the name of the party to whom the materials were furnished, or for whom the work or labor was done or performed. Such notice shall be served by mailing the same by registered mail or certified mail, postage prepaid, in an envelope addressed to the Principal, Owner or Surety, at any place where an office is regularly maintained for the transaction of business, or served in any manner in which legal process may be served in the state in which the aforesaid project is located, save that such service need not be made by a public officer.

b) After the expiration of one (1) year following the date on which Principal ceased work on said Contract, it being understood, however, that if any limitation embodied in this bond is prohibited by any law controlling the construction hereof such limitation shall be deemed to be amended so as to be equal to the minimum period of limitation permitted by such law.

c) Other than in a state court of competent jurisdiction in and for the county or other political subdivision of the state in which the project, or any part thereof, is situated, or in the United States District Court for the district in which the project, or any part thereof, is situated, and not elsewhere.

4. The amount of this bond shall be reduced by and to the extent of any payment or payments made in good faith hereunder, inclusive of the payment by Surety of mechanics' liens which may be filed of record against said improvement, whether or not claim for the amount of such lien be presented under and against this bond.

Signed and sealed this .... day of .... A.D. 195....

IN THE PRESENCE OF:

.... (Principal) (Seal)

.... (Title)

.... (Surety) (Seal)

.... (Title)

Labor and material payment bond (reverse)

Cas. Co. v. *Dallas,* 175 S.W. 1098, Tex., 1915.) The rule is also stated in *Knapp* v. *Swaney,* 23 N.W. 162, Mich., 1885, that "Public property cannot be the subject of such a lien (mechanic's lien) unless the statute shall expressly so provide; it is by implication excepted from lien statutes, as much as from general tax laws and for the same reasons."

In some states there is authorization for filing a lien against the money earned by a public construction contractor, and in these cases all parties concerned should be fully informed of the particular local procedures that govern the filing and the foreclosure of such a lien. (*Hazard* v. *Board of Education,* 75 At. 237, N.J., 1910.)

Generally, government agencies have the statutory power to require a surety bond for public construction work; however, the omission of statutory authorization does not prohibit a government body from requiring a surety bond for such a contract. In a leading case on this subject, *Knapp* v. *Swaney,* 23 N.W. 162, Mich., 1885, it was stated: "A county may go to great pains and great expense to make its courthouse unquestionably safe, that individual citizens may not suffer injuries consequent upon its construction; but, if it may do this, it would be very strange if it were found lacking in authority to stipulate, in the contract for the building, that the contractors, when calling for payment shall show that they are performing their obligations to those who supply the labor and materials, and that the county is not obtaining the building at the expense of a few of its people. We cannot think such is the case."

In *Philadelphia* v. *Stewart,* 45 At. 1056, Pa., 1900, the court approved the judgment of the representatives of a municipality to require a surety bond to protect workmen and suppliers of material and commented: "There is nothing *ultra vires* (beyond the scope of the powers of the municipality) or contrary to public policy in this condition, it is the right as well as the interest of the city to secure good work upon its contracts for public improvements, and there is no better policy towards that end than to satisfy honest and competent workmen that they can rely on being paid. There being no right of mechanic's lien against public works, the workmen and materialmen are, to that extent, in the contractor's power as to pay, and that fact has a natural tendency to produce skimped work and inferior materials by the class of men willing to run that risk. Against this risk the city is entitled to protect itself by exacting assurance from the contractor that he will pay his honest debts incurred in doing the city's work."

The recognized purpose of a payment bond on public construction projects is to protect and to secure labor and material men in the same manner that mechanic's liens, which are available on private construction work, do. Statutes that require such surety bonds have the same objectives. (*Lingler* v. *Andrews,* 10 N.E. 2d 1021, Ohio, 1926; *Cavanaugh* v. *Globe Indemnity Co.* 44 P. 2d 216, Kan., 1935; *U.S.F. & G. Co.* v. *Tafel Electric Co.,* 91 S.W. 2d 42, Ky., 1935; *Pneucrete Corp.* v. *U.S.F. & G. Co.,* 46 P. 2d 1000, Cal., 1935.)

Most jurisdictions have statutes related to those described above, which define the terms and prescribe the conditions of performance surety bonds on public contracts. The performance bond is for an entirely different purpose than the payment bond, which is for the protection of those who perform labor on, or furnish material for a construction job. In some jurisdictions, a single bond is required covering both performance and payment of labor and materials, although the obligations may be independent of each other. Because it is not unusual for a contractor to be in difficulty in both the performance of his work and the payment of labor and material bills arising from the work, it is obvious that such a situation creates serious problems as to priority of rights of the persons covered by a single bond. *Fosmire* v. *National Surety Co.*, 229 N.Y. 44, 1920 was a case involving a single bond of the type described in this paragraph. The court pointed out that the purpose of the bond is to protect the interest of the State with relation to the completion of the contract; the bond is not intended to make the contractor's employees the beneficiaries of a cause of action to be enforced in hostility to the State's benefit nor is it intended that the security under the bond should be exhausted at the instance and for the benefit of persons other than the State.

Formerly, federal statutes provided for a single bond on government construction contracts to cover both completion and labor and materials. However, in 1935, upon the enactment of the Federal Miller Act, the requirements were changed and since that time federal construction contracts call for a separate surety bond to protect the government on performance of the contract and another surety bond for the protection of workmen and material suppliers. The premium cost of the bond protection is not increased by furnishing two bonds for the purposes described in this paragraph.

Although surety bonds on public construction work have come into general use, the forms have not been standardized nationally, because they must comply with the requirements of specific statutes, which differ from state to state. Governmental units that contract for substantial construction work have developed their own bond forms that recognize the statutory requirements governing each particular situation; to this extent the federal government and many states and municipalities have standardized bond forms. The performance bond and labor and material payment bond forms shown are among those obtainable nationally from surety companies and are acceptable for public construction contracts, provided there is no statutory control of the express conditions and forms of such bonds.

It is significant that the federal government and almost every state require surety bonds for labor and material on certain public construction contracts. An acceptable justification for the requirement of a surety bond for the protection of material of materialmen, is that the charge for their product will be increased when no surety bond is provided. The costs of collection and the bad debt loss are factors that must be considered by the supplier of material in an unbonded transaction, as contrasted with the protected account when a surety bond is required. If a supplier's prices are loaded with unprotected credit costs, the government agency will eventually pay for them as part of the total contract price.

7.5 Meaning of statutory bond

The term "statutory bond," when used in connection with public construction contracts, means a bond that is required by statute and that contains all the conditions required by law to be contained in such a bond. (*City of Knoxville* v. *Melvin E. Burgess,* 175 S.W. 2d 548, Tenn., 1943.) For example, where a local statute called for a surety bond that would obligate a public construction contractor to pay for labor and materials, the surety company was held liable for them, despite an indemnification in the bond, for the benefit of the public agency only, for damages arising from the contractor's default in performance. The statutory requirement and the remedy were read into the surety bond. (*Camdenton, etc., ex rel. W. H. Powell Lumber Co.* v. *New York Casualty Co.,* 104 S.W. 2d 319, Mo., 1937.)

The beneficiary of a statutory bond is entitled to all the remedies and procedures that are granted by statute. The statutory bond is distinguished from a common law bond, which is a type of voluntary agreement or contract, and the beneficiary must abide by the terms of the bond or the rules of common law for the purpose of enforcing the bond. (*40 Words and Phrases 102.*)

In *Philip Carey Co.* v. *Maryland Casualty Co.,* 206 N.W. 808, Iowa, 1926, the terms of the bond that was filed by the contractor did not conform to the requirements of the statute, which required a surety bond for any public construction contract in excess of $1,000. The surety company claimed that the document was not intended as a statutory bond, was solely for the benefit of the public agency, and was not for the benefit of the supplier of material to the contractor. The court decided that "where the situation is such as to require a statutory bond, and the bond given by the contractor conforms in material and essential respects to the requirements of the statute, the parties will be held to have intended to make a statutory bond, notwithstanding the omission from the bond of other conditions required by the statute, or the inclusion of stipulations contrary to the statute. We think the statute itself so provides. 'If the law has made the instrument necessary, the parties are deemed to have had the law in contemplation when the contract was executed.' *Fogarty* v. *Davis,* 264 S.W. 879."

The importance of determining the extent of the guarantee, if any, for payment of labor and materials is emphasized in *Warner* v. *Hallyburton,* 121 S.E. 756, N.C., 1924. The contract was for the construction of a schoolhouse, and the statutory requirement as to public contracts was for "a bond before beginning the work, and payable to said county, city, etc., conditioned for payment of all labor done, or material and supplies furnished for said work; that said bond may be put in suit by any laborer or material and supply man having a valid claim; and further that if the official or said county, city or town or other municipal corporation fails to require this bond, he shall be guilty of a misdemeanor, etc." The contract had no express provision that the contractor, who was the principal upon the bond, was to pay labor or the materialmen. The bond, in this case, was held to be one of indemnity to protect the municipality from liens and claims that might be made effective against it. If the statute had provided "that any bond taken in such cases should inure to the benefit of laborers and materialmen, this might be construed as constituting a part of contracts to which the statute applied." The court ruled that the contract and bond were designed to secure satisfactory completion of the project, and nothing else, and because the contract contained no stipulation binding the contractor to pay labor and materialmen, the surety company was not liable on the bond. The statute involved in *Warren* v. *Hallyburton* was later amended, so that now in North Carolina every surety bond for public construction is presumed to have been taken in accordance with the amended statute, and the provisions of the statute are presumed to be written in the bond. This amendment was so interpreted in *Standard Supply Co.* v. *Vance Plumbing, etc. Co.,* 143 S.E. 248, N.C., 1928.

7.6 Liability of public official for failure to require surety bond

The general, but not universal, theory of the liability of a public official who does not obtain a surety bond for labor and materials that is expressly required by an applicable statute is that negligence in the performance of a public duty creates no liability on the public agency or its representative. A statute expressly creating such a liability is not affected by the stated theory, however. (*Johns-Manville* v. *Lander County,* 240 P. 925, Nev., 1925; *Harnbach* v. *Ward,* 125 P. 140, Wash., 1912.) In *E. I. Dupont de Nemours & Co.* v. *Glenwood Springs,* 19 F. 2d 225, Colo., 1927, it appeared that the city failed to require the statutory bond, and upon the contractor's default, the city took over the work, but permitted the contractor to remain in charge and control of the work. The parties in the lawsuit did not dispute the power of the city to enter into the contract, or that the city had failed to require the bond provided for by the statute. The court held that "in the absence of a statutory requirement that a city pay claims for material furnished to a contractor, or that it withhold funds to insure payment of such claims, no enforceable liability on the part of the city exists therefor." In commenting on the reason for its decision, the court stated: "A rule of law which would require a municipal corporation to pay for material furnished to a person to whom it lets a contract, because of the failure of its officers to require a sufficient bond, or because of their failure to exercise the privilege of withholding from the contractor payment of money due him, until claims for material are paid, would be against sound public policy. As was said in the case of *Merwin* v. *Chicago,* 45 Ill. 133, 'A municipal corporation is a part of the government. Its powers are held as a trust for the common good. It should be permitted to act only with reference to that object, and should not be subjected to duties, liabilities or expenditures merely to promote private interest or private convenience.'" The same rule applies in many states to the public officers upon the ground that the negligence of the agents of the public body in the failure to require the bond is the negligence of the governmental unit, not of the officers as individuals. (*Blanchard* v. *Burns,* 162 S.W. 63, Ark., 1913; *Hydraulic Press Brick Co.* v. *School Dist.,* 79 Mo. App. 655, 1899; *Fore* v. *Feimster,* 88 S.E. 977, N.C., 1916; *Pidgeon Thomas Iron Co.* v. *LeFlore County,* 99 So. 677, Miss., 1924.)

The rule differs in some jurisdictions. In those places, if the affected workman or the supplier of material can show that he believed that a statutory bond had been provided to protect his interests and that he cannot collect from the contractor, the government unit and its agents or representatives may be held liable. Inasmuch as liens cannot be filed against public property as a remedy for unpaid labor or materials, the rule in some states gives the workman or the supplier the right to sue the negligent public body or its officers. (*Owen* v. *Hill,* 34 N.W. 649, Mich., 1887; *Northwest Steel Co.* v. *School Dist.,* 148 P. 1134, Oregon, 1915; *Hardison* v. *Yeaman,* 91 S.W. 1111, Tenn., 1905.)

7.7 Subrogation and surety rights

When a public construction contractor defaults in the performance of the work, the obligation of the surety (one who has become legally liable for the debt, default, or miscarriage of another) on its performance bond becomes effective. The rights of several parties are then affected, particularly as to money earned by the contractor before his default, but not paid to him by the government agency.

If, when the contractor defaults, his surety undertakes to complete the work for which the bond was issued, the surety becomes entitled to the remedies that the government unit has against the contractor under the contract. The term "subrogation" describes such situations; it is an equitable rule based on principles of justice, where the surety substitutes for the public agency as far as the latter's rights and remedies against the defaulting contractor under the contract are concerned.

A surety or indemnitor on the bond of a contractor who assumes and completes the work after the contractor abandons or defaults on his contract is subrogated to all of the rights which the public agency might have enforced against the contractor if the public agency had declared the contract forfeited and had completed the work by its own methods. (*Southern R. Co.* v. *Bretz,* 104 N.E. 19, Ind., 1914.)

A leading case on the subject of the subrogation of a surety is *Prairie State Natl. Bank* v. *United States,* 164 U.S. 227, 1896. The contract was for the erection of a custom house for the federal government at Galveston, Texas. The dispute in the lawsuit was between the bank and Hitchcock who had the prior rights to money retained from the progress payments to the contractor as security against a breach of the contract. The contractor gave a representative of the bank a power of attorney to receive the final payment under the contract; the consideration for the power of attorney was money advanced and to be advanced by the bank to the contractor. The federal official rejected the power of attorney, but consented to mail the check for the final payment to the contractor at the address of the bank's representative. The bank and the contractor agreed upon this arrangement, and, on that basis, the bank advanced about $6,000 to him. Hitchcock claimed the money on the ground that he was the surety for the contractor who had defaulted in the performance of the contract. Hitchcock completed the job at a cost of about $15,000 in excess of the current payments by the federal government without knowledge of the rights claimed by the bank. The court in awarding the retention money to Hitchcock said:

"Thus the respective contentions are as follows: the Prairie Bank asserts an equitable lien in its favor, which it claims originated in February, 1890, and is therefore paramount to Hitchcock's lien, which it is asserted, arose only at the date of his advances. The claim of Hitchcock, on the other hand, is that his equity arose at the time he entered into the contract of suretyship, and therefor his right is prior in date and paramount to that of the Bank. . . .

"That Hitchcock, as surety on the original contract, was entitled to assert the equitable doctrine of subrogation is elementary. That doctrine is derived from the civil law, and its requirements are, as stated in *Aetna L. Ins. Co.* v. *Middleport,* 124 U.S. 534: '(1) That the persons seeking its benefits must have paid a debt due to a third party before he can be substituted to that party's rights; and (2) that in doing this he must not act as a mere volunteer, but on compulsion to save himself from loss by reason of a superior lien or claim on the part of the person to whom he pays the debt, as in cases of sureties, prior mortgages, etc. The right is never accorded in equity to one who is a mere volunteer in paying a debt of one person to another.'

"As said by Chancellor Johnson in *Gadson* v. *Brown,* Speers, Equity 38, 41 (quoted and referred to approvingly in the opinion in *Aetna L. Ins. Co.* v. *Middleport,* previously quoted), 'the doctrine of subrogation is a pure unmixed equity, having its foundation in the principles of natural justice, and from its very nature never could have been intended for the relief of those who were in any condition in which they were at liberty to elect whether they would or would not be bound and, as far as I have been able to learn its history, it never has been so applied. If one with the perfect knowledge of the facts will part with his money, or bind himself by his contract in a sufficient consideration, any rule of law which would restore him his money or absolve him from his contract would subvert the rules of social order. It has been directed in its application exclusively to the relief of those that were already bound who could not but choose to abide the penalty.

"Under the principles thus governing subrogation, it is clear whilst Hitchcock was entitled to subrogation the bank was not. The former in making his payments discharged an obligation due by Sundberg for the performance of which he, Hitchcock, was bound under the obligation of his suretyship. The Bank, on the contrary, was a mere volunteer, who lent money to Sundberg on the faith of a presumed agreement and of supposed rights acquired thereunder. The sole question, therefore, is whether the equitable lien which the Bank claims it has, without reference to the question of its subrogation, is paramount to the right of subrogation which unquestionably exists in favor of Hitchcock. In other words, the rights of the parties depend upon whether Hitchcock's subrogation must be considered as arising from and relating back to the date of the original contract, or as taking its origin solely from the date of the advance by him."

The decision emphasized that Hitchcock was subrogated to the right to claim the money and the remedies that were available to the contractee, the United States; he was not subrogated to the rights of the contractor in default, who had no rights whatever upon default. The surety's rights included the 10 per cent retained by the federal government, which would have been available to the federal government in case it had been compelled to complete the work. The retention of a percentage of contract payments is as much for the indemnity of the guarantor or surety as it is for the contractee, and the surety is released from his obligation if the contractee disregards the provisions relating to the 10 per cent retention.

In concluding the opinion in the *Priarie Bank* case, the court said: "Sundberg & Company (the contractor) could not transfer to the bank any greater rights in the fund than they themselves possessed. These rights were subordinate to those of the United States and the sureties. Depending, therefore, solely upon rights claimed to have been derived in February, 1890, by express contract with Sundberg & Company, it necessarily results that the equity, if any, acquired by the Prairie Bank in the 10 per cent fund then in existence and thereafter to arise was subordinate to the equity which had, in May, 1888, arisen in favor of the surety Hitchcock."

The decision in the *Prairie Bank* case was cited by the court in its opinion in support of the case of *Henningsen* v. *U.S.F. & G. Co.,* 143 F. 810, 1906; affirmed 208 U.S. 404. In this lawsuit, Henningsen and Clive, as partners, contracted with the United States to construct some buildings; the surety bond was a guarantee of performance and the payment of labor and materials for $11,000. The buildings were constructed, but the contractors did not pay for $15,000 worth of labor and materials. Upon the default of the contractor, the surety sued the contractor and everyone with claims for labor and materials. The judgment held the surety liable for the claims and awarded payment to the creditors of the sum guaranteed, upon a pro rata basis. During the progress of the work on the contract, the contractor assigned all payments then due and to become due to one S. in trust for the National Bank of Commerce of Seattle to secure a loan of $3,500 and subsequent loans, and, at the same time, gave as further security an order addressed to the U.S. Quartermaster requesting him to deliver all payment checks to S. This lawsuit was by the surety to restrain a payment to S. under the assignment and order. The court, in affirming the decision of the lower court, ruled as follows:

"Is its (the surety's) equity superior to that of one who simply loaned money to the contractor to be by him used as he saw fit, either in the performance of his building contract or in any other way? We think it is. It paid the laborers and materialmen and thus released the contractor from his obligations to them, and to the same extent released the government from all equitable obligations to see that the laborers and supply men were paid. It did this not as a volunteer, but by reason of contract obligations entered into before the commencement of the work." In further comment on the subrogation of the surety to the rights of the government because the surety was compelled to make good the default of the contractor, the court pointed out: "Upon precisely the same principle the surety is entitled to be subrogated to the rights of the laborers and materialmen where, as in the present case, it is compelled by reason of the obligation of the bond to pay them for labor and material because of the default of its principal. That right of subrogation relates back, as was held by the Supreme Court in *Prairie State Natl. Bank* v. *United States, supra,* to the time the contract of suretyship was entered into."

Generally, the surety who completes a defaulted contract and has incurred a loss in fulfilling the condition of the surety bond has a source of reinbursement by the state or municipality, which includes money earned by the contractor and not yet paid to him, as well as money to be earned. (*First National Bank* v. *City Trust, S.D. & Surety Co.,* 114 F. 529, 1902; *Prairie State Natl. Bank* v. *United States,* 164 U.S. 227, 1896.)

Provisions in public construction contracts that require the government unit to retain a certain percentage of the contract payments and to require estimates, certificates from architect or engineers, or receipted bills for labor and materials, are, generally, for the protection of the surety as well as for the government. (*Fort Worth Independent School Dist.* v. *Aetna Casualty & Surety Co.,* 48 F. 2d 1, 1931; *Commercial Casualty Ins. Co.* v. *Durham County,* 128 S.E. 469, N.C., 1925.)

The contractor who abandons his contract loses the right to compel the government unit to pay any money for the work he had performed, nor does an assignee of the contractor have any right to such funds. (*Labbe* v. *Bernard,* 82 N.E. 688, Mass., 1907.)

A surety that pays labor and materialmen after the contractor has defaulted is entitled to payments earned by the contractor, even without a formal assignment of such money by the contractor to the surety. The same right of the surety also applies to money unearned at the time of the default, but which is due later because the surety has completed the contract. (*Massachusetts Bonding & Insurance Co. and ano.* v. *State of New York,* 259 F. 2d 33, N.Y., 1958.)

A variation must be considered in connection with the general rule of the source of reimbursement to the surety. For example, in *North Pacific Bank* v. *Pierce County,* 167 P. 2d 454, Wash., 1946, the county entered into a contract with N. who agreed to produce and furnish to the county 10,000 cubic yards of crushed stone to improve a county road. On the same date the contractor furnished a surety bond for the performance of the contract. The application for the bond included a conditional covenant of assignment by N. to the surety corporation of all money due or to become due under the contract; however, the county was not notified of the assignment until five months after the contract was made. About two months after the contract had been made, N. gave an assignment to the bank of "all monies now due me or to become due me under a certain contract with" the county, and authorized the delivery of all checks to the bank with power to endorse them. The bank loaned N. $4,026 about one month after the contract date. Five months after the contract date, the contractor defaulted and the surety completed the contract. The court, ruling on one point in the case, said: "It is now the accepted rule in this jurisdiction that where a public improvement contract does not *clearly require* the state or municipality to withhold funds applicable to the payment of a contract, the contractor may effectually assign such funds." In reviewing the decisions of the courts in the State of Washington, the following expression was quoted from *Pacific Coast Steel Co.* v. *Old National Bank of Spokane,* 235 P. 947, Wash., 1925; "The sum total of all these decisions seems to be that where a contract provides for a specific amount of the contract price to be retained by the municipality for the payment of labor and material, the balance of the estimates may be assigned by the contractor, and that the assignee is entitled to receive all the assigned amounts, if '(1) there is no provision in the contract that the municipality may, in addition to the percentage reserved, retain out of the balance amounts sufficient to cover labor and material claims.'" The contract in the *North Pacific Bank* case did not include such a provision, and the court ruled against the surety. The opinion also pointed out: "As between the bonding company and the contractor, the indemnity agreement could, upon breach of the contract, relate back to its original date, but as to the bank, which in the meantime had taken its assignment, had given notice thereof to the county, and advanced funds on the security thereof, the indemnity agreement could operate only from the time of default and proper notice thereof and, even then, only in such a way as would not prejudice the rights of the bank under its assignment. *Fidelity & Deposit Co.* v. *Auburn,* 272 P. 34, Wash.; *Hall and Olswang* v. *Aetna Casualty & Surety Co.,* 296 P. 162, Wash." The conclusions stated in the *North Pacific Bank* case were approved and were cited in support of the decisions in *Daneis* v. *M. DeMatteo Const. Co.,* 102 F. Supp. 874, N.H., 1952, and *National Surety Corp.* v. *Fisher,* 317 S.W. 2d 334, Mo., 1958.

7.8 Surety bonds, labor, and materialmen

The protection offered by surety bonds to benefit labor and suppliers of materials on public construction contracts must be determined by an examination of the local laws, which will disclose whether the bond is governed by a statute or by the stipulations in the specifications or the contract. The extent of the benefits that are available to labor and materialmen is to be determined from the language of the statute governing the surety bond requirement, and, too, the surety bond and the contract must be read as one document. (*Standard Gas & Power Corp* v. *N.E. Casualty Co.,* 101 At. 281, N.J., 1917; *Los Angeles Stone Co.* v. *National Surety Co.,* 173 P. 79, Cal., 1918.) "Surety contracts of this character, with a paid surety, are regarded as in the nature of insurance contracts, and will be most strongly construed against the surety" (*Leslie Lumber & Supply Co.* v. *Lawrence, et al.,* 11 S.W. 2d 458, Ark., 1928). A statute that provides for surety bonds for labor and materialmen is usually construed liberally in order to accomplish the objects of such a law in the protection of the class of workmen and materialmen. (*Arnold* v. *United States,* 280 F. 338, 1922; *Lane* v. *State,* 43 N.E. 244, Ind., 1895; *Baumann* v. *West Allis,* 204 N.W. 907, Wis., 1925.)

Since statutes do not give labor and materialmen a lien on public work for their services and products, the terms of this type of surety bond are the one protection they have for payment. (*Holcomb* v. *American Surety Co.,* 42 S.W. 2d 765, Ark., 1931.) The consideration to the contractor and his surety for their obligation under the labor and material bond is found both in the award of the contract for public work and the agreement of the contractor to pay these charges; and the workman and the supplier of material have the right to sue on the bond for payment of their bills for services and materials. (*Lyman* v. *Lincoln,* 57 N.W. 531, Neb., 1894; *National Surety Co.* v. *Hall-Miller Decorating Co.,* 61 So. 700, Miss., 1913; *Williams* v. *Markland,* 44 N.E. 562, Ind., 1896.) The protection of the surety bond is available to them even though they may not have known of the existence of the bond. (*Toner* v. *Long,* 111 At. 311, N.H., 1920; *Knight & J. Co.* v. *Castle,* 87 N.E. 976, Ind., 1909; *Montgomery* v. *Rief,* 50 P. 623, Utah, 1897.)

The definition of labor as used in a particular statute, which gives a remedy to obtain payment for labor performed for contractors on public construction projects, includes not only the labor in its ordinary sense, but also includes mental effort by the skilled superintendent. *U.S.F. & G. Co., for use of Reedy* v. *American Surety Co. of N.Y.,* 25 F. Supp. 28, 1938. "The word 'work' has a more comprehensive meaning than the term 'labor', and has been defined to mean to exert one's self for a purpose, to put forth effort for the attainment of an object, to be engaged in the performance of a task, duty, or the like; and as thus defined, covers all forms of physical or mental exertions, or both combined, for the attainment of some object other than recreation or amusement; the word 'work' including within its scope something more than mere physical exertion which falls under the protection of the word 'labor' as used in a statute" (*24 Words and Phrases 21*).

The meaning of labor as intended by the statutes relative to public work is construed by the courts to give obvious import to the language of the particular statute; this liberal construction of the law by the courts preserves protective measures for the benefit of a particular class of people. Judicial decision is guided by the language of the particular statute, however, whether the words are general or restrictive. When a statute provided that it was intended for: "All persons who perform labor, or furnish labor, . . . for the construction and improvement of any . . . railroad . . . by contract . . . shall have a lien," it was decided that the language intended to provide for the purpose for which the labor was furnished and not the kind of labor. The court ruled that it included a civil engineer who actually superintended and directed the construction of the work. In commenting on this subject, it has been judicially stated that: "It is somewhat difficult to draw the line between the kind of work and labor which is entitled to a lien, and that which is mere professional or supervisory employment, not fairly to be included in those terms." (*Mining Co.* v. *Collins,* 104 U.S. 179, 1881; *Central Trust Co.* v. *Richmond, N. I. & B. R. Co.,* 54 F. 723, Ky., 1892.)

The statutory language that expressly provided the lien protection for "A person performing labor for . . . a contractor . . . for the construction of a public improvement," was held to bar a claim by an insurance company for premiums that it paid for workmen's compensation insurance for the employees of a defaulted contractor. (*The Travelers Insurance Company* v. *Village of Ilion and others,* 126 Misc. 275, N.Y., 1925.) Also, *In Re Zaephel & Russell,* 49 F. Supp. 709, Ky., 1941, a statute that authorized a lien in favor of persons who "furnished labor, materials, or supplies for the construction" of a public improvement, did not intend a valid lien for public liability insurance premiums.

It was held in *McCormick* v. *Los Angeles City Water Co.,* 40 Cal. 185, 1868, that a person was not entitled to a lien for the value of his services rendered in cooking for the laborers employed on a public improvement under construction, notwithstanding the cooking was done on the ground as the work progressed.

The Supreme Court of New Hampshire, in *Perrault* v. *Shaw,* 69 N.H.

180, 1895, held that one does not perform labor or furnish materials for making brick, within the meaning of the statute, by furnishing board under a contract with the manufacturer, to the workmen employed in the brick yard.

In *Rara Avis Gold & Silver Mining Co.* v. *Bouscher,* 9 Colo. 385, 1886, the plaintiff was denied a lien for labor performed as an accountant and disbursing agent, although such services were necessary in the performance of the contract.

Labor is protected by the surety bond for payment when the workmen are engaged in removing stone from a quarry for the purpose of building a breakwater located at another place. (*U.S. F. & G. Co.* v. *United States,* 189 F. 339, 1911.)

A surety bond for all claims for labor and material that are furnished for the work is liable to the supplier of lumber used by the contractor for the false work in connection with the construction of a bridge. (*Empire State Surety Co.* v. *Des Moines,* 132 N.W. 837, Iowa, 1911.)

The question of enforceable claims against a surety bond for trucking, towage, and freight is answered differently in various courts. In *Alpina ex rel. Beaudrie* v. *Murray Co.,* 123 N.W. 1128, Mich., 1909, it was held that the surety bond for labor and material is not liable for transporting of lumber, coal, machinery, and other supplies to and from a dredge used by the contractor in performing his work. Also, in *United States, to use of Sabine & E. T. Ry. Co.,* v. *Hyatt, et al.,* 92 F. 442, 1899, the payment bond was held not liable for charges for railroad freight on materials that were loaded and unloaded by the contractor. The regular lien for freight charges, however, was available to the carrier, which, in this particular case, was waived by the railroad. In *American Surety Co. of New York* v. *Lawrenceville Cement Co., et al.,* 110 F. 717, 1901, the bond was held liable for charges by truckmen who moved materials from a certain place to the site "although the distance may be somewhat considerable." Waterborne transportation "carried on by the servants of the contractor, or for short distances without the aid of steam or a fully equipped vessel," was also held to be protected by the surety bond.

A rental charge for a boat chartered to a contractor at a per diem rate and used to carry materials is not covered by a surety bond for labor and material. (*United States, to use of Thomas Laughlin Co.,* v. *Morgan,* 111 F. 474, 1900.)

It is difficult, at times, to distinguish between a materialman who is protected by the payment bond and one who is not so protected. The language of the applicable statutes that are intended to protect the supplier of materials is of vital importance. The general criterion is whether or not the item in question became a part of the work; although in some states the lienable material must be tangible or visible as a part of the work to be included in the term. In *Red Wing Sewer Pipe Co.* v. *Donnelly,* 113 S.W. 1, Minn., 1908, it was held that where a supplier furnished materials for a construction project that were suitable for the purpose intended, it was not necessary for the supplier to prove, in an action on the surety bond to recover the price, that the material he delivered was actually incorporated in the construction work.

A lien for material furnished was claimed for the rental and moving of a concrete mixer that was used in the construction of a street improvement. The supplier attempted to support the validity of the lien by the fact that he also furnished the operator for the mixer. The court ruled that the rental of the concrete mixer was not classified as "material and labor," but was considered to be "appliances or equipment." "The claim is for the rental of a concrete mixer and the cost of moving the same, and not for human labor furnished by them" (*Cincinnati Quarries Co.* v. *Hess, et al.,* 162 N.E. 686, Ohio, 1927).

The judicial interpretation of material varies in different states, and items like gasoline and oil for the operation of contractors' equipment provide significant illustrations of the difference in formal court opinions. In *Mid-Continental Petroleum Corp.* v. *Southern Surety Co.,* 9 S.W. 2d 229, Ky., 1928, the surety bond was "for all labor performed or furnished, and for all materials used in carrying out of said contract." The dispute in the lawsuit was over a bill of $3,050.47 for gasoline, which was consumed by excavating and hoisting machinery and by gasoline pumps, which were used to remove water from trenches. The following comment in the decision discloses the variation in availability of a lien in different states: "Coming now to the question sharply in issue on this appeal, viz., whether the gasoline and oil furnished by appellant were lienable materials, it appears that a number of courts have held that coal and gasoline, used to generate power, and oil, used to lubricate machinery, employed by the contractor in construction work, are not lienable articles. The precise question has never been before this court, but the opinion in *Fidelity & Deposit Co.* v. *Hegewald,* 139 S.W. 975, Ky., is strongly persuasive that under our construction of the mechanic's and materialmen's statute, gasoline and oil when so furnished and used are lienable materials. The cases holding to the contrary are from states where the lien statute is strictly construed. *Schultz* v. *C. H. Quereau Co.,* 210 N.Y. 257; *Sampson* v. *Commonwealth,* 202 Mass. 326; *Barker Lumber Co.* v. *Marathon Paper Mills Co.,* 146 Wis. 12. On the other hand, dynamite and blasting powder when used in construction work are generally held lienable materials. *Schagticoke Powder Co.* v. *Greenwich & J. Ry. Co.,* 183 N.Y. 306; *Hercules Powder Co.* v. *Knox-*

ville, L. & J. R. Co., 113 Tenn. 382; *Giant Powder Co.* v. *Oregon Pac. R.R. Co.,* 42 F. 470." The court approved the lien for gasoline and followed the reasoning in *Smith* v. *Oosting,* 203 N.W. 131, Mich., 1925 that, as powder and dynamite used for blasting have been regarded as "materials," the use of gasoline has also contributed to and enhanced the value of the highways. Although neither the blasting material or the gasoline are physically incorporated into the highway, "both are wholly consumed in aid of the work."

The surety on a payment bond is not liable for the purchase price of machinery or materials for a contractor's plant, whether or not the plant is wholly or partly consumed in the work of the construction project. (*Nye-Schneider-Fowler Co.* v. *Bridges, H. & Co.,* 151 N.W. 942, Neb., 1915; *Beals* v. *Fidelity & Deposit Co.,* 76 App. Div. 526, N.Y., 1902; *United States Rubber Co.* v. *American Bonding Co.,* 149 P. 706, Wash., 1915; *Miller* v. *American Bonding Co.,* 158 N.W. 432, Minn., 1916.)

Among those considered to be materialmen are: (1) a manufacturer of cement blocks made in steel molds furnished by a sewer contractor; (2) a dealer in reinforcing steel kept in stock until requisitioned for delivery on a public construction project; (3) a supplier of finishing hardware consisting of standard stock goods of his trade; (4) one who contracts with the principal contractor to furnish the stone, cut and ready to set, for a city school, but has no part in the placing of the stone or in the construction of the building; (5) one who sells steam radiators to the principal contractor, but performs no work on the building; (6) a dealer in brick and other building material who sold brick made especially for the project to a building contractor for use in the construction of a city school building; (7) a stone dealer who furnished the curbstone of a paving contract; (8) an owner of a stonecrushing plant who sold different sizes of crushed stone to the principal contractor, but had no supervision over the construction work; and (9) the licensee of a patented process of laying pavement who contracts to furnish to a paving contractor the special equipment incident to the process.

Subcontractors have been found to be: (1) a woodmill operator who contracts with one making additions to and alterations in a city hall to furnish interior trim, installed by the latter, in accordance with the specifications; (2) a firm agreeing to furnish to a company contracting for the construction of a municipal plant, materials to be in strict accordance with the engineer's plans and specifications, and to be delivered as directed by the contractor's representative on the job; (3) one contracting to furnish the interior finish of a city school in conformity with the plans and specifications; (4) one who entered into a contract with another who had contracted with the state highway commission to construct a highway, to furnish all the sand at a certain rate per square yard of pavement (examples in this and preceding paragraph from *43 Am. Jur. sec. 148*).

7.9 Sub-subcontractors and payment bonds

The question of the rights of a sub-subcontractor under the provisions of a surety bond for the payment of labor and material has been before both federal and state courts on many occasions. The judicial determination of a claim in such cases depends upon the language of the related statute. The express intention of the statute under which a claim of a sub-subcontractor may be asserted is not always clearly stated, yet the success or failure of such a claimant in collecting his bill is conditioned upon the provisions of the governing statute. No guarantee exists, and none should be presumed, of uniformity of the terms of related laws in different states; therefore, the parties concerned must know the statutory rights of a creditor who is faced with the prospect of a loss because his protection was not within the limits of the law. The fate of certain claims that are discussed in the following paragraphs emphasize the importance of knowing and understanding the language in related federal and state statutes and the effect of words when labor services and materials have been supplied on public construction contracts.

One of the claims is at the state level and is found in *Wynkoop* v. *People,* 1 App. Div. 2d 620, N.Y., 1956, affirmed 4 N.Y. 2d 892, where the statutory provision in section 5 of the Lien Law of the State of New York provided: "A person performing labor for or furnishing materials to a contractor, his subcontractor or legal representative, for the construction or demolition of a public improvement pursuant to a contract by such contractor with the State or a public corporation, shall have a lien for the principal and interest of the value or agreed price of such labor or materials upon the moneys of the state or of such corporation applicable to the construction or demolition of such improvement, to the extent of the amount due or to become due on such contract." In this lawsuit, the prime contractor for a parkway project ordered material from the first subcontractor, a lumberman, who, in turn, ordered the material from a second subcontractor. The second subcontractor ordered part of the work from a third subcontractor, who ordered material from a foundry, the fourth subcontractor. The foundry delivered the finished material directly to the lumberman who paid the second subcontractor, and thereafter the lumberman was later paid by the prime contractor. The second subcontractor did not pay the third subcontractor, the plaintiff Wynkoop, who filed a notice of lien against the contractor's money in the hands of the parkway authority; this lawsuit was to fore-

A.I.A. DOCUMENT NO. A-310
(1958 Edition)

BID BOND

This document approved and issued by The American Institute of Architects
1735 New York Avenue, N. W., Washington 6, D. C.

KNOW ALL MEN BY THESE PRESENTS,

That we, ...

.... (hereinafter called the "Principal"),

as Principal, and the ....

...., of ....,

a corporation duly organized under the laws of the State of ....,

(Hereinafter called the "Surety"), as Surety, are held and firmly bound unto ....

....

.... (Hereinafter called the "Obligee"),

in the sum of .... Dollars

($), for the payment of which sum well and truly to be made, the said Principal and the said Surety, bind ourselves, our heirs, executors, administrators, successors and assigns, jointly and severally, firmly by these presents.

WHEREAS, the Principal has submitted a bid for ....

....

....

....

A.I.A. DOCUMENT NO. A-310
(1958 Edition)

Bid bond (face)

close that lien. The court pointed out that the chain of qualified lienors ended with the second subcontractor, and the decision added: "Under the Lien Law (N.Y. section 2, subdiv. 10) anyone who enters into a contract with a contractor or subcontractor for a public improvement or with a person who has contracted with or through the contractor

NOW, THEREFORE, if the Obligee shall accept the bid of the Principal and the Principal shall enter into a contract with the Obligee in accordance with the terms of such bid, and give such bond or bonds as may be specified in the bidding or contract documents with good and sufficient surety for the faithful performance of such contract and for the prompt payment of labor and material furnished in the prosecution thereof, or in the event of the failure of the Principal to enter such contract and give such bond or bonds, if the Principal shall pay to the Obligee the difference not to exceed the penalty hereof between the amount specified in said bid and such larger amount for which the Obligee may in good faith contract with another party to perform the work covered by said bid, then this obligation shall be null and void, otherwise to remain in full force and effect.

Signed and sealed this .... day of .... A.D. 195......

.... } .... (Seal)
(Principal)

....
(Title)

.... } .... (Seal)
(Surety)

....
(Title)

Bid bond (reverse)

acquires the status of 'subcontractor'. . . . The chain (as disclosed by the facts in this case) could continue indefinitely. If it be within the power of remote subcontractors effectively to file liens against moneys due the contractor, then the contractor to protect himself would be required to pay no one until expiration of the time within which liens might be filed, namely, 30 days after completion of the public improvement (N.Y. Lien Law section 12). The Legislature has limited the class of subcontractors entitled to file liens against the moneys due the contractor of a public improvement to 'A person performing labor for or furnishing materials to a contractor, *his* subcontractor or legal representative.' (N.Y. section 5, Lien Law). The restriction in the statute to *his,* as distinguished from a subcontractor, qualifies only those subcontractors who are no more remote from the contractor than persons performing labor for or furnishing materials to one who has contracted with the contractor." Wynkoop was not within the restricted field and his lien was nullified by the court.

The requirements of surety bonds for performance and for labor and material in contracts for public buildings and public works projects that are undertaken by the federal government are prescribed by the Miller Act in *40 U.S.C. Ann. sec. 270a.* The statute provides that before a contract exceeding $2,000 in amount for a public improvement is awarded, the contractor shall furnish two bonds; a performance bond for the protection of the United States, and a payment bond "for the protection of all persons supplying labor and material in the prosecution of the work provided for in said contract for the use of each such person." Section 270b of the same statute specifies who the beneficiaries of the payment bond are and gives them the right to sue on that bond. The leading case on the subject of claims of subcontractors and sub-subcontractors on federal contracts is *MacEvoy* v. *United States,* 322 U.S. 102, 1943, where the federal government made a contract with the MacEvoy Company to furnish materials and to do the necessary work for construction of housing units on a cost-plus-fixed-fee basis. The payment bond was in pursuance of the Miller Act, was issued by a surety corporation, and was conditioned upon payment by MacEvoy "to all persons supplying labor and material in the prosecution of the work provided for in said contract." MacEvoy bought building material for the project from Miller, who obtained the material from the Tomkins Company. Miller did not pay Tomkins about $12,000 of the bill for the material. Miller did not engage in any work of construction on the project, but MacEvoy paid Miller in full for the material. Tomkins notified MacEvoy and the surety of its claim and then sued on the payment bond. The lawsuit involved whether under the Miller Act the supplier of materials to a materialman of a government contractor can recover on the payment bond an unpaid balance of the account for the materials. The court decided against the Tomkins Company. The court ruled that the practical considerations and the express statutory language show no legislative intention to impose the claim of "remote and undeterminable liabilities incurred by an ordinary materialman, who may be a manufacturer, a wholesaler or a retailer. Many such materialmen are usually involved in large projects; they deal in turn with innumerable submaterialmen and laborers. To impose unlimited liability under the payment bond to those submaterialmen and laborers is to create a precarious and perilous risk on a prime contractor and his surety. To sanction such a risk requires clear language in the statute and in the bond so as to leave no alternative. Here the proviso of section 2a of the Act forbids the imposition of such a risk, thereby foreclosing Tomkin's right to sue on the payment bond." The court pointed out that the prime contractor knows who his subcontractors are, and he can secure himself against their creditors for labor and materialmen by requiring his subcontractors to give him a surety bond for that purpose. To extend the obligation of payment on the contractor's payment bond beyond labor and materialmen of the subcontractor who performs a part of the original contract is not the intent of the Miller Act.

Briefly stated, the provisions of the Miller Act do protect those who supply labor or materials directly to the prime contractor and also those who directly render or supply labor or furnish material to a subcontractor who has a direct contractual relationship with the prime contractor. The statute does not afford protection for all persons who perform or supply services of labor or furnish material used on the project or in the prosecution of the work.

An important decision on the subject of sub-subcontractors and their rights under the payment bond is *Elmer* v. *U.S.F. & G. Co.,* 174 F. Supp. 437, Miss., 1959, affirmed 275 F. 2d 89. The Tyler-Hyde Construction Co. was the prime contractor; they subcontracted the work of grading and paving to Scholes, who sublet the asphalt paving to Acme, who, in turn, subcontracted with Testing Laboratories to supervise the asphalt plant. Testing Laboratories was not paid the sum of $1,966 for its services, and they sued on the payment bond. The court ruled against the Testing Laboratories, and said:

"The legal question to be answered is: Do the facts, especially the contractual relations of plaintiff with the subject matter of the contract between the United States and Tyler-Hyde, above set forth, meet the requirements of the Miller Act, as interpreted by the Supreme Court in *MacEvoy* v. *United States,* 322 U.S. 102, or is it afflicted with the same infirmities in this case?

"Of course, in the MacEvoy case, it was found that Tomkins not only had no direct dealings with either the prime or the subcontractor, but furnished materials to a materialman, and the opinion states that he did not work on the job, while in the present case, Elmer, the plaintiff under contract with Acme, whose contract was with Scholes, who was a subcontractor to Taylor-Hyde, went upon the premises and rendered, apparently valuable and satisfactory services, to the knowledge of all concerned. However, if we accept the Supreme Court's statement in the MacEvoy case, then *the right to bring suit on a payment bond is limited to (1) 'those* . . . who deal directly with the prime contractor,' and (2) those of the same classes, 'who, lacking express or implied contractual relationship with the prime contractor, *have direct contractual relationships with a subcontractor,* and who give the statutory notice of their claims to the prime contractor,' it is hard to see how plaintiff can sue upon the bond.

"Finally, the court, itself, supplied the meaning of 'subcontractor' in cases such as MacEvoy and here as follows: 'In a broad, generic sense a subcontractor includes anyone who has a contract to furnish labor or material to the prime contractor. In that sense Miller was a subcontractor. But under the more technical meaning, as established by usage in the building trades, *a subcontractor is one who performs for and takes from the prime contractor a specific part of the labor or material requirements of the original contract, thus excluding ordinary laborers and materialmen.* To determine which meaning Congress attached to the word in the Miller Act, we must look to the Congressional history of the statute as well as to the practical considerations underlying the act.'

"This court, therefore, feels bound by this interpretation, and plaintiff's (Elmer's) demand must be rejected." The decision was affirmed on appeal in 275 F. 2d 89.

7.10 Release of surety from liability

The general principles that govern the release or discharge of a surety are founded upon the confidence and the good faith of each party to the suretyship transaction. In recognizing the rights of the surety, the public body must be cautious in its transactions so that there will be no act or omission that will materially increase the surety's responsibility or risk, unless the surety consents. Without that consent, any resulting material injury or prejudice to the surety will release the surety from its liability to the state or municipality to that extent. The usual material variations from the express requirements of a public construction contract are found in the time and the amount of progress payments, the amount of the retained percentage, and the demand for certificates or receipted bills; any or all of these departures from the contract provisions formed a basic representation that induced the surety to issue the bond.

Public construction contracts generally provide that the plans and specifications may be changed or modified. If the variation is of a minor nature and does not greatly extend the scope of the contract, the surety is considered to have consented to the change, in advance. However, this conclusion is not correct where extensive and material modifications are made in the contract.

The doctrine of *strictissimi juris* is from the common law, and in its application to the liability of the surety, the doctrine means that the express obligations of the surety are not to be extended beyond their strict meaning. *(40 Words and Phrases '60, p.p. 65.)* This doctrine does not apply to a corporation organized to conduct the business of issuing surety bonds, for which it receives compensation. Such a corporation is an insuror, and if the surety bond provisions have been drafted by the surety, they will be construed in favor of the obligee or the government unit. On the other hand, a voluntary surety is entitled to insist on the very letter of his contract of surety. (*State ex rel. City of Beckley* v. *Roberts,* 40 S.E. 2d 841, W. Va., 1946; *Ohio County* v. *Clemens,* 100 S.E. 680, W. Va., 1919.)

When a surety has knowledge of the facts that create a right of release from the bond, and by an affirmative act actually waives a ground for discharge by implication the surety loses its right to be released. In *Williams* v. *Pacific Coast Casualty Co* 140 P. 74, Wash., 1914, the surety of the subcontractor knew that the general contractor had advanced money to the subcontractor for payroll and materials, and then the surety consented, in writing, to payments on a different basis from that originally agreed upon. The court ruled that the surety had waived the breach of its bond agreement.

In another case, a contract expressly permitted "any additions to, or alterations or deviations from said drawings and specifications without invalidating this agreement." The original plans and specifications were canceled during the progress of the work, and new ones were substituted with an increased cost of about $14,000. The surety bond provided "that any alterations which may be made in the terms of the contract, or in the work to be done under it, or the giving by the owner of any extension of time for performance or contract, or any other forbearance on the part of either the owner or principal to the other shall not in any way release the principal and the surety or sureties, or either of them, their heirs, executors, etc. from liability hereunder, notice to the surety or sureties of any such alteration, extension or forbearance being hereby waived." The surety contended that the substitution of new plans and specifications was not within the scope of the construction contract and its bond. The court, however, denied the claim of the surety and said that the contract as modified was, in effect, the same as the original contract and that there was no radical variation effected by the substituted plans and specifications. The court also pointed out that the surety was protected against the increased cost of the contract by the fact that it was entitled to an additional premium charge that had previously been demanded by the broker who placed the bond. The demand for the additional premium charge was considered by the court as the practical construction of the modification provisions by an experienced insurance underwriter. The decision stated the applicable rule as: "it is well settled law that, even though a construction contract provides that alterations may be made during progress of the work, radical or revolutionary changes—changes not fairly within the contemplation of the parties at the time the contract was made – changes constituting a material departure from the original undertaking – do not come within its terms and will therefore release a non-consenting surety upon the usual performance bond. It is also well settled that material changes not so extensive as to constitute a departure from the original undertaking—changes which are fairly within the scope of the original undertaking—changes which may reasonably be said to have been originally contemplated by the parties as permissible alterations—are covered by a construction contract permitting alterations, and the making of such changes does not release the surety on the ordinary performance bond." (*Mass. Bonding Co.* v. *John R. Thompson Co.* 88 F. 2d 825, 1937.)

The protection of labor and materialmen as generally provided by the payment bond was discussed in *Equitable Surety Co.* v. *McMillan* 234 U.S. 448, 1914. The contract provisions did not cover changes, and the original plans were modified to relocate the front of a school from one street to another, and, as a result, a considerable amount of grading that had been done was wasted and new grading was required. In a decision in favor of labor and materialmen, the court said that the surety had notice that labor and materialmen would rely upon the payment bond, and the opinion added: "If the changes were so great as to amount to an abandonment of the contract and the substitution of a substantially different one, so that persons supplying labor and materials would necessarily be charged with notice of such abandonment, a different question would be presented, but in the case of such a change as was here made—a mere change of position and location of the building, without affecting its general character; involving changes in grading but having nothing to do with the furnishing of the materials upon which the action is based—it seems to us that the responsibility of the surety to the materialman remains unaffected."

The general rule on deviations from the contract or bond provisions with respect to conditions, amounts, and times of payment is that the surety is released to the extent that it has been injured but is not discharged from the entire bond obligation. (*Prudence Co.* v. *Fidelity and Deposit Co.,* 2 F. Supp. 454, 1933; *Burr* v. *Gardella,* 200 P. 493, Cal., 1921; *Museum of Fine Arts* v. *American Bonding Co.* 97 N.E. 633, Mass., 1912; *Crouse* v. *Stanley,* 154 S.E. 40, N.C., 1930; *Maryland Casualty Co.* v. *Eagle River Union Free High School District,* 205 N.W. 926, Wis., 1925.) In considering this subject, it is significant that surety bonds, in certain cases, contain provisions against discharge where premature payments and retentions are concerned. (*Guttenberg* v. *Vassel,* 65 At. 994, N.J., 1907; *Zang* v. *Hubbard Bldg. Realty Co.,* 125 S.W. 85, Tex., 1910.)

When a contract for constructing a ship provided that the owner would withhold a part of the contract price until he had inspected and had accepted the ship as satisfactory, the payment of the withheld money to the contractor without notice to the surety, released the surety because of the material violation of the terms of the contract and bond. The court said, in *Island Nav. Co.* v. *American Surety Co. of New York,* 227 S.W. 809, Ky., 1921, that:

"The rule is well settled in this jurisdiction that a surety, even for pay, is not bound if the principal obligee (the owner) failed to comply with the material and substantial provisions of the contract, the faithful performance of which is guaranteed.

"This is the generally accepted rule throughout the country, though there are a few exceptions. It is fundamental that any agreement or dealing between the creditor and the principal in an obligation of a debt, which essentially varies the terms of a contract, without the consent of the surety, will release the surety from liability. . . . When his contract is changed without his knowledge or authority, it becomes a new contract and is invalid, because it is deficient in the essential element of consent. And it has been held that it is not of any consequence that the alteration of the contract is trivial or even that it is clearly for the advantage of the surety, if it appears that it varied his responsibility, and was without consent. It is the surety's right to determine for himself what is, or is not, for his benefit. . . . Many decisions in stating the general rule employ the word 'material' and hold that any material change in the contract made by the principal parties to it, without the assent of the parties to it, without the assent of the surety, discharges the latter.... It is a general principle that any material alteration in a building contract will release nonconsenting sureties given to guarantee the faithful performance of the contract, and to protect the owner against any claims or liens for labor or materials used in the construction of the building.... So, generally, the courts declare broadly that the owner's failure to retain, until completion of the building, a specified percentage of the price of labor and material, as required by the contract, discharges the surety absolutely, and not merely to the extent of the premature payment."

In *United States* v. *Edward J. Freel, etc.,* 186 U.S. 309, 1902, the surety was released from his bond on a contract for construction of a drydock, because of a change without his consent. The revision was in the location of the drydock, which required the contractor to do additional excavating and to make connections with the water at a higher cost, and the time of performance was extended. These changes, the court held, were not contemplated by the contract and were beyond the terms of the undertaking of the surety. The court cited the following case as one of the illustrations within the principle of its decision, and quoted: "The supreme court of Indiana, in *Zimmerman* v. *Judah,* 13 Ind. 286, held that a supplementary agreement to put an additional story on a house released the surety for the contractor in the original contract."

Although the language of the contract and the bond governs the responsibility of the participating parties, the government disbursing officer is bound to recognize the rights of the surety when a notice of lien is filed against a contractor, prior to the payment of an estimate due the contractor. When conflict exists on the point in the courts of different jurisdictions, there is the risk of the operation of a discharge of a surety bond when payment is made by the owner to the contractor after a notice of lien has been filed. (*Lucas County* v. *Roberts,* 49 Iowa 159, 1878; *Shelton* v. *American Surety Co.* 127 F. 736, 1904; *Harris* v. *Taylor,* 129 S.W. 995, Mo., 1910; *Silberstein* v. *Kittrick,* 169 P. 250, Cal., 1917.) Decisions to the contrary are found in *Hayden* v. *Cook,* 62 N.W. 165, Neb., 1892; *Carson Opera House Assn.* v. *Miller,* 16 Nev. 327, 1881; *Foster* v. *Gaston,* 23 N.E. 1092, Ind., 1890; *Massachusetts Bonding & Ins. Co.* v. *Realty Trust Co.,* 73 S.E. 1053, Ga., 1912; *McCrum* v. *Love,* 58 Pa. Sup. Ct. 404, 1914; *Flint* v. *Chicago Bonding & S. Co.,* 168 N.W. 528, Mich., 1918.

chapter 8

Liquidated damages and delay

The term "liquidated damages" or "penalty" is used in contracts when the parties agree that damages must be paid for a default if the work is not completed within a certain number of days or not later than a stated time or date.

8.1 Liquidated damages vs. penalties in contracts

Liquidated damages is defined as "a specific sum stipulated or agreed upon in the contract as the amount to be paid to a party who alleges and proves a breach of it." (Bouvier's Law Dict.). Penalty is a term "mostly applied to a pecuniary punishment." (Bouvier's Law Dict.). In case of a penalty the injured party is entitled to recover from the other party the damages actually sustained, not the sum stated in the penalty. However, when a contract provides for liquidated damages, the amount recoverable is the stated sum (Anson on *Law of Contract,* p. 334). The question as to whether liquidated damages or a penalty is to be paid depends upon the intention in the entire contract.

Contracting parties may agree to an estimated figure for damages that result from doing or not doing a particular act or for not complying with a contract provision. Particularly when it is difficult to establish actual damages, prior agreement to pay stipulated liquidated damages for a breach is appropriate. A penalty is in the nature of security for the amount of any damages actually sustained, and a defaulting party can be relieved from that forfeit. Liquidated damages is a proper term when a sum is fixed in a contract for a breach in a matter of uncertain value, for example, a stated sum per day for delay in a construction contract. The term penalty applies to the payment of a sum that is in excess of the value of the matter of the breach, and may be considered by the courts as punishment and not as compensation for a loss.

Many decisions have been written by the courts on various provisions in contracts for liquidated damages in cases of breach of contract, especially in connection with delay in completion. A comprehensive summary of the rules is found in the following quotation from section 339 of *Restatement of the Law—Contracts*:

"Liquidated Damages and Penalties

(1) An agreement, made in advance of breach, fixing the damages therefor, is not enforceable as a contract and does not affect the damages recoverable for the breach unless

- (a) the amount so fixed is a reasonable forecast of just compensation for the harm that is caused by the breach, and
- (b) the harm that is caused by the breach is one that is incapable or very difficult of accurate estimation.

(2) An undertaking in a penal bond to pay a sum of money as a penalty for nonperformance of the condition of the bond is enforceable only to the extent of the harm proved to have been suffered by reason of such nonperformance, and in case for more than the amount named as a penalty with interest."

The modern position of the courts in practically all states was expressed in *Priebe & Sons* v. *United States,* 332 U.S. 407, 1947, as follows: "Today the law does not look with disfavor upon 'liquidated damages' provisions in contracts. When they are fair and reasonable attempts to fix just compensation for anticipated loss caused by breach of contract, they are enforced. *Wise* v. *United States,* 249 U.S. 361; *Sun Printing and Pub. Assn.* v. *Moore,* 183 U.S. 642; *Restatement Contracts,* section 339; *Kothe* v. *Taylor Trust,* 280 U.S. 224. They serve a particularly useful function when damages are uncertain in nature and amount or are unmeasurable as is the case in many government contracts. *United States* v. *Bethlehem Steel Co.,* 205 U.S. 105; *United States* v. *Walkof,* 144 F. 2d 75." The fact that the damages suffered are shown to be less than the damages contracted for is not fatal. These provisions are to be judged as of the time of making the contract.

When a contract for digging a drainage ditch and placing a concrete tube through a highway where a bridge had to be removed provided for payment of $40 per day as liquidated damages in case the completion date was not observed, the contractor did not finish his work for 30 days after the agreed completion date. In a lawsuit by the contractor to recover the balance he claimed was due for the work he performed, the county counterclaimed for $1,200 liquidated damages for the contractor's delay. The court granted the contractor the amount of his claim, less the $1,200 due the county under the liquidated damages clause. The court said: "The court rules as a matter of law that the provision in the contract for the payment of liquidated damages in the event that the project was not completed on or before July 1, 1955, was a lawful provision and that it ought to be enforced within reason. This was a public contract and it was highly proper for the board of supervisors to provide for liquidated damages in the event of failure to complete it on time. It is, of course, practically impossible to determine in advance the actual damages which may be sustained to the public on account of a failure to timely complete a public work of this nature. The court finds that the provision for $40 per day is a fair approximation of such damage, and that the provision in the contract is therefore legitimate and lawful and is not be interpreted as a penalty. The burden is on the contractor to show otherwise" (*Korshoj Constr. Co. Inc.* v. *Mills County*, 156 F. Supp. 138, Iowa, 1957).

8.2 Liquidated damages

Public policy approves liquidated damage clause: The United States Supreme Court has held that a liquidated damage clause in a public construction contract is not contrary to public policy, and it is a proper means "of inducing due performance, or of giving compensation, in case of failure to perform; and the courts give it effect in accordance with its terms" (*Robinson* v. *United States*, 261 U.S. 486, 1922).

In *United States* v. *Kanter*, 137 F. 2d 828, Mo., 1943, the contract stated that the actual damage for default in timely delivery would be impossible to determine, and therefore, the parties agreed to $10 per calendar day. The opinion stated that the courts look with favor upon agreements which provide for reasonable liquidated damages, and the provision will be enforced consistently, as it was written and agreed upon. If the amount stated for liquidated damages bears no reasonable relationship to the actual damages anticipated from a default, however, the courts are not inclined to enforce such claims, because such provisions are regarded to be "contracts for unenforceable penalties." When the liquidated damage agreement is reasonable, it will be enforced "although the party complaining of a failure of performance neglects or is unable to prove actual damages by reason of a breach of contract." If the party who seeks to recover on such a provision has contributed to part of the default or is chargeable with all of the delay or default, the provision will not be enforced. Where both parties have failed, the delay will not be apportioned unless the contract so provides. The *Robinson* case, described in a previous paragraph, was distinguished from the *Kanter* case in this paragraph, because the contract in the former case provided for completion date extension of one day for each day of delay caused by the government, which assumes apportioning the delay between the parties.

The specifications of a contract for two federal buildings provided that "time for the completion of the work shall be considered as of the essence of the contract, and that for the cost of all extra inspection and for all amounts paid for rents, salaries, and other expenses entailed upon the United States by delay in completing the contract, the United States shall be entitled to the fixed sum of $200 as liquidated damages, computed, estimated and agreed upon, for each and every day's delay not caused by the United States." There was a delay of 101 days beyond the contract period, and the government deducted $20,200 which the contractor's representative sought to recover by court action. The contractor argued that the deduction was a penalty which requires proof of actual damage, and a single sum of liquidated damages was specified without regard to whether one or both buildings should be delayed in completion. The decision approved the deduction of $20,200 as liquidated damages, stated that as the result of modern decisions, courts will try to determine the intention of the parties when they make an agreement that includes a stipulation to pay a designated sum for the specified breach of contract. "When that intention is clearly ascertainable from the writing, effect will be given to the provision, as freely as to any other, where the damages are uncertain in nature or amount, or are difficult of ascertainment, or where the amount stipulated for is not as extravagant or disproportionate to the amount of property loss, as to show that compensation was not the object aimed at, or as to imply fraud, mistake, circumvention, or oppression. There is no sound reason why persons competent and free to contract may not agree upon this subject as fully as upon any other, or why their agreement, when fairly and understandably entered into with a view to just compensation for the anticipated loss, should not be enforced" (*Wise* v. *United States,* 249 U.S. 361, 1919).

Suggested format: A liquidated damage clause in a public construction contract usually anticipates the difficulty of establishing actual damages; the inconvenience to the public because of the delay is not the subject of measurement in dollar value. Judicial approval exists, however, for reasonable liquidated damages when the contracting parties expressly agree to the amount and state that the actual loss is not ascertainable. When public authorities and commissions that have the legal right to impose tolls construct toll roads and bridges, there is usually a measurement of dollar loss based upon estimated collections on each day when the toll facility is open to users. Whether the contract is for a free or a toll facility, the liquidated damage clause should state either the basis for computing the sum to be paid by the contractor, or that the sum is reasonable and proportionate and is not computed upon a definite basis because of the difficulty in establishing one. The clause must also clearly disclose an express intention of the applicability of the clause to the particular contract.

An example of a reasonable sum and an express agreement of the contracting parties as to the basic reasons and the intention of the liquidated-damage clause is disclosed by the following excerpt from the specifications of contract for the construction of piers to accommodate, within a prescribed time, the superstructure of a toll bridge:

> "Date of Completion, Liquidated Damages.
>
> The contractor further agrees that he will begin the work herein embraced within ten days of the date hereof, unless the consent of the State, in writing, is given to begin at a later date, and that he will prosecute the same so that the several piers and the entire contract shall be completed in accordance with the following schedule:
>
> Pier M3 ready for steel erection July 1, 1958.
> Pier M4 ready for steel erection September 1, 1958.
> Pier M1 ready for steel erection December 1, 1958.
> Pier M6 ready for steel erection December 1, 1958.
> Pier M2 ready for steel erection December 1, 1958.
> Pier M5 ready for steel erection December 1, 1958.
>
> Complete all work on the contract on or before May 1, 1959.
>
> The term 'ready for steel erection' as used in the foregoing paragraph shall mean the earliest date on which the superstructure contractor can start his erection of superstructure metal work on the pier and proceed without interruption by operations under this contract.
>
> In case the contractor shall fail to complete any one pier 'ready for steel erection' within the time fixed in the foregoing, or within the time to which such completion may have been extended *without assessment*, the contractor shall pay to the State the sum of Five Hundred Dollars ($500.00) for each and every calendar day that the time consumed in completing such pier exceeds the time allowed therefor.
>
> In case the contractor shall fail to have completed two or more of the piers 'ready for steel erection' on the dates specified, or within the time to which such completion may have been extended *without assessment*, the contractor shall pay to the State the sum of One Thousand Dollars ($1,000.00) for each and every calendar day that two or more piers remain not 'ready for steel erection' and the time consumed in such completion exceeds the time allowed therefor.

In case the contractor shall fail to complete the entire contract within the time specified, following completion of all piers 'ready for steel erection', the contractor shall pay to the State the sum of Five Hundred Dollars ($500.00) for each and every calendar day that the time consumed in final completion exceeds the time allowed therefor.

Completion of the work under this contract on the specified time schedule is necessary and vital in the maintenance of a suitable schedule for the completion of the entire project. Other contractors will be dependent upon the facilities to be provided hereunder for the progress of their work, and failure to complete the project within the overall project schedule will result in loss of toll revenue to an extent not readily computable at the time of the letting hereof. The liquidated damages herein prescribed are established as a reasonable approximation of the loss in revenues as well as the additional cost of engineering expenses which will be required to be paid by the State.

Said sum, in view of the difficulty of accurately ascertaining the loss which the State will suffer by reason of delay in completion, is hereby fixed and agreed by the parties hereto as the liquidated damages that will be suffered by reason of such delay, and not as a penalty. The State will deduct and retain out of the monies which may become due hereunder, the amount of any such liquidated damages, and in case the amount which may become due hereunder shall be less than the amount of liquidated damages suffered, the contractor shall be liable to pay the difference upon demand by the State.

Liquidated Damages, Extensions of Time.

When the work embraced in the contract is not completed on or before the date specified therein, liquidated damages may be imposed as provided in the contract. The contractor may present to the State, in writing, any claim or claims for additional extensions of contract time for completion due to causes beyond his control, such as extra work or supplemental contract work added to the original contract, fires, strikes, floods, accidents, or unreasonable delays in receiving ordered materials and equipment. Such claims shall be presented to the district engineer within ten (10) calendar days after the occurrence of the claimed delay, accompanied by all necessary supporting data, and if based on valid grounds, will be considered by the State and such extensions of time granted as may seem to it to be fair and reasonable. However, no claims will be considered when based on delays caused by conditions existing at the time bids were received and of which the contractor might be reasonably expected to have full knowledge at the time of bidding, or upon delays caused by failure on the part of the contractor to anticipate properly the requirements of the work contracted for as to materials, labor and equipment, or as to time lost by any conditions of weather, whatsoever. Delays caused by non-availability of the site, in whole or in part, will not be considered unless they extend beyond the date contained in the special notes of this proposal relating to right-of-way availability. Further, delays caused by late delivery of steel will not be considered under any circumstances, notwithstanding the above, unless caused by act of God casualty or strike occurring subsequent to the award. Further, if the original completion date, as extended in accordance with these contingencies, is nevertheless exceeded and work still remains to be done, a final extension of time will be granted at the time of the completion of all work, only because the State's fiscal agent will not make final payment without such a document, but this final so-called extension of time will be granted solely because of the auditor's rules and will in no event relieve the contractor from liquidated damages otherwise assessable hereunder." (Copied from Ogdensburg Bridge Authority's contract OBA 57-1)

Bonus or premium not required: It has been determined judicially that an agreement that fixes, in advance, the amount of any damages that may be caused is valid; therefore, the basis of the liquidated damage clause is a written understanding of an anticipated loss resulting from a breach of contract. The amount fixed by the contracting parties is considered to be a reasonable attempt to state a just compensation, and it is not necessary that it be accompanied by another provision for a bonus or premium if there is no breach of contract or if the contract is completely performed earlier than a certain date of completion.

Waiver of liquidated damage clause: In *United States* v. *United Eng. & C. Co.*, 234 U.S. 236, 1914, the specifications provided that the government had the right to deduct $25 per day for each "calendar day after and exclusive of the date within which completion was required up to and including the date of completion and acceptance of the work, said sum being specifically agreed upon as the measure of damages to the United States by reason of delay in the completion of the work; and the contractor shall agree and consent that the contract price, reduced by the aggregate of damages so deducted, shall be accepted in full satisfaction for all work done under the contract." The government had deducted $6,000 for 240 days delay. The contractor and the engineer in charge disputed the design and construction of the floor of a pump well, and the interpretation of other requirements, all of which resulted in delay and cessation of work "without the fault of the claimant (contractor)." In a second supplemental contract made after the date of completion, which was stipulated in the prime contract, nothing was said as to time of completion or as to the delays under prior contracts. The Supreme Court of the United States stated: "We think the better rule is that when the contractor has agreed to do a piece of work within a given time, and the parties have stipulated a fixed sum as liquidated damages, not wholly disproportionate to the loss for each day's delay, in order to enforce such payment the other party must not prevent the performance of the contract within the stipulated time; and that where such is the case and thereafter the work is completed, though delayed by the fault of the contractor, the rule of the original contract cannot be insisted upon, and liquidated damages measured thereby are waived." The opinion commented that one party to a contract must not do anything that will interfere with the other party who is trying to complete his work within a stated time. The Court pointed out that the representatives of the government caused certain delays and, later on, the contractor was chargeable with other periods of delay. This situation is a basis for damages "measured by the actual loss sustained" and "even where both parties are responsible for the delays beyond the fixed time, the obligation for liquidated damages is annulled, and, in the absence of a provision substituting another date, it cannot be revived, and the recovery for subsequent delays must be for actual loss proved to have been sustained." This ruling by the court was followed in *United States* v. *John Kerns Const. Co.*, 140 F. 2d 792, Ark., 1944, in which the defense of the contractor in a suit by the government to recover liquidated damages was the delay of the public engineers which contributed to the late completion of the contract. If the other party to the contract "engages in conduct which prevents the doing of the thing contracted for in the specified time, such action amounts to a waiver of the claim for liquidated damages for delay."

Liquidated damages vs. actual damages: An interesting decision, which sustained a provision for liquidated damages over actual and greater damages, is found in *Trans-World Airlines Inc.* v. *Travelers Indemnity Co.*, 262 F. 2d 321, Mo., 1959. The contract stipulated liquidated damages of $50 per day to the city as owner of the property, and several hundred dollars per day to the airline company as lessee of specified areas of the property. The lawsuit included a claim by the company of $3,480,000 for delay damages not provided for in the contract. The court held: "Unquestionably the great weight of authority, it seems to me, almost unchallenged, is to the effect that where a contract and bond provides for the penalties to be paid in the event of a failure to comply with the terms of a contract, are clearly binding upon all the parties, and that special or unliquidated damages may not be collected in addition to liquidated damages which are clearly provided for and spelled out in the contract." The court, in approving on appeal the decision of the lower court added:

"We think it cannot be disputed that the general rule enforced by the courts of the State of Missouri and elsewhere is to the effect that where the parties especially provide or stipulate for liquidated damages, such liquidated damages take the place of any actual damages suffered and that any recovery for breach is limited to the amount so agreed upon. The rule is stated in *15 Am. Jur. § 264, Damages,* p. 697, as follows:

'The effect of a clause for stipulated damages in a contract is to substitute the amount agreed upon as liquidated damages for the actual damages resulting from breach of the contract, and thereby prevent a controversy between the parties as to the amount of damages. If a provision is construed to be one for liquidated damages, the amount named forms, in general, the measure of damages in case of a breach, and the recovery must be for that amount. No other or greater damages can be awarded, even though the actual loss may be greater or less.' "

In *Comey* v. *U.S. Surety Co.,* 217 N.Y. 268, 1916, the surety bond provided for a fixed sum per day as liquidated damages for any delay, and when the contractor refused to complete the work, the owner, with the approval of the surety, engaged another contractor to finish the project. Increased cost resulted, and the court approved a judgment for both the excess cost and the liquidated damages for delay. This precedent was cited in *McKegney* v. *Illinois Surety Co.,* 180 App. Div. 507, N.Y., 1917, both for the amount actually paid to complete the contract and the liquidated damages for delay. The opinion states: "The plaintiff is entitled to that compensation which will leave him as well off as he would have been if the contract had been fully performed. This includes not only the cost of completion, but also any special loss by reason of delays, etc."

Apportionment of delay: Although the courts in most states follow the rule that liquidated damages for delay will not be allowed when both parties to the contract are mutually at fault, variations in the application of the rule exist in Massachusetts and Texas.

The case of *Wallis et al.* v. *Inhabitants of Wenham,* 90 N.E. 396, Mass., 1910 involved a contract that called for speedy and continuous work to meet a specified completion date, with liquidated damages of $10 per day for delay, and included an agreement by the contractor to coordinate his work with the owner and with other contractors on the site. The provisions of the contract authorized extensions of time for obstruction or delay "in the prosecution or completion of the work by the neglect, delay or default of any other contractor or by any damage which may happen" by fire, strike of workmen, or action of the elements. The contractor claimed that because he was delayed by the municipality, nullifying the time limit for completion, he could be held liable only for actual damages and not for the stated liquidated damages. The Supreme Judicial Court of Massachusetts pointed out that whereas some jurisdictions support that contention (*Dodd* v. *Churton,* 12 Q.B. 562, 1897; *Holme* v. *Guppey,* 3 M & W 387; *Willis* v. *Webster,* 1 App. Div. 301; *Graveson* v. *Tobey,* 75 Ill. 540), it is universally agreed "that under such an agreement as this the owner cannot hold the contractors liable in the amount of the stipulated damages for any delays which have been due to his own fault. *Russell* v. *DiBandeira,* 13 C. B. N.S. 149; *Kenny* v. *Monahan,* 169 N.Y. 591, affirming 53 App. Div. 421; *Home Bank* v. *Drumgoole,* 109 N.Y. 63; *Marsh* v. *Kauff,* 74 Ill. 189; *Palmer* v. *Stockwell,* 9 Gray 237; *Amoskeag Manuf. Co.* v. *United States* 17 Wall. 592. But in many of the decisions in which contractors have been completely exempted from such liquidated damages for a failure to finish the whole work within the stipulated time it has been either assumed or found as a fact that the whole of the delay was due to the fault of the owner or of persons for whose conduct the owner was responsible. *Ludlum* v. *Vail,* 166 N.Y. 611, affirming 53 App. Div. 628; *Perry* v. *Levenson,* 82 App. Div. 94; *Boden* v. *Maher,* 105 Wis. 539; *Weber* v. *Collins,* 139 Mo. 501; *White* v. *Fresno Bank,* 98 Cal. 166; *Erickson* v. *United States,* 107 F. 204; *Altoona Electric Co.* v. *Kittaning Street Rwy.,* 126 F. 559; *District of Columbia* v. *Camden Iron Works,* 15 App. Cas. D.C. 198; *Dunavan* v. *Caldwell & Northern R. R.,* 122 N.C. 999. So in *Cornell* v. *Standard Oil Co.,* 91 App. Div. 345, the contractor finished the work as soon as the owner allowed him to do it. This was the principle applied in *Champlain Constr. Co.* v. *O'Brien,* 177 F. 271 in which the owner was found to be principally at fault for the delay which had occurred, but it was impossible to apportion the responsibility between him and the contractor." The court concluded that: "The builder is not relieved from his contract, but the owner cannot recover for delays which have been caused by himself or by those for whose conduct he was responsible. The parties are taken to have understood that the contractor's time limit was extended by the amount of such delays."

The Texas rule is found in *Bedford-Carthage Stone Co. et al.* v. *Ramey et al.,* 34 SW. 2d 387, Tex. 1930. The plaintiff had contracted to erect a college building by July 1, 1925, and he agreed to pay the college liquidated damages of $500 per day for delay. The plaintiff made a contract with the defendant stone company to quarry and deliver Leuders stone as rapidly as needed for the erection of the building. This contract provided for stipulated dates of shipment of the stone and $100 per day liquidated damages for delay. The stone shipments were considerably delayed; the stone company defended on the ground that the contractor did not pay for the material as he had agreed to do, and it also alleged that cold weather contributed to the delays as an act of God. The Court of Civil Appeals of Texas stated that the "better rule" is, that upon good and satisfactory evidence or by standards agreed to by the parties, damages resulting from delays caused by the default of the respective parties will be apportioned between them.

Unenforceable penalties: A clarifying discussion of a liquidated damage clause in *Steffen* v. *United States,* 213 F. 2d 266, 1954, included the following: "A provision in a contract for liquidated damages will be enforced by the court provided it is in reality liquidated damages and not a penalty. If such provision is in fact a penalty it will not be enforced and the injured party will be entitled to recover the actual damages suffered. *Restatement of Contracts, sec. 339; Fidelity & Deposit Co. of Maryland* v. *Jones,* 256 Ky. 181. . . ." Also "As a general rule, whether the stipulated sum is liquidated damages or a penalty depends upon the intention of the parties, to be determined by a consideration of the entire contract and the circumstances under which it was executed. *Kothe* v. *Taylor Trust,* 280 U.S. 224, 226. . . . An agreement to pay a fixed sum regardless of the nature of the breach or the extent of the damage, or which has no reasonable relation to the probable damage which may follow a breach, will be treated as a penalty and will not be enforced. If the actual damage sustained would be wholly uncertain and incapable or very difficult of being ascertained, except by conjecture, the courts are inclined to look to the measure of damages fixed by the contract. But where the actual damage can be correctly ascertained by the application of a proper measure of damages to the actual facts, the courts are inclined to treat the provision as a penalty and leave the parties to their proof of actual damages suffered."

In the *Steffen* case, a contractor for demolition of a building was unable to complete within the stipulated time, despite several extensions of time; a new contractor was brought in, and he also failed to complete the job. A third contractor was engaged to finish the work, for which he was to be paid more than $7,000, instead of $1,111 for the first contractor, and $850 for the second contractor. The government sued the first contractor for breach of contract and for more than $5,000 damages, the additional cost of the work. The first contractor had deposited with his proposal a surety bond of $1,000 as a guarantee of performance, which the surety paid to the government. That money was applied as a credit against the additional cost of more than $6,000. It was found that the contract with the completion contractor contained terms and conditions that differed considerably from those in the first contract. The case was ordered returned to correct an erroneous instruction to the jury concerning items to be considered in their verdict, but it appears from the opinion cited that it is not correct to measure damages by comparing contracts that call for widely different specifications or conditions.

8.3 Delay

Causes and relative principles: Subject to the provisions in a particular contract, a government unit is generally held answerable to its contractors for money damages for the delay it causes them in situations usually created by one or more of the following conditions:

1. Failure to provide the right of way
2. Failure to complete the preliminary arrangements to make the site available
3. Change in plans or spectifications
4. Act or omission of the engineer or architect who represents the government unit
5. Act or omission of others contractors at the site
6. Failure to make progress payments
7. Suspension of the work by order or by the act of the governmental unit.

For a contractor to recover damages for delay caused by the state or municipality, it must appear that:

1. The contractor's work was delayed.
2. He suffered damages because of the delay.
3. The contractee was responsible for the act or omission that caused the delay.
4. That act or omission caused the delay in the contractor's work.

If reasonable delay was anticipated when the specifications disclosed a specific situation, the contractor has no ground for a claim for special damages if the delay occurs. For example, where the specifications stated that the contractor must await the availability of the site of the public work, by purchase or by condemnation, the contractor has no valid claim for delay if he starts his work without waiting for the acquisition of the site. (*Normile, Fastaburd & McGregor* v. *United States,* 239 U.S. 344, 1915; *Crook Co.* v. *United States,* 270 U.S. 4, 1926.)

Another example of such a case was found in *Connolly* v. *State,* 120 Misc. 854, N.Y., 1923, where the contractor knew that the town through which his contract ran was laying watermains in the bed of the affected right of way. His claim for damages against the State for the delay of the town in completing its installation was denied because he had prior knowledge of the attending circumstances.

In *State* v. *Feigel,* 178 N.E. 435, Ind., 1931, the State failed to secure the right of way, and the contractor sued for damages for a breach of contract on account of the delay. The court held that there was implied in the contract a covenant by the State that the right of way was available to the contractor. The State, as one of the contracting parties, was held liable for the failure of its representatives to get the right of way. In support of this reasoning, the court's opinion referred to *Schunnemunk Constr. Co.* v. *State of New York,* 116 Misc. 770, N.Y., 1918, where the contractor's work was interrupted by the State's failure to provide the necessary right of way, and the latter decision stated: "The issue is whether the State has violated the contract by interfering with and interrupting the claimant in its performance of the obligation. Such interference between individuals or corporations confers a right of action.... If an individual in good faith believed that he was the owner of land, and contracted with a builder for the construction of a house upon it, and during the progress of the work discovered that he had not title, and interrupted the performance of his builder in erecting the structure, he would be liable for the damage caused.... The situation is no different here." The general acceptance of this principle by the courts in other jurisdictions is disclosed by: *Brennan Const. Co.* v. *State of New York,* 117 Misc. 816, N.Y., 1921; *Carr, Auditor, etc.* v. *State,* 127 Ind. 204, 1871; *City of Indianapolis* v. *Indianapolis Water Co.,* 113 N.E. 369, Ind., 1916; *Hartman* v. *Greenhow,* 102 U.S. 672, 1880; *Grogan* v. *San Francisco,* 18 Cal. 590, 1861; *Chapman* v. *State,* 104 Cal. 690, 1894; *Mansfield* v. *N.Y.C. & H.R.R. Co.,* 102 N.Y. 205, 1886.

The responsibility of a government unit to its contractor was discussed in *Shore Bridge Corpn.* v. *State of New York,* 186 Misc. 1005 N.Y., 1946, affirmed in 271 App. Div. 811. The claimant sued the State for damages caused by delay due to errors by the State in the alignment and design as well as the driving of test piles. For one month of the alleged delay, the contractor's plant, machinery, and equipment were idled at the site and could not be released for use elsewhere. The contract was, nevertheless, completed on time, and one of the questions before the court was whether or not damages were recoverable regardless of the timely completion of the job. The contractor alleged that the State breached the contract by unreasonable delay and that his own progress schedule had been arranged to complete his work one month before the specified date of completion. Time of completion was of paramount importance, and the contractor had the right to concentrate upon his performance to carry on the work to his best advantage and with economy. The court ruled that one party to a contract does not have the power to destroy the mutual obligations of the agreement by delaying tactics and thereby force the other party into a default because of nonperformance. The owner has the duty to do his part to facilitate the work of his contractor. (*Mansfield* v. *N.Y.C. & H.R.R. Co.,* 102 N.Y. 205, 213.) In every contract is implied an obligation that neither party will interfere with or impede the progress of the other party in carrying out the work to completion. (*Patterson* v. *Nieverhofer,* 204 N.Y. 96.) "Whether a delay is so great," said the court, "as to be beyond the contemplation of the parties to the contract is a question of fact to be determined in the light of all the surrounding circumstances."

Another instance of interference by a government unit is found in the case of *American Bridge Co.* v. *State of New York,* 245 App. Div. 535, N.Y., 1935, where the foundation contractor encountered serious difficulties, which delayed the claimant who had the contract for the bridge superstructure. The State ordered the claimant to fabricate the steel, although it was obvious that the delay in erecting the superstructure would be of long duration. The steel was stored out-of-doors, and before it could be incorporated in the work, it was necessary that the material be repainted, which involved considerable additional expense for the claimant. This action by the State was held to be interference, rather than merely a matter of delay for which a recovery was allowed to the claimant.

In *Guerini Stone Co.* v. *P. J. Carlin Const. Co.,* 248 U.S. 334, 1918, it was held: "It is sufficiently obvious that a contract for the construction of a building even in the absence of an express stipulation upon the subject, implies as an essential condition that a site shall be furnished upon which a structure may be erected." And, in *Great Lakes Const. Co.* v. *Republic Creosoting Co.,* 139 F. 2d 456, 1943, there is the statement: "And it is equally obvious that a contract to construct the flooring in a building implies timely provision of the situs for its location."

Obligations with multiple contractors and one site: Although the state or the municipality has the obligation, both express and implied, to facilitate the work of a contract and not to hinder its progress, there is a distinction in the principle when several contractors are engaged on the same site of a construction project. In *United States* v. *Blair,* 321 U.S. 730, 1944, the specifications required the general contractor to cooperate with the other contractors on the site and not to interfere with them. The contract provided for liquidated damages in case the general contractor delayed the completion of the work, except unforeseen situations created by the government. The court held that the government had no obligation to assist the general contractor to complete the contract before the stated date, and the general contractor was not entitled to damages because the government failed to force accelerated performance by another contractor.

When more than one contractor is engaged in work upon the same public site, the government unit is not automatically liable if delay by one of its contractors causes another to be delayed. Responsibility for delay may be charged to the state or to the municipality if: (1) the specifications are misleading; (2) if the specifications require substantial changes in the procedures; (3) if the public officers or their representatives interfere with the progress of the work or fail to advance it reasonably; or if they take or require any action other than that expected by the parties to the contract. (*Endes Plumbing Corp.* v. *State of New York,* 198 Misc. 546, N.Y., 1950.)

In *Stehlin-Miller-Henes Co.* v. *City of Bridgeport,* 117 At. 811, Conn., 1922, the contractor agreed to do electrical work and to install heating and ventilation; liquidated damages were stipulated, respectively, for delay. The contract provided that the work to be done "as fast as the building is in condition," and the contractor was to avoid delaying the general contractor. A three-months delay ensued in the work of the general contractor, and the electrical and heating contractor could not complete his work on time. He charged the municipality with failure to provide a building in which he could perform his work, and asserted that the actual completion date of the structure was almost one year beyond the stipulated date. The municipality had to engage another contractor to take the job over from the original general contractor to expedite the completion of the school building, and the time lost caused damage to the electrical and heating contractor. The court said: "The rule is undoubted in circumstances such as were present in this case that an implied contract arose on the part of the defendant (municipality) to keep the work on the building, whether done by itself or other contractors, in such a state of forwardness as would enable the plaintiff (electrical and heating contractor) to complete its contracts within the time limited."

The rule in the *Bridgeport* case was cited in support of a decision in *Byrne* v. *Bellingham Consol. School Dist. #301,* 108 P. 2d 791, Wash., 1941. The damage clause in the contract in the *Byrne* case provided for a claim against any party causing damage by a wrongful act or neglect. The building was delayed 10½ months in completion, requiring 18 months for the contractor to complete the electrical work, instead of 7½ months as stipulated in the contract. The architect issued extensions of time, but the court held the school district responsible to this contractor, even though the delay may have been caused by the general contractor. The decision stated: "The rule announced by the courts in practically every state in the Union, including this state (Washington) is that, in the absence of any provision in the contract to the contrary, a building or construction contractor who has been delayed in the performance of his contract, may recover from the owner of the building damages for such delay if caused by the default of the owner. *Hetherington-Berner Co.* v. *Spokane,* 135 P. 484. Such right of recovery is predicated upon a breach of what we have already stated is an implied obligation on the part of an owner to furnish to the contractor a building in a state of forwardness sufficient to enable the contractor to complete the contract within the time limit."

Unavailability of right-of-way: During a contract to construct a state highway, when the contractor was notified by the state's chief engineer to proceed, he assembled his equipment and completed preparations to begin his operations. The right-of-way was not available because of acts and omissions of the state, and the delay caused the contractor damages of more than $20,000, for which he sued the state in the case of *State* v. *Feigel,* 178 N.E. 435, Ind., 1931. The state contended that: (1) the specifications required the contractor first to examine the site; (2) the contract empowered the chief engineer to suspend the work of the contractor when, in his opinion, the conditions warranted that action; (3) the delay was fully compensated by an extension of time as provided in the contract; and (4) the contractor started his work knowing that the right-of-way was not available. The contractor was paid under the terms of the contract, but the court ruled that the contractor also had the right to sue for breach of the contract. As to the state's claim that the damage suit for delay was not valid because of the issuance of the extension of time, the decision stated: "This might be true if the delay was caused without the fault of either party, but it certainly would not be true where the delay was caused by the public body without right."

Work methods disputes: A common cause of delay in the progress of government construction contracts is disputes over whether the method preferred by the contractor or the one preferred by the engineer in charge should be used to produce the end result. In *Meads & Co.* v. *City of New York,* 191 App. Div. 365, N.Y., 1920, the contract provided that the contractor had the option of either sheathing from the grade levels to the bottom of the excavation, or of pitching back the banks at such an angle that the adjoining material could not fall into the excavation and sheathing from two feet above the base of the banks to the bottom of the excavation, providing that the tops of the banks should in no case come within five feet of the property line. He decided to follow the method of pitching back the banks. While the work was in progress, a slide occurred and the architect ordered the contractor to install trussing, sheathing and shoring instead of the method the contractor had selected. The contractor protested that the order was a change in method, but performed as directed. The court held that the architect's right to issue the order was debatable, and upon all the facts and circumstances, concluded that the right did not exist. The contractor's method met the requirements of the specifications, and the court held that "The law is that so long as a contractor produces work which satisfies the specifications, he can, in the interest of economy, choose his own methods. This is not only law, but common sense; for when a contractor bids, his estimates, which influence the bid, are necessarily based on his own methods of work, so long as those methods are not controlled by the specifications. *Horgan* v. *The Mayor,* 160 N.Y. 516."

The principle in the *Meads & Co.* case was applied also in *Baker Co.* v. *State of New York,* 267 App. Div. 712, N.Y., 1944, where the state let a contract for construction of a powerhouse and concrete tunnels to one contractor, and let another contract to the plaintiff, Baker & Co., for installing a piping system and heating equipment. The plaintiff assumed that the tunnel work would be continuous, which was vital to his schedule of operations. The schedule was completely disrupted when the state, over the claimant's protest, permitted the other contractor to build the powerhouse and the tunnel in a disjointed manner, and, too, the claimant suffered water damage when the other contractor overran his completion date by eleven months. The state was held liable to the claimant because it permitted the other work to be done in the manner stated, which compelled the claimant to perform his work by the same procedure. This was ruled by the court to be interference with the claimant's work and prevented him from using his usual economical methods.

Subcontractor's rights: A general contractor is obligated to his subcontractor to make good all losses caused by delay in the work not attributable to the subcontractor. The basis of the rule is that the general contractor, in control of the work, might have prevented the delay. As long as he expects the subcontractor to fulfill his obligation, he has corresponding obligation, whether express or implied, to make good any losses of the subcontractor, unless the contract expressly exempts him or permits an intention by inference that the responsibility is not his. (*Norcross* v. *Wills,* 198 N.Y. 336, 1910.)

8.4 No damages for delay clause and its application

Generally, public construction contracts contain provisions that are intended to prevent the contractor from making a claim for damages because of an act or omission of the governmental unit. The provisions include an agreement by the contractor that he will be compensated for any such damages if the state or municipality grants an extension of time for the delay periods. The courts construe the "no damages" provision strictly, according to the language to which the parties agreed; however, it is a settled rule that a contractor is not stopped from making a claim for damages if the government unit fails to perform its contractual obligations. (*Moran Bros. Co.* v. *United States,* 61 Ct. Cl. U.S. 73, 1925; *Kelly* v. *United States,* 31 Ct. Cl. 361, 1896; *Selden Breck Const. Co.* v. *Regents of Univ. of Michigan,* 274 F. 982, 1921.) Whereas judicial decisions seek to avoid, if possible, any harshness by enforcing the stop provisions, the clear and unambiguous statements in a contract must be given the construction that was intended, even though the result may be harsh. (*Erickson* v. *Edmond School Dist.,* 125 P. 2d 275, Wash., 1942.) Each case depends upon its own facts, therefore, the answer to the question of the contractor's right to recover damages for delay is to be found in the limitations, restrictions, and circumstances in the particular case. (*Cauldwell-Wingate Co.* v. *State of New York,* 276 N.Y. 365, 1938; *Wells Bros.* v. *United States,* 254 U.S. 83, 1920; *Orlando* v. *Murphy,* 84 F. 2d 531, 1936.)

A public construction contract provision that the contractor is not entitled to damages for delay, but only to an extension of time, when the delay is occasioned by act or omission of the municipality, was discussed in *Manerad* v. *City of Eugene,* 124 P. 662, Oregon, 1912. The provision stipulated: "The contractor shall not be entitled to damages on account of delay, but if such delay be occasioned by the city, the contractor shall be entitled to an extension of time in which to complete the work, to be determined by the engineer." The contract was made for the construction of a canal for a power plant for the city in accordance with plans and specifications. The city reserved the right to make any changes in the plans thought necessary. At the beginning, a delay occurred because of a lawsuit by an adjacent property owner, and the city gave an extension of time to the contractor. In the suit against the city, the contractor charged that the city caused delays by refusing for an unreasonable time to secure space to place waste material, and for borrow pits. The engineer in charge, the contractor alleged, failed to cross-section the earth construction as is customary in such cases, and he also omitted to stake out the side lines of the earth construction. The contractor also complained that much time was lost in operations because no municipal engineering representative was regularly assigned to the site to furnish the necessary information and direction for the progress of the work, and because the city made many mistakes in giving lines and grades, the correction of which delayed the regular work of the contractor. During the fall a great rainfall and flood made it impossible to do any work until the following spring. The suspension of the work was agreed upon between the contractor and the municipality, but the municipality took over the contract and completed the work. The contractor sued for damages, and the city charged him with abandonment of the work, forcing the city to complete the project; therefore, the city claimed reimbursement for the completion cost. The court disallowed the contractor's claim and stated: "under its (the contract's) terms mere delay of the city constituted no ground for damage. It only extends the time for the plaintiff (contractor) to complete the work within the discretion of the engineer." As to the contractor's claim of custom of his trade as supporting his position, the court added: "If custom is to be pleaded as an element of plaintiff's case, it must be shown to be, among other things, general in the business involved, and that it was either known to the party of such general notoriety as to presumption that it was thus understood." No proof of custom or of knowledge on the part of the city was offered by the contractor, and the claim of custom by the contractor was also disallowed.

A leading case on the subject of the "no damage" clause is *Cauldwell-Wingate Co., et al.* v. *State of New York,* 276 N.Y. 365, 1938. The lawsuit involved a contract for the superstructure of a state building; the foundation was supposed to have been completed and ready within three weeks. A delay of about nine months occurred, however, because of an error in the specifications as to subsurface conditions. The plans showed a cross-section of the site that did not disclose some old masonry foundations built on wooden piles over a swamp. The contractor sued for increased cost because of the erroneous plans. The court ruled that since the contract was contingent on the performance of the foundations contractor, the superstructure contractor assumed the risk of delay in the foundation work, but did not assume such a risk due to the misrepresentations of the plans and specifications. The contract for the superstructure provided for extensions of time for delay caused by any act or neglect of the state, and the contractor would make no claim for damages for delays from any cause. It was decided in the lawsuit that the contract obligated the superstructure contractor to wait a reasonable time for the completion of the work of the foundation, and the "no damage" clause did not apply to delays because of direct interference and misrepresentations of the state.

In *A. Kaplen & Son, Ltd.* v. *Housing Authority,* 126 At. 2d 13, N.J., 1956, the contract was for public housing, and it was necessary to remove the occupants of the buildings before the new construction could be started. A delay of about five months occurred, and the contractor sued to recover damages of about $160,000. The contract provided that no payment of any kind, except extensions of time, be made to the contractor "because of hindrances or delays from any cause in the progress of the work, whether such hindrances or delays be avoidable or unavoidable, and the Contractor agrees that he will make no claim for compensation, damages, or mitigation of liquidated damages for any such delays, and will accept in full satisfaction for such delays said extension of time." The court ruled that the quoted provision was agreed to by the contracting parties and was binding against the contractor. It was also decided that the "no damages" clause is not against public policy, but is a proviso "obviously conceived in the public interest in protecting public agencies contracting for large improvements on the basis of fixed appropriations or loan commitments against vexatious litigation based on claims, real or fancied, that the agency has been responsible for unreasonable delays."

A contractor sued a state university for damages due to delay caused by the university representatives, principally because of late delivery of the site. The "no damage" clause in the contract attempted to excuse the university from responsibility for damages through the fault of any other contractor on the site, or through delay caused by the university; the extension of time allowance was stated to be the recognized and accepted equivalent of the lost time. The court said in the case of *Selden Breck Const. Co.* v. *Regents of U. of Michigan,* 274 F. 982, Mich., 1921, that the act of the contractor in completing the project after the alleged breach of contract was not a waiver of any right to recover damages. The decision pointed out: "Consideration of the subject satisfies me that the correct rule is that upon breach of a building contract by the failure of the owner to perform his obligations under such contract, which delays the contractor in completing his work thereunder, the latter is not obliged to abandon such work, but may elect to continue therewith after such breach and, upon performance of the contract on his part, is entitled to recover the damages sustained by him as a result of the delay caused by such owner. *W. H. Stubbings Co.* v. *World's*

Columbian Exposition Co., 110 Ill. App. 210; *Allanion* v. *Albany,* 43 Barb. 33, N.Y.; *Florence Oil & Refining Co.* v. *Reeves,* 56 P. 674."

The "no damages" clause because of delay of the contractor was discussed in *Psaty & Fuhrman* v. *Housing Authority,* 68 At. 2d 32, R.I., 1949. The contract called for completion of construction work in 300 days from the date of the authority's order to start, with landscaping to be finished in an additional 200 days. The contract also provided for liquidated damages of $250 for each day of delay in construction work and $25 per day for landscaping. The "no damages" clause stated: "No payment or compensation of any kind shall be made to the contractor for damages because of hindrance or delay from any cause in the progress of the work, whether such delays be avoidable or unavoidable." Extensions of time could be made available under the same clause. The contractor charged the authority with bad faith, and contended that the "no damages" clause excused the authority for reasonable delay only. The court held that such a construction of the clause would practically always subject the authority to an inquiry as to whether a reasonable person would have acted the same way or in a different manner. The reason for the "no damages" clause is to avoid just that type of inquiry, and, also, the clause is valid and binding unless there is evidence of malicious intent, fraud, or bad faith on the part of the governmental unit. These conclusions, the court said, are in accord with the decision of *Mach* v. *State of New York,* 122 Misc. 86, N.Y., 1923, affirmed 211 App. Div. 825.

In *Taylor-Fichter Steel Constr. Co.* v. *Niagara Frontier Bridge Commision,* 261 App. Div. 288, N.Y., 1941, affirmed 287 N.Y. 669, the contractor claimed a breach of the contract for the construction of steel superstructures for two bridges over the Niagara River. Another contractor was to erect the supporting substructures consisting of abutments and piers. The claim was based on unlawful interference by the commission, so that the work the contractor could have completed continuously in 14 weeks, was spread over 37 weeks, which added to the contractor's expense. The specifications included a schedule of completion dates of the different parts of the substructure, and also provided that the contractor would have no claim, other than extension of time to complete, if the substructure contractor did not meet the schedule of completion dates. The claimant-contractor asked the commission for a postponement of the commencement of its work because the substructure operations were hampered by winter weather conditions and also by unexpected conditions below the surface, yet the superstructure was completed ahead of time. The court held that the commission's engineers had the right to exercise their judgment as to the time when the substructure would be ready, and while their judgment turned out to be wrong, there was no interference with the progress of the claimant-contractor's work; the provisions of the contract estopped the claimant-contractor from a damage claim for delay, since the engineers had taken all reasonable steps to expedite the work.

When a state or municipality fails to furnish temporary heat, as agreed, for interior construction work, the "no damages" clause is not a defense to a claim for delay caused a contractor. It has been held in such a situation that the omission to furnish heat, as agreed, was active interference with the contractor. *DeRiso Bros.* v. *State,* 161 Misc. 934, N.Y., 1937.

If a contract expressly provides that the governmental unit shall have complete discretion to suspend performance or to change the work or materials and allows additional time because of such suspension, no damages for delay are recoverable, particularly when the provisions of the contract are that "no claim shall be made or allowed to the contractor for any damages which may arise out of any delay caused by" the contractor. (*Wells Bros.* v. *United States,* 254 U.S. 83, 1920.)

8.5 Contractor's rights upon delay in progress payments

When an action was brought under a federal statute to recover money due a subcontractor from the contractor and the surety, the subcontractor claimed that progress payments were not made for such a long time that he could not proceed with the work. However, an oral agreement had been made under which the subcontractor agreed to complete the work with the arrangement that payment would be made when the contract was completed. The subcontractor's lawsuit was to recover the reasonable value of his work instead of the agreed price because the contractor had interefered with and had allegedly prevented performance of the subcontractor's work. The court denied the claim of the subcontractor for payment based on reasonable value and said that when one party to a contract does not fulfill his obligation, the other party "may cease performance, and rather than sue for damages, i.e. his outlay to the date of the breach, plus lost profits . . ." he may rescind the contract and make the document void and annulled, and seek to restore the status quo, that is, as far as possible, place the contracting parties in the same position as before making the contract, in which case the amount recoverable is not limited by the contract price, but is measured by the reasonable value of the work performed. (*United States* v. *Behan,* 110 U.S. 338, La., 1884; *Bucholz* v. *Green Bros. Co.,* 172 N.E. 101, Mass., 1930; *Connolly* v. *Sullivan,* 53 N.E. 143, Mass., 1899.) It is clear from the following decisions, too, that the party who claims the contract has been breached or broken by the other party has the option to sue for the price of the contract or to claim the reasonable value of his work. (*Schwasnick* v. *Blandin,* 65 F. 2d 354, Vt., 1933.)

The rule is explained in *Krotts* v. *Clark Const. Co.* 249 F. 181, Ill., 1918. In this case the owner refused to pay the contractor on the monthly progress estimate of the architect as required by the lump-sum contract, and by this evidence of renunciation of the contract by the owner, the contractor was justified in abandoning the contract and in suing to recover, as the minimum measure of damages for the breach of the contract, the outlay reasonably incurred by him, without regard to loss or profit on full performance of the work.

In a contract for the construction of the approaches for the superstructure of a bridge, monthly progress payments were to be made upon estimates approved by the chief engineer of the District, less 12½ per cent to be retained until completion and acceptance of the work. When about seventy per cent of the work had been completed and paid for on the monthly progress estimate basis, payments of more than $33,000 were not made because of the lack of funds; the contractor suspended the work and refused to continue until the past due estimates were paid. A judgment for the amount claimed was obtained by the contractor, which was paid several months later from the sale of bonds which were issued for the purpose. Thereafter, with sufficient funds available, the parties made a supplemental agreement, the contractor completed the work, and he was paid the amount of the original contract. The lawsuit *Underground Constr. Co.* v. *Sanitary Dist.,* 11 N.E. 2d 361, Ill., 1937, was started to collect special damages for ten months' delay for lack of funds, and the supplemental agreement disclosed that the contractor reserved his right to claim and sue for damages due to the delay. Three questions appeared during the trial:

1. Did the judgment recovered for the two unpaid installments of more than $33,000 decide the claims of the contractor in this lawsuit for special damages for delay?

2. Did the contractor have but one of three courses to pursue when the District breached the contract by failing to pay the progress estimates, that is

(a) to treat the contracts as rescinded or canceled and to sue for the work he had completed, or

(b) to continue the work and sue for payment on each installment or progress payment as it became due, or

(c) to complete the work and sue for the unpaid balance of the entire contract price.

3. Could the contractor suspend the work until the installments were paid and then sue for special damages because of the delay?

The court ruled that the judgment for the sum of more than $33,000, which the contractor had collected from the District, had not decided the claims for delay, because, at the time the progress payments were due but were not paid, no one could tell how long the delay or the suspension of the work would continue.

Concerning the effect of installment payments under a contract, the court said: "Generally, payment for so much work done, though the amount of the installment is generally so measured, is, in contemplation of the parties, but a part payment of the whole sum to be paid under the contract. Professor Williston in his work on contracts expresses the opinion that the contractor, on nonpayment of an installment of the contract price, might refuse to perform further until such payment was made, and, if delayed for a long and unreasonable time, might refuse to go on with the work altogether, though a day's delay in payment would not justify permanent cessation of the work. *2 Williston on Contracts,* page 1626. . . . This court recognized the justice of such a principle in *Chicago* v. *Duffy,* 75 N.E. 912, in which case it was held that failure on the part of the city to pay installments when due, prevented it from claiming damages from resulting delay in completion of the contract." The decision of the court was that the contractor, Underground Construction Co., was not limited to a choice of one of the three remedies outlined in (a), (b), and (c) described in the second preceding paragraph, but he had the right to suspend the work until the accrued estimates were paid and then sue for special damages due to the delay caused by the District.

A contractor was allowed interest, not general damages, when his payments were delayed. The court ruled that he elected to complete the work when he had the right of rescission or of annulling the contract. The decision in *Ryan* v. *New York,* 159 App. Div. 105, N.Y., 1913, held that there is "no legal ground for awarding plaintiff general damages for any loss occasioned by delay on the part of the city in making its payment to him on contract time. For such delays, legal interest is presumptively sufficient compensation, and that the plaintiff has already been awarded. While a failure by the city to pay as agreed might have afforded plaintiff ground for rescinding the contract, he did not elect so to do. After having proceeded with the contract to completion it would certainly be a novelty in the law if he could recover, beyond interest, damages for the city's failure to pay as and when agreed."

The *Ryan* decision was based upon the judicial holding in *Wharton*

& Co. v. *Winch,* 140 N.Y. 287, 1893, which involved the construction of a street railway. When the payments were not made at definite times as agreed, the contractor stopped the work and sued for a breach of the contract. The court held that under the provisions of the contract "the mere failure of the defendant to make punctual payment of an installment due according to the provisions of the contract, was not such a breach of the entire contract as to permit the plaintiff to refuse to proceed further under it, and recover damages for the profits which he would have earned had the contract been fully performed on his part." The court pointed out: "It is undoubtedly true that the defendant's failure to pay the installment was such a breach of the contract as absolved the plaintiff from all obligation to further perform on his part while the default continued. Nor was he bound to grant the defendant any indulgence and wait for any period of time in order to enable him to make good his broken promise. The obligation of the plantiff (contractor) to proceed under the contract depended upon a condition precedent. If it was not fulfilled, one of two courses was open to the plaintiff. He might at once rescind the contract and refuse to go on, and immediately recover for the materials furnished and services rendered under it; or he might proceed with the performance of the contract on his part, and at the same time, if he chose, bring suit to recover the past due installment. In the present case the trial court has found that the plaintiff elected to rescind the contract, and as he had already received more than sufficient to compensate him for the work done under it, that he can recover nothing in this action."

chapter 9

Federal construction contracts

The right and capacity of the federal government to enter into a contract within its constitutional powers is an incident to the general right of its sovereignty. In discussing the related principles of law that apply to the federal government, the Supreme Court of the United States stated in the case of *Cooke, et al.* v. *United States,* 91 U.S. 389, 1875, that "If it (the United States) comes down from its position of sovereignty, and enters the domain of commerce, it submits itself to the same laws that govern individuals there." The fact that the federal government is a party to a contract does not affect the rules of construction and interpretation that apply in the case of contracts between private parties. (*Hollerbach* v. *United States,* 233 U.S. 165, 1914.)

"When the United States, with constitutional authority, makes contracts, it has rights and incurs responsibilities similar to those of individuals who are parties to such instruments. There is no difference, said the Court in *United States* v. *Bank of the Metropolis,* 15 Peters 392, except that the United States cannot be sued without its consent." *Perry* v. *United States,* 294 U.S. 330, 1934. The decision in the *Perry* case declared: "The fact that the United States may not be sued without its consent is a matter of procedure which does not affect the legal and binding character of its contracts. While the Congress is under no duty to provide remedies through the courts, the contractual obligation still exists and, despite infirmities of procedure, remains binding upon the conscience of the sovereign. *Lynch* v. *United States,* 292 U.S. 571."

In a discussion of the general extent of the power of the federal government under the Constitution, the opinion in the case *United States* v. *Butler,* 297 U.S. 1, 1935, pointed out that: "It hardly seems necessary to reiterate that ours is a dual form of government; that in every state there are two governments, the state and the United States. Each state has all government powers save such as the people, by their Constitution, have conferred upon the United States, denied to the states, or reserved to themselves. The federal union is a government of delegated powers. It has only such as are expressly conferred upon it and such as are reasonably to be implied from those granted. In this respect we differ radically from nations where all legislative power, without restriction or limitation, is vested in a parliament or other legislative body subject to no restrictions except the discretion of its members."

9.1 General principles of federal contracts and jurisdiction of Court of Claims

An interesting discussion of the principles that govern the relationship between the federal government and its construction contractors is contained in the court's opinion in *United States* v. *A. Bentley & Sons Co.*, 293 F. 229, Ohio, 1923. The contract was on a cost-plus basis for the construction of an army camp. The lawsuit was to recover from the contractor an alleged overpayment of $5,000,000. The federal government claimed that a fiduciary or trust relationship existed, under which the government was a beneficiary of a trust and the contractor was the trustee. The court stated that the federal government has awarded cost-plus contracts for the expenditure of enormous sums of money, and that although such a contract changes the usual manner of computing payment, it does not convert the independent contractor into an employee, and there is no trust or fiduciary relationship created by a cost-plus contract. Public officers and agents act as guardians for the government when they are engaged in their official duties, and persons who transact business with the representatives of a government agency "are in duty bound to inquire as to and to take notice of the extent of such officers' and agents' authority and power, and are to be held to a recognition that such officers and agents must observe fairness and good faith as between themselves and the government. Both are expected to exercise honesty and common sense. But no cases have been found which hold that a person who is not an officer or a representative of the government occupies in dealing with it the position of or a position akin to that of a guardian or trustee. When the government enters into a contract with an individual or corporation, it divests itself of its sovereign character as to the particular transaction and takes that of an ordinary citizen and submits to the same law as governs individuals under like circumstances."

In *United States* v. *Standard Rice Co.,* 323 U.S. 106, 1944, the court declared, with reference to the interpretation of federal contracts generally, that: "Although there will be exceptions, in general the United States as a contractor must be treated as other contractors under analogous situations. When problems of the interpretation of its contracts arise the law of contracts governs. *Hollerbach* v. *United States,* 233 U.S. 165; *United States* v. *Bethlehem Steel Corp.,* 315 U.S. 289. We will treat it like any other contractor and not revise the contract which it draws on the ground that a more prudent one might have been made. *United States* v. *American Surety Co.,* 322 U.S. 96." In *Priebe & Sons* v. *United States,* 332 U.S. 407, 1947, it was stated: "It is customary, where Congress has not adopted a different standard, to apply to the construction of government contracts the principles of general contract law."

However, in transacting business with the federal government, the contractor "takes the risk of having accurately ascertained that he who purports to act for the Government stays within the bounds of his authority. The scope of his authority may be explicitly defined by Congress or be limited by delegated legislation, properly exercised through the rule-making power. And this is so even though . . . the agent himself may have been unaware of the limitations upon his authority." (*Federal Crop Ins. Corp.* v. *Merrill,* 332 U.S. 380, 1947.)

The federal government may conduct its business affairs through the administrative agencies or specially created authorities or commissions, and while the representatives of the United States are so engaged, the federal government "is neither bound nor estopped by acts of its officers or agents in entering into an agreement to do or cause to be done what the law does not sanction or permit" *Utah Power & Light Co.* v. *United States,* 243 U.S. 389, 1916.

Public construction contractors are expected to know that they are bound by notice of the restrictions stated in section 11 of Title 41 of the United States Code, which prohibit the making of a contract unless it is authorized by law or by an adequate appropriation for the work. Certain exceptions and limitations are expressed in the statute as applying to the Departments of the Army, Navy, and Air Force, for clothing, subsistence, forage, fuel, quarters, transportation, or medical and hospital supplies. The purpose of the statute is to prevent commitments of federal government expenditures unless funds for the proposed construction work are available.

Where a contract was made according to federal statutes for improving a harbor channel, it appeared that the amount of money appropriated was adequate to pay for the estimated quantity of material to be excavated. The inspector who was charged with keeping a record of the work done "within the limits of available funds," made a mistake. So much work had already been done, that payment at the agreed unit prices would call for a far greater amount than the appropriation available. The engineer stopped the work, and the contractor sued to recover at the unit price for the excavation already performed and also for the cost of blasting rock, which was not removed because of the order to cease work. In the case of *Sutton* v. *United States,* 256 U.S. 575, 1921, the court decided that the statutes that authorized the work did not confer any authority on the contracting officer to spend more than the amount then appropriated for the purpose and pointed out that those who dealt with him "must be held to have had notice of the limitations upon his authority." The court also said that the mistake by the inspector in keeping his records is answered by the fact "that since no official of the government could have rendered it liable for this work (i.e., the excess over the available funds) by an express contract, none can by his acts or omissions create a valid contract implied in fact." The subsequent use of the excavation by the government did not "imply a promise to pay for it at any time thereafter Congress should appropriate money to be applied in completing the improvement." The court also commented as follows: "There is no necessity to consider what may be the equitable rule when there is a claim of unjust enrichment through work done upon the land of another under a mistake of fact. See *Bright* v. *Boyd,* 1 Story 478; *Williams* v. *Gibbes,* 20 How. Pr. 535; *Canal Bank* v. *Hudson,* 111 U.S. 66; *Armstrong* v. *Ashley,* 204 U.S. 272. Nor need we consider whether the doctrine is ever applicable to transactions with the government. For the right to sue the United States in the Court of Claims, here involved, must rest upon the existence of a contract express or implied in fact. *United States* v. *North American Transp. & Trading Co.,* 253 U.S. 330."

The principles of contract law as they apply to certain rights of a contractor are discussed in *Albert & Harrison* v. *United States,* 68 F. Supp. 732, 1946. The lawsuit was over a claim for extra work, additional pay to bricklayers, and also for liquidated damages, which were withheld by the Government in connection with a contract for the renovation of three buildings at an arsenal. The contractor, as the low bidder, was asked by the contracting officer to meet him at the arsenal and sign the contract. Upon visiting the arsenal, the contractor discovered that there were 116 more openings to be bricked up than he had expected, and he asked that new bids be requested. After the Government's representative threatened to forfeit the bid bond unless the contractor signed the contract under protest, he completed the work. The contractor renewed his claim for extra compensation to the contracting officer, who denied the claim. The same denial was made by the Secretary of War, who had jurisdiction of the project; the General Accounting Office also rejected the contractor's claim. The court found in favor of the contractor because he was misled by the specifications, and it was pointed out in the court's opinion that the contractor "was coerced by these circumstances (the threat of forfeiture of the bid bond) into signing the contract when the Government had no right to a contract based upon the plaintiff's (the contractor's) bid. The situation is quite different from that in the *Massman* case – *Massman Construction Co.* v. *United States,* 102 Ct. Cl. U.S. 699. There the Government was in no way at fault or responsible for the plaintiff's mistake, and the plaintiff was probably not entitled to withdraw his bid. It was, therefore, not improperly coercive for the agent of the Government to threaten to enforce the bid bond. Here the plaintiff's dilemma was the result of the Government's misleading documents."

The contractor's claim for damages against the government was allowed in *Struck Construction Co.* v. *United States,* 96 Ct. Cl. U.S. 186, 1942, when he was compelled to use more cement and to spend more for labor because the contracting officer decided that the concrete mixed in accordance with the specifications was unsatisfactory. The court said that the contractor was subjected to oppressive conduct that amounted to coercion when the Government's representative refused to approve a sample wall that was built according to the specifications, but did approve a sample wall which he knew contained extra cement and also was provided at an additional labor cost. In commenting on the principles that apply to construction contracts in such cases, the court pointed out that: "Coercion sufficient to avoid a contract need not, of course, consist of physical force or threats of it. Social or economic pressure illegally or immorally applied may be sufficient. See *Hartsville Oil Mill* v. *United States,* 271 U.S. 43, and *Hazelhurst Oil Mill Co.* v. *United States,* 70 Ct. Cl. U.S. 335." It was also stated judicially that: "It is difficult to apply terms with moral implications, such as 'good faith,' to impersonal legal entities such as corporations or governments especially in situations where they act on one matter through a number of agents. Some of the moral qualities may be lost or diluted as the decision passes from one agent to another. But the test of good faith should be the same for an entity which must act through agents as for an individual acting for himself. If the aggregate of the actions of all of the agents would, if all done by one individual, fall below the standard of good faith, the entity for whom the various agents acted should be held to have violated that standard. It is the responsibility of the entity, the principal, to so coordinate the work of its agents that the aggregate of their actions will conform to required legal standards."

In considering the rights of a public construction contractor in a claim against the federal government, the original doctrine of immunity of the sovereign exempted the government from suit on the ground that "there can be no legal right as against the authority that makes the law on which the right depends." (*Kawananvakoa* v. *Polyblank,* 205 U.S. 349, 1906) In England, the procedure with relation to a claim under a contract with the Crown, is by a "Petition of Right," but, as there is no sovereign officer in the United States to whom such a petition could be directed, the Federal Act of March 3, 1797, Chapter 20, section 4, authorized contractors to petition Congress for redress. However, in 1855, the Court of Claims was created and the provisions as to its jurisdiction are now found in section 1491 of the *United States Code.* Generally, the Court of Claims is empowered to render a judgment upon a contract with the United States "or for liquidated or unliquidated damages in cases not sounding in tort." The establishment of the Court of Claims and its jurisdiction constitute the assent in advance and also the consent of the government for a judicial determination of a contract claim within the scope of the powers of the Court of Claims. Suits against the United States for claims upon contract for damages were authorized by the *Tucker Act* of March 3, 1887, and by this enactment the liability of the federal government for breach of contract was clearly expressed in the statute law.

The United States may be sued on a contract claim in the United States Court of Claims. A Chief Judge and four Associate Judges are appointed by the President, by and with the advice and consent of the Senate, as provided in Chapter 7 of Part I of Title 28 of the *United States Code.* The jurisdiction of the court is expressed in section 1491 of Part IV of Title 28, and provides: "The Court of Claims shall have jurisdiction to render judgment upon any claim against the United States founded either upon the Constitution, or any Act of Congress, or any regulation of an executive department, or upon any express or implied contract with the United States, or for liquidated or unliquidated damages in cases not sounding in tort." Certain exceptions of jurisdiction are stated, such as Tennessee Valley Authority and treaties with foreign countries.

The procedure of the United States Court of Claims is governed by Chapter 165 of Part VI of Title 28 of the *United States Code,* and the general provisions as to the time for filing suit are stated in section 2501 as follows: "Every claim of which the Court of Claims has jurisdiction shall be barred unless the petition thereon is filed within six years after such claim first accrues."

When a claimant institutes an action against the federal Government in the United States Court of Claims, the matter is brought to the formal attention of the court by the filing of a petition in the office of the Clerk of the Court, and he furnishes copies to the office of the Attorney-General. The government then files an answer, but if there is an allegation of fraud or of a counterclaim in the answer, the claimant is entitled to file a reply. The court does not hear the trial. Under the provisions of section 792, Title 28 of the *United States Code,* the court is authorized to appoint up to seven commissioners who are empowered to hear and receive evidence, administer oaths to witnesses, and then make a report of findings of fact and a recommendation of law to the court. The rules of the Court of Claims do not provide for a trial by jury.

9.2 Local regulations and federal construction contracts

State and municipal regulations of building construction are usually enacted for the purposes of establishing standards of operations and materials for the safety and convenience of the community where the respective regulations are applicable. When a contractor engages to build a government structure he must determine beforehand if the regulations of the state or the municipality in which the site is located, apply to the public project that he has agreed to construct.

Congress has power, under Article 1, section 8, clause 17 of the U.S. Constitution, to exercise exclusive jurisdiction over "all Places purchased by the consent of the Legislature of the State in which the Same shall be, for the Erection of Forts, Magazines, Arsenals, Dock-Yards, and other needful buildings; . . .". The term "other needful buildings" has been defined to include whatever structures are necessary in the performance of the function of the United States. (*Janus* v. *Dravo Contracting Co.,* 302 U.S. 134, 1937.)

If the federal government does not own the land upon which the work is to be done, then the local regulations will apply to the contractor, unless they are inconsistent with the policy expressed in the related federal statute. In *Public Housing Administration* v. *Bristol Township,* 146 F. Supp. 859, Pa., 1956, an electrical contractor on a government housing project was held to be subject to the provision in the building code of the township, which required a contractor to secure a work permit as a protection of public safety. The court ruled that the Act of

Congress that authorized the project did not have any intention of exempting from the local building code any electrical contractors who perform work necessary to sell housing units to private parties. The court, in commenting on the federal statute, said: "In fact Congress expressed its intent that projects under this Act 'shall, so far as may be practicable, conform in location and design to local planning and tradition.'—42 U.S.C. Ann. 1545—and that government ownership of the project shall not 'deprive any State or political subdivision thereof, including any Territory or possession of the United States, of its civil and criminal jurisdiction in and over such property, or impair the civil rights under the State or local law of the inhabitants on such property.' 42 U.S.C. Ann. 1547."

A building corporation that was legally transacting business in Virginia contracted with the United States to construct a post office building on lands owned by the federal government. The city in which the site was located required a contractor's license tax of $250 for the privilege of engaging in the contracting business in Virginia. A permit to utilize the sidewalks adjoining the property, a necessary requirement to perform the work, was granted by the city to subcontractors, even though all pedestrian traffic was blocked. The contractor contended in court, in the case of *Sollitt & Sons Constr. Co.* v. *Com.,* 172 S.E. 290, Va., 1934, that the imposition of the license tax extended the jurisdiction of the State over property owned by the federal government and also infringed on the due process clause in the fourteenth amendment of the United States Constitution. The State of Virginia conceded in the lawsuit that the land was owned by the United States but that the contract could not be carried out without the exclusive use of the sidewalks; therefore, the State claimed that the license tax was valid. The court disallowed the claim of the contractor and upheld the right of the State of Virginia to impose the tax. It was pointed out in the decision that the states have the original power to tax property, business, and persons within their borders. The opinion stated: "The tax imposed is a mere bagatelle, when compared with the material benefit derived by the company in the exclusive use of the State's property." The basis of the tax was the appropriation of the street area. The contractor was not engaged on federal government property exclusively, and was not an instrument of the government and exempt from taxation by a state. In the opinion of the court in the *Sollitt* case, the following was quoted from the decision in *Thomson* v. *Union Pacific Railroad,* 9 Wall. 579, 1869: "We perceive no limits to the principle of exemption which the claimants seek to establish. It would remove from the reach of State taxation all the property of every agent of the government. Every corporation engaged in the transportation of mails, or of government property of any description, by land or water, or in supplying materials for the use of the government or in performing any service of whatever kind, might claim the benefit of the exemption."

The decision in *Miller* v. *Arkansas,* 352 U.S. 187, 1956 was in favor of the contractor who entered into a contract to construct facilities at an Air Force Base in Arkansas, over which the federal government had not yet acquired jurisdiction. The contractor commenced his work without obtaining a license under Arkansas law for such activity. It was provided in the Armed Services Procurement Act of 1947 – 41 *United States Code,* section 152 – that awards will be made, after advertisement, to the responsible bidder whose proposal will be more advantageous to the government, price and other factors considered. The Act also authorized procurement regulations that contained guiding considerations in defining who is a responsible bidder. The Arkansas licensing law sets forth similar factors to guide the Contractors' Licensing Board, which, the court said, were in conflict with the federal law and regulations. In the opinion it was held: "Subjecting a federal contractor to the Arkansas contractor license requirements would give the State's Licensing Board a virtual power of review over the federal determination of 'responsibility' and would thus frustrate the expressed federal policy of selecting the lowest responsible bidder. In view of the federal statute and regulations, the rationale of *Johnson* v. *Maryland,* 254 U.S. 51, 57, is applicable: 'It seems to us that the immunity of the instruments of the United States from state control in the performance of their duties extends to a requirement that they desist from performance until they satisfy a state officer upon examination that they are competent for a necessary part of them and pay a fee for permission to go on. Such a requirement does not merely touch the government servants remotely by a general rule of conduct; it lays hold of them in their specific attempt to obey orders and requires qualifications in addition to those that the Government has pronounced sufficient. It is the duty of the Department to employ persons competent for their work and that duty it must be presumed has been performed.' "

Where a municipal code required a permit by any contractor to construct a building, it also required that the applicant for the permit must file copies of the specifications and a diagram of the site, as well as a survey of the lines. The court in *Birmingham* v. *Thompson,* 200 F. 2d 505, Ala., 1952 declared this local requirement to be invalid against a contractor who engaged to construct a United States Veterans'

Hospital on land that had been ceded to the federal government by the State. The decision said that the State had not reserved the right to regulate the construction of buildings on the site, and no such right was vested in the city, in view of the constitutional power of Congress to exercise exclusive jurisdiction over all property purchased by the United States by the consent of the state for the purpose of erecting needful buildings.

The Act of Congress (*40 U.S.C. Ann. 421*) relative to the acquisition of land for low-cost housing or slum clearance projects expressly provides that this acquisition must not deprive any state or political subdivision of its criminal and civil jurisdiction over the property; the muncipality where the work is to be done has no right to force a contractor for the project to procure a license, give bonds, and submit to inspection. The court said in *Oklahoma City, et al.* v. *Sanders,* 94 F. 2d 323, Okla., 1938, "The cession of exclusive jurisdiction over premises acquired by the United States government, included the power of regulation and control in such matters as ordinarily fall within the police power of the State."

Where a federal contract was drawn for constructing a post office on a site purchased in New York City with the consent of the State, the question arose over the State Labor Law requirement in effect before the purchase of the property to board over all open steel tiers for the protection of the workmen. The contract provided that "State or Municipal Building Regulations do not apply to work inside the Government's lot lines." The court ruled that the State law requirement was valid, particularly because the independent contractor does not share any immunity with the contractee – the United States. The decision held that the contract provision which purported to make the regulation inapplicable "is intended to relieve the contractor from provisions as to types of material, fire hazards and the like, which are covered by the New York City Building Code"; whereas the State Labor Law requirement "is effective in the federal area, until such time as the Congress may otherwise provide." *James Stewart & Co.* v. *Sadrakula,* 309 U.S. 94, 1939. The opinion pointed out:

"Does the acceptance of sovereignty by the United States have the effect of displacing this subsection of the New York Labor Law? We think it did not. The subsection continues as a part of the laws of the federal territory.

"It is now settled that the jurisdiction acquired from a state by the United States whether by consent to the purchase or by cession may be qualified in accordance with agreements reached by the respective governments. The Constitution does not command that every vestige of the laws of the former sovereignty must vanish. On the contrary, its language has long been interpreted so as to permit the continuance until abrogated of those rules existing at the time of the surrender of sovereignty which govern the rights of the occupants of the territory transferred. This assures that no area however small will be left without a developed legal system for private rights."

9.3 Standard federal construction contract documents

Before 1921 each department of the federal government had its own type of construction contracts, and variations existed among bureaus in the same department. Form No. 23 was established in 1926 by an interdepartmental board, and it was used as the standard form in most Government construction contracts until 1953 when it was revised.

A public construction contract is the written intention of what is required by the government and what the contractor can expect for performance of the work under the contract. The combination of these two concepts constitutes the respective obligations and the objectives of an agreement. The intention must be distinctly stated, with express legal rights and duties that disclose the promise and the consent of each of the contracting parties. Even though the law of contracts does not prescribe the form of the writing, the current standard federal forms show the professional skill that understands the necessity of providing for the rights of the contracting parties. There is no substitute for experience in matters of this kind, and progress in the formulation of contract documents indicates affirmative action by responsible public officials who are alert to judicial expressions and comments on the subject. The maintenance of a proper balance of the equities of the interested parties will undoubtedly safeguard the rights of the public, and the contractor will have no cause to complain of oppression or coercion.

Although dispute and controversy are inevitable in many instances, the informed contractor who first examines the documents that show the mutual obligations of the contracting parties can eliminate much contention. He will find that each form is a separate part of the written agreement that, together, form the complete contract. The standard federal construction contract forms are contained in the codification of federal government regulations that apply to public contracts and are found in Title 41 of the *Code of Federal Regulations* (CFR), which was revised and reorganized on January 1, 1960. Also, Chapter I of Title 41 of the *United States Code* establishes (as of March 10, 1959) a new

Federal Procurement Regulations System. The introduction to the System states (see volume 24 Federal Register, page 1933), in part:

"The System consists of two major elements: (a) procurement policies and procedures for Government agencies prescribed by the Administrator of General Services in the *'Federal Procurement Regulations'* (FPR), and (b) the publication of agency implementing regulations in succeeding chapters of Title 41, including the gradual transfer to such succeeding chapter of existing agency regulations affecting procurement. Ultimately, government procurement regulations will be contained in the same title of the *Code of Federal Regulations* for convenient utilization by both the public and the Government personnel. Greater consistency in agency procurement regulations is anticipated under this System.

"Background. The System is an outgrowth of recommendations of the President's Cabinet Committee on Small Business and the work of the interagency Task Force for Review of Government Procurement Policies and Procedures, which was established by the Administrator of General Services pursuant to a Presidential directive of September 26, 1956. The system is intended to promote the objectives of eliminating needless inconsistencies between agency procurement regulations, minimizing complexities and inequities in procurement policies and procedures, and making agency procurement regulations more readily available to businessmen and others concerned."

Paragraph 5 of the Introduction to *Federal Procurement Regulations* provides: "While the Federal Procurement Regulations are applicable to all executive agencies, they are not made mandatory on the Department of Defense, except with respect to matters concerning standard government forms and clauses, Federal Specifications and Standards, and except as directed by the President, Congress, or other authority. Therefore, the extent of implementation of FPR by the Department of Defense and participation in the System will be determined by that Department. However, coordination with that Department in the development and issuance of these regulations in contemplated."

The provisions of Chapter 137, which is entitled "Procurement Generally" (10 U.S.C. Ann. 2301, etc.) apply to purchases and services by the Departments of the Army, Navy, and Air Force, as well as by the Coast Guard and The National Aeronautics and Space Administration.

The Department of Defense issues the *Armed Services Procurement Regulation,* in coordination with the Departments of the Army, Navy and Air Force. This *Regulation* applies to all purchases and contracts made by the United States "for the procurement of supplies or services which obligate appropriated funds (including available contract authorizations), unless otherwise specified herein (but see 1-109.4)." Paragraph 1-109.4 provides that in "the event of any specific conflict between the provisions of this *Regulation* and any treaty or executive agreement to which the United States is a party, such treaty or agreement will govern; *provided,* however, if such conflict affects a provision of this regulation which is based on the requirements of law, such conflict shall, in accordance with Departmental procedures, be referred to the ASPR (*Armed Services Procurement Regulation*) Committee for consideration. Any procurement action which constitutes a deviation from an ASPR provision based on a requirement of law shall be held in abeyance pending consideration by the ASPR Committee."

The *Armed Services Procurement Regulation* has a considerable number of sections and appendixes. Some sections relate to special methods of procurement, and also contain standard forms with contract clauses that pertain to labor, patents, termination of contracts, and principles and procedures of contract cost. Comprehensive manuals are included in the *Regulation* for the control of government property in the possession of contractors, and also in the possession of nonprofit research and development contractors. A board of contract appeals is created by the *Regulation,* with appropriate power to deal with related matters.

Section 1-1.213 of the *Federal Procurement Regulations* states that:

" 'Construction contractor' means a person (or firm) who, before being awarded a contract satisfies the contracting officer that he qualifies as one:

(a) Who owns, operates, or maintains a place of business, regularly engaged in the construction, alteration, or repair of buildings, structures, communication facilities, or other engineering projects, including the furnishing and installing of necessary equipment; or

(b) Who, if newly entering into a construction activity, has made all necessary prior arrangements for personnel, construction equipment, and required licenses to perform the construction work."

"Procurement" is defined in section 1-1.209 of the *Federal Procurement Regulations* as "the acquisition (and directly related matters), from non-Federal sources, of personal property and nonpersonal services (including construction) by such means as purchasing, renting, leasing, contracting, or bartering, but not by seizure, condemnation, donation, or requisition." And "procurement item" is defined in section 1-1.220, to mean "any personal property or nonpersonal service, including construction, alteration, repair, or installation which is the object of procurement."

The methods of "procurement" anticipate a competitive situation, usually by formal advertising and an award to the "lowest responsible bidder," with the express "objective to use that method of procurement which will be most advantageous to the Government, price, quality, and other factors considered." (section 1-1.301, *Federal Procurement Regulations.*)

The term "contracting officer," which is used frequently throughout most of the standard federal contract forms, "means the person executing this contract on behalf of the Government, and any other officer or civilian employee who is a properly designated Contracting Officer; and the term includes, except as otherwise provided in this contract, the authorized representative of a Contracting Officer acting within the limits of his authority."

The standard federal construction contract forms are:

1. Standard Form 19: Invitation, Bid, and Award (Construction, Alteration, or Repair – January 1959
2. Standard Form 19A: Labor Standards, Provisions Applicable to Contracts in Excess of $2,000 – January 1959
3. Standard Form 20: Invitation for Bids (Construction Contract) – March 1953
4. Standard Form 21: Bid Form (Construction Contract) – January 1959
5. Standard Form 22: Instructions to Bidders (Construction Contracts) – March 1953
6. Standard Form 23: Construction Contract – March 1953
7. Standard Form 23A: General Provision (Construction Contracts) – March 1953
8. Standard Form 24: Bid Bond – November 1950
9. Standard Form 25: Performance Bond – November 1950
10. Standard Form 25A: Payment Bond – November 1950
11. Standard Form 27: Performance Bond, Corporate Co-Surety Form – November 1950
12. Standard Form 27A: Payment Bond, Corporate Co-Surety Form – November 1950
13. Standard Form 27B: Continuation Sheet, Corporate Co-Surety Bond – November 1950
14. Standard Form 28: Affidavit of Individual Surety – November 1950.

Standard Forms 23 and 23A are reproduced in this chapter and the others in Appendix B. Originals may be inspected at General Services Administration regional offices; the forms are obtainable from the Superintendent of Documents, U.S. Government Printing Office, Washington 25, D.C.

Additional terms, conditions, and provisions that are not inconsistent with those contained in the standard forms may be included in the contract documents by the head of a department, but deviations are expected to be at a minimum for the sake of maintaining uniformity. The printed standard forms cannot be changed to provide for a deviation; changes are shown in the specifications or in a continuation sheet. In case of deviations in classes of cases, the General Services Administration has the right to consider them jointly with the agency desiring the change.

The use of the standard form is optional and is to be adapted to the procedure of the particular agency for negotiated contracts, including those which are made on a competitive basis, but without full compliance with formal advertisement for bids (see section 1-16.403 of the *Federal Procurement Regulations*).

9.4 Prescribed use of Standard Forms

Section 1-16.402-2 of the *Federal Procurement Regulations* provides that: "Standard Forms 19 and 19A shall be used for contracts estimated to exceed $2,000 but not to exceed $10,000. Standard Form 22 also may be used." Section 1-16.402-3 provides: "Standard Forms 20, 21, 22, 23, and 23A shall be used for contracts estimated to exceed $10,000." The use of the standard forms is prescribed for construction contracts by formal advertising. Standard Form 19 consists of a single sheet on which bids are invited or solicited by the Government, submitted by the bidder, and accepted by an award by a contracting officer. The *Federal Procurement Regulations* (section 1-16.402-4) direct attention to the fact that: "General Provisions on the reverse (side) are adequate for contracts of $2,000 or less and become adequate for contracts in excess of that amount when Standard Form 19A is added." The General Provisions contain eleven clauses, all of which are made a part of the contract by a reference to them in the contractor's "Bid" section on page 1 of Standard Form 19.

Standard Form 19A, "Labor Standards Provisions Applicable to Contracts in Excess of $2,000," is used as a supplemental part of Standard Form 19 when the contract cost is estimated to exceed $2,000.

Standard Form 20, "Invitation for Bids (Construction Contract)," is applicable to contracts in excess of $10,000, and is intended to inform prospective contractors of the time and the place where the sealed bids will be received and publicly opened, and also furnishes information on the bidding material, bid deposit or guarantee, and the required surety bonds. The form contains a general description of the work and states that the specifications include information as to the applicable contract provisions. Bids must be furnished on the forms furnished by the issuing office or copies.

Standard Form 21, "Bid Form (Construction Contract)," includes, on page 1, specific information on price and other related matters, as well as an agreement by the bidder that within the the number of days specified, he will execute the "Construction Contract" (Standard Form 23) and will also give the standard federal forms of surety bonds for performance and payment, if required. Page 2 of the "Bid Form" states the time of commencement of the work and its completion and also includes an acknowledgement of the receipt by the bidder of any addendum that must be identified in the bid form in the space provided for that purpose.

Although the two-page Standard Form 22, "Instructions to Bidders (Construction Contracts)," is not incorporated in the contract, the knowledgeable bidder will carefully consider the instructions to make certain that he has complied with the preliminary administrative steps leading to a contract.

Standard Form 23, "Construction Contract," is a single-sheet or streamlined version of a formal public document, with an absolute minimum of provisions that make up a mutual agreement between contracting parties. Whereas there are certain noncontroversial provisions for the amount of the contract, a general statement of the work to be done, and the correct execution of the form, the clause of special importance, as far as the obligations of the interested parties are concerned, is the incorporation in the contract documents of the "General Provisions (Standard Form 23A), specifications, schedules, drawings, and conditions all of which are made a part hereof."

9.5 Standard Form 23A, "General Provisions (Construction Contracts)"

The following paragraphs will examine the contract provisions contained in the *cardinal four* clauses of Standard Form 23A:

3. Changes
4. Changed Conditions
5. Termination for Default – Damages for Delay – Time Extensions
6. Disputes.

These four clauses, which frequently require judicial interpretation and construction, are of prime importance to the contracting parties.

Each of this group of clauses makes successive reference to the other, and these references emphasize the necessity of clearly understanding the significant provisions in the component parts of the Federal Construction Contract.

Clause 3 – Changes: Clause 3 of Standard Form 23A, is entitled "Changes." Its provisions constitute an express authorization by the contractor to the government to change the drawings, the specifications, or both, within the general scope of the contract. Equitable adjustment is also authorized for any increase or decrease in the cost of any changes or in the time required for performance by the contractor. When the contracting officer anticipates a change in the work, it is usual for him to precede the issuance of the order to change by a stop order. The government is entitled to a reasonable time in which to decide whether or not to make the changes, and in proper situations, to provide for the nature and extent of the changes within the general scope of the project. The inevitable delay is recognized, as a rule, by an extension of time for completion of the work, this delay having been anticipated by the parties when the contract was made. If the contractor believes that he is entitled to compensation for accomplishing the change as directed in writing, and he is not satisfied with the denial of compensation or the adequacy of the compensation, he must assert his claim in writing within 30 days from the date he receives the change order. The contracting officer is authorized to consider and adjust such a claim at any time before the final settlement of the contract, if he feels the facts warrant such action. If the contractor and the contracting officer are unable to agree upon the adjustment, then the provisions of Clause 6, "Disputes," must be observed, but the work must be continued by the contractor. No allowance for extra work or material is permitted by Clause 3, except as provided by the terms of the clause; the payment of compensation for an increase in the work is not considered to be damages caused to the contractor as a breach of contract, but is considered an equitable adjustment for additional work. Whereas reasonable periods of delay by the government are considered in the contract provisions, increased expenses incurred by the contractor due to unreasonable delay must be supported by proof that the costs were directly brought on by the excessive time loss, although the reasonableness or unreasonableness of delay is always a question of fact. (*Ross Engineering Co. Inc.* v. *United States,* 92 Ct. Cl. U.S. 253, 1940.)

In the case of *Great Lakes Construction Co.* v. *United States,* 95 Ct. Cl. U.S. 479, 1942, the contract was for the construction of a federal penitentiary. During the progress of the work, 109 changes were made, but the urgent need for the availability of the building did not permit

STANDARD FORM 23
REVISED MARCH 1953
GENERAL SERVICES ADMINISTRATION
GENERAL REGULATION NO. 13

CONSTRUCTION CONTRACT
(See instructions on reverse)

CONTRACT NO.

DATE OF CONTRACT

NAME AND ADDRESS OF CONTRACTOR

CHECK APPROPRIATE BOX
☐ Individual
☐ Partnership
☐ Incorporated in the State of

DEPARTMENT OR AGENCY

CONTRACT FOR *(Work to be performed)*

PLACE

AMOUNT OF CONTRACT *(Express in words and figures)*

ADMINISTRATIVE DATA *(Optional)*

THIS CONTRACT, entered into this date by the United States of America, hereinafter called the Government, represented by the Contracting Officer executing this contract, and the individual, partnership, or corporation named above, hereinafter called the Contractor, witnesseth that the parties hereto do mutually agree as follows:

Statement of Work. The Contractor shall furnish all labor, equipment, and materials and perform the work above described for the amount stated above in strict accordance with the General Provisions (Standard Form 23a), specifications, schedules, drawings, and conditions all of which are made a part hereof and designated as follows:

WORK SHALL BE STARTED	WORK SHALL BE COMPLETED

16—17413-8

Standard form 23 (face)

adequate time for preparation of complete plans and specifications, and the bid was prepared in spite of certain omissions. Shortly after the contract was made, a corrected set of plans was delivered to the contractor, which showed 700 additions and corrections. Disputes arose between the contracting parties as to the allowances for changes, but

Alterations. The following changes were made in this contract before it was signed by the parties hereto:

In witness whereof, the parties hereto have executed this contract as of the date entered on the first page hereof.

THE UNITED STATES OF AMERICA

By ______________________________

____
(Official title)

CONTRACTOR

____
(Name of Contractor)

By ______________________________
(Signature)

____
(Title)

INSTRUCTIONS

1. This form shall be used, as required by GSA regulations, for contracts for the construction, alteration, or repair of public buildings or works.

2. The full name and business address of the Contractor must be inserted in the space provided on the face of the form. The Contractor shall sign in the space provided above with his usual signature and typewrite or print name under all signatures to the contract and bonds.

3. An officer of a corporation, a member of a partnership, or an agent signing for the Contractor shall place his signature and title after the word "By" under the name of the Contractor. A contract executed by an attorney or agent on behalf of the Contractor shall be accompanied by two authenticated copies of his power of attorney, or other evidence of his authority to act on behalf of the Contractor.

16—17412-3 U. S. GOVERNMENT PRINTING OFFICE

Standard form 23 (reverse)

the work was continued. In the action for damages caused by delay, the court ruled that Clause 3 of the contract gave the government the right to make changes. Although increases or decreases were provided by the contracting officer through supplemental agreements, the judicial conclusion was that the delay, because the necessary changes were con-

STANDARD FORM 23-A
MARCH 1953 EDITION
GENERAL SERVICES ADMINISTRATION
FED. PROC. REG. (41 CFR) 1-16.401
23-202

GENERAL PROVISIONS
(CONSTRUCTION CONTRACTS)

1. DEFINITIONS

(a) The term "head of the department" as used herein shall mean the head or any assistant head of the executive department or independent establishment involved, and the term "his duly authorized representative" shall mean any person authorized to act for him other than the Contracting Officer.

(b) The term "Contracting Officer" as used herein, shall include his duly appointed successor or his authorized representative.

2. SPECIFICATIONS AND DRAWINGS

The Contractor shall keep on the work a copy of the drawings and specifications and shall at all times give the Contracting Officer access thereto. Anything mentioned in the specifications and not shown on the drawings, or shown on the drawings and not mentioned in the specifications, shall be of like effect as if shown or mentioned in both. In case of difference between drawings and specifications, the specifications shall govern. In any case of discrepancy either in the figures, in the drawings, or in the specifications, the matter shall be promptly submitted to the Contracting Officer, who shall promptly make a determination in writing. Any adjustment by the Contractor without this determination shall be at his own risk and expense. The Contracting Officer shall furnish from time to time such detail drawings and other information as he may consider necessary, unless otherwise provided.

3. CHANGES

The Contracting Officer may at any time, by a written order, and without notice to the sureties, make changes in the drawings and/or specifications of this contract and within the general scope thereof. If such changes cause an increase or decrease in the amount due under this contract, or in the time required for its performance, an equitable adjustment shall be made and the contract shall be modified in writing accordingly. Any claim of the Contractor for adjustment under this clause must be asserted in writing within 30 days from the date of receipt by the Contractor of the notification of change: *Provided, however,* That the Contracting Officer, if he determines that the facts justify such action, may receive and consider, and adjust any such claim asserted at any time prior to the date of final settlement of the contract. If the parties fail to agree upon the adjustment to be made the dispute shall be determined as provided in Clause 6 hereof. But nothing provided in this clause shall excuse the Contractor from proceeding with the prosecution of the work as changed. Except as otherwise herein provided, no charge for any extra work or material will be allowed.

4. CHANGED CONDITIONS

The Contractor shall promptly, and before such conditions are disturbed, notify the Contracting Officer in writing of: (1) subsurface or latent physical conditions at the site differing materially from those indicated in this contract, or (2) unknown physical conditions at the site, of an unusual nature, differing materially from those ordinarily encountered and generally recognized as inhering in work of the character provided for in this contract. The Contracting Officer shall promptly investigate the conditions, and if he finds that such conditions do so materially differ and cause an increase or decrease in the cost of, or the time required for, performance of this contract, an equitable adjustment shall be made and the contract modified in writipg accordingly. Any claim of the Contractor for adjustment hereunder shall not be allowed unless he has given notice as above required; provided that the Contracting Officer may, if he determines the facts so justify, consider and adjust any such claim asserted before the date of final settlement of the contract. If the parties fail to agree upon the adjustment to be made, the dispute shall be determined as provided in Clause 6 hereof.

5. TERMINATION FOR DEFAULT—DAMAGES FOR DELAY—TIME EXTENSIONS

(a) If the Contractor refuses or fails to prosecute the work, or any separable part thereof, with such diligence as will insure its completion within the time specified in this contract, or any extension thereof, or fails to complete said work within such time, the Government may, by written notice to the Contractor, terminate his right to proceed with the work or such part of the work as to which there has been delay. In such event the Government may take over the work and prosecute the same to completion, by contract or otherwise, and the Contractor and his sureties shall be liable to the Government for any excess cost occasioned the Government thereby, and for liquidated damages for delay, as fixed in the specifications or accompanying papers, until such reasonable time as may be required for the final completion of the work, or if liquidated damages are not so fixed, any actual damages occasioned by such delay. If the Contractor's right to proceed is so terminated, the Government may take possession of and utilize in completing the work such materials, appliances, and plant as may be on the site of the work and necessary therefor.

(b) If the Government does not terminate the right of the Contractor to proceed, as provided in paragraph (a) hereof, the Contractor shall continue the work, in which event he and his sureties shall be liable to the Government, in the amount set forth in the specifications or accompanying papers, for fixed, agreed, and liquidated damages for each calendar day of delay until the work is completed or accepted, or if liquidated damages are not so fixed, any actual damages occasioned by such delay.

(c) The right of the Contractor to proceed shall not be terminated, as provided in paragraph (a) hereof, nor the Contractor charged with liquidated or actual damages, as provided in paragraph (b) hereof because of any delays in the completion of the work due to unforeseeable causes beyond the control and without the fault or negligence of the Contractor, including, but not restricted to, acts of God, or of the public enemy, acts of the Government, in either its sovereign or contractual capacity, acts of another contractor in the performance of a contract with the Government, fires, floods, epidemics, quarantine restrictions, strikes, freight embargoes, and unusually severe weather, or delays of subcontractors or suppliers due to such causes: *Provided,* That the Contractor shall within 10 days from the beginning of any such delay, unless the Contracting Officer shall grant a further period of time prior to the date of final settlement of the contract, notify the Contracting Officer in writing of the causes of delay. The Contracting Officer shall ascertain the facts and the extent of the delay and extend the time for completing the work when in his judgment the findings of fact justify such an extension, and his findings of fact thereon shall be final and conclusive on the parties hereto, subject only to appeal as provided in Clause 6 hereof.

6. DISPUTES

Except as otherwise provided in this contract, any dispute concerning a question of fact arising under this contract which is not disposed of by agreement shall be decided by the Contracting Officer, who shall reduce his decision to writing and mail or otherwise furnish a copy thereof to the Contractor. Within 30 days from the date of receipt of such copy, the Contractor may appeal by mailing or otherwise furnishing to the Contracting Officer a written appeal addressed to the head of the department, and the decision of the head of the department or his duly authorized representatives for the hearings of such appeals shall, unless determined by a court of competent jurisdiction to have been fraudulent, arbitrary, capricious, or so grossly erroneous as necessarily to imply bad faith, be final and conclusive: *Provided,* That, if no such appeal to the head

Standard form 23A

of the department is taken, the decision of the Contracting Officer shall be final and conclusive. In connection with any appeal proceeding under this clause, the Contractor shall be afforded an opportunity to be heard and to offer evidence in support of its appeal. Pending final decision of a dispute hereunder, the Contractor shall proceed diligently with the performance of the contract and in accordance with the Contracting Officer's decision.

7. PAYMENTS TO CONTRACTORS

(a) Unless otherwise provided in the specifications, partial payments will be made as the work progresses at the end of each calendar month, or as soon thereafter as practicable, or at more frequent intervals as determined by the Contracting Officer, on estimates made and approved by the Contracting Officer. In preparing estimates the material delivered on the site and preparatory work done may be taken into consideration.

(b) In making such partial payments there shall be retained 10 percent on the estimated amount until final completion and acceptance of all work covered by the contract: *Provided, however,* That the Contracting Officer, at any time after 50 percent of the work has been completed, if he finds that satisfactory progress is being made, may make any of the remaining partial payments in full: *And provided further,* That on completion and acceptance of each separate building, public work, or other division of the contract, on which the price is stated separately in the contract, payment may be made in full, including retained percentage thereon, less authorized deductions.

(c) All material and work covered by partial payments made shall thereupon become the sole property of the Government, but this provision shall not be construed as relieving the Contractor from the sole responsibility for all materials and work upon which payments have been made or the restoration of any damaged work, or as a waiver of the right of the Government to require the fulfillment of all of the terms of the contract.

(d) Upon completion and acceptance of all work required hereunder, the amount due the Contractor under this contract will be paid upon the presentation of a properly executed ~~and duly certified~~ voucher therefor, after the Contractor shall have furnished the Government with a release, if required, of all claims against the Government arising under and by virtue of this contract, other than such claims, if any, as may be specifically excepted by the Contractor from the operation of the release in stated amounts to be set forth therein. If the Contractor's claim to amounts payable under the contract has been assigned under the Assignment of Claims Act of 1940, as amended (41 U. S. C. 15), a release may also be required of the assignee at the option of the Contracting Officer.

8. MATERIALS AND WORKMANSHIP

Unless otherwise specifically provided for in the specifications, all equipment, materials, and articles incorporated in the work covered by this contract are to be new and of the most suitable grade of their respective kinds for the purpose and all workmanship shall be first class. Where equipment, materials, or articles are referred to in the specifications as "equal to" any particular standard, the Contracting Officer shall decide the question of equality. The Contractor shall furnish to the Contracting Officer for his approval the name of the manufacturer of machinery, mechanical and other equipment which he contemplates incorporating in the work, together with their performance capacities and other pertinent information. When required by the specifications, or when called for by the Contracting Officer, the Contractor shall furnish the Contracting Officer for approval full information concerning the materials or articles which he contemplates incorporating in the work. Samples of materials shall be submitted for approval when so directed. Machinery, equipment, materials, and articles installed or used without such approval shall be at the risk of subsequent rejection. The Contracting Officer may in writing require the Contractor to remove from the work such employee as the Contracting Officer deems incompetent, careless, insubordinate, or otherwise objectionable, or whose continued employment on the work is deemed by the Contracting Officer to be contrary to the public interest.

9. INSPECTION

(a) Except as otherwise provided in paragraph (d) hereof all material and workmanship, if not otherwise designated by the specifications, shall be subject to inspection, examination, and test by the Contracting Officer at any and all times during manufacture and/or construction and at any and all places where such manufacture and/or construction are carried on. The Government shall have the right to reject defective material and workmanship or require its correction. Rejected workmanship shall be satisfactorily corrected and rejected material shall be satisfactorily replaced with proper material without charge therefor, and the Contractor shall promptly segregate and remove the rejected material from the premises. If the Contractor fails to proceed at once with the replacement of rejected material and/or the correction of defective workmanship the Government may, by contract or otherwise, replace such material and/or correct such workmanship and charge the cost thereof to the Contractor, or may terminate the right of the Contractor to proceed as provided in Clause 5 of this contract, the Contractor and surety being liable for any damage to the same extent as provided in said Clause 5 for terminations thereunder.

(b) The Contractor shall furnish promptly without additional charge, all reasonable facilities, labor, and materials necessary for the safe and convenient inspection and test that may be required by the Contracting Officer. All inspection and tests by the Government shall be performed in such manner as not unnecessarily to delay the work. Special, full size, and performance tests shall be as described in the specifications. The Contractor shall be charged with any additional cost of inspection when material and workmanship are not ready at the time inspection is requested by the Contractor.

(c) Should it be considered necessary or advisable by the Government at any time before final acceptance of the entire work to make an examination of work already completed, by removing or tearing out same, the Contractor shall on request promptly furnish all necessary facilities, labor, and material. If such work is found to be defective or nonconforming in any material respect, due to fault of the Contractor or his subcontractors, he shall defray all the expenses of such examination and of satisfactory reconstruction. If, however, such work is found to meet the requirements of the contract, the actual direct cost of labor and material necessarily involved in the examination and replacement, plus 15 percent, shall be allowed the Contractor and he shall, in addition, if completion of the work has been delayed thereby, be granted a suitable extension of time on account of the additional work involved.

(d) Inspection of material and finished articles to be incorporated in the work at the site shall be made at the place of production, manufacture, or shipment, whenever the quantity justifies it, unless otherwise stated in the specifications; and such inspection and written or other formal acceptance, unless otherwise stated in the specifications, shall be final, except as regards latent defects, departures from specific requirements of the contract, damage or loss in transit, fraud, or such gross mistakes as amount to fraud. Subject to the requirements contained in the preceding sentence, the inspection of material and workmanship for final acceptance as a whole or in part shall be made at the site. Nothing contained in this paragraph (d) shall in any way restrict the Government's rights under any warranty or guarantee.

10. SUPERINTENDENCE BY CONTRACTOR

The Contractor shall give his personal superintendence to the work or have a competent foreman or superintendent, satisfactory to the Contracting Officer, on the work at all times during progress, with authority to act for him.

2

Standard form 23A (cont.)

11. PERMITS AND RESPONSIBILITY FOR WORK, ETC.

The Contractor shall, without additional expense to the Government, obtain all licenses and permits required for the prosecution of the work. He shall be responsible for all damages to persons or property that occur as a result of his fault or negligence in connection with the prosecution of the work. He shall also be responsible for all materials delivered and work performed until completion and final acceptance, except for any completed unit thereof which theretofore may have been finally accepted.

12. OTHER CONTRACTS

The Government may undertake or award other contracts for additional work, and the Contractor shall fully cooperate with such other contractors and Government employees and carefully fit his own work to such additional work as may be directed by the Contracting Officer. The Contractor shall not commit or permit any act which will interfere with the performance of work by any other contractor or by Government employees.

13. PATENT INDEMNITY

Except as otherwise provided, the Contractor agrees to indemnify the Government and its officers, agents and employees against liability, including costs and expenses, for infringement upon any Letters Patent of the United States (except Letters Patent issued upon an application which is now or may hereafter be, for reasons of national security, ordered by the Government to be kept secret or otherwise withheld from issue) arising out of the performance of this contract or out of the use or disposal by or for the account of the Government of supplies furnished or construction work performed hereunder.

14. ADDITIONAL BOND SECURITY

If any surety upon any bond furnished in connection with this contract becomes unacceptable to the Government, or if any such surety fails to furnish reports as to his financial condition from time to time as requested by the Government, the Contractor shall promptly furnish such additional security as may be required from time to time to protect the interests of the Government and of persons supplying labor or materials in the prosecution of the work contemplated by this contract.

15. COVENANT AGAINST CONTINGENT FEES

The Contractor warrants that no person or selling agency has been employed or retained to solicit or secure this contract upon an agreement or understanding for a commission, percentage, brokerage, or contingent fee, excepting bona fide employees or bona fide established commercial or selling agencies maintained by the Contractor for the purpose of securing business. For breach or violation of this warranty the Government shall have the right to annul this contract without liability or in its discretion to deduct from the contract price or consideration the full amount of such commission, percentage, brokerage, or contingent fee.

16. OFFICIALS NOT TO BENEFIT

No member of or Delegate to Congress, or Resident Commissioner, shall be admitted to any share or part of this contract, or to any benefit that may arise therefrom; but this provision shall not be construed to extend to this contract if made with a corporation for its general benefit.

17. BUY AMERICAN ACT

The Contractor agrees that in the performance of the work under this contract the Contractor, subcontractors, material men and suppliers shall use only such unmanufactured articles, materials and supplies (which term "articles, materials and supplies" is hereinafter referred to in this clause as "Supplies") as have been mined or produced in the United States, and only such manufactured supplies as have been manufactured in the United States substantially all from supplies mined, produced, or manufactured, as the case may be, in the United States. Pursuant to the Buy American Act (41 U. S. C. 10a–d), the foregoing provisions shall not apply (i) with respect to supplies excepted by the head of the department from the application of that Act, (ii) with respect to supplies for use outside the United States, or (iii) with respect to the supplies to be used in the performance of work under this contract which are of a class or kind determined by the head of the department or his duly authorized representative not to be mined, produced, or manufactured, as the case may be, in the United States in sufficient and reasonably available commercial quantities and of a satisfactory quality, or (iv) with respect to such supplies, from which the supplies to be used in the performance of work under this contract are manufactured, as are of a class or kind determined by the head of the department or his duly authorized representative not to be mined, produced, or manufactured, as the case may be, in the United States in sufficient and reasonably available commercial quantities and of a satisfactory quality, provided that this exception (iv) shall not permit the use in the performance of work under this contract of supplies manufactured outside the United States if such supplies are manufactured in the United States in sufficient and reasonably available commercial quantities and of a satisfactory quality.

18. CONVICT LABOR

In connection with the performance of work under this contract, the Contractor agrees not to employ any person undergoing sentence of imprisonment at hard labor.

19. NONDISCRIMINATION IN EMPLOYMENT

In connection with the performance of work under this contract, the Contractor agrees not to discriminate against any employee or applicant for employment because of race, creed, color, or national origin; and further agrees to insert the foregoing provision in all subcontracts hereunder except subcontracts for standard commercial supplies or for raw materials.

20. DAVIS-BACON ACT (40 U. S. C. 276a–a(7))

(a) All mechanics and laborers employed or working directly upon the site of the work will be paid unconditionally and not less often than once a week, and without subsequent deduction or rebate on any account (except such payroll deductions as are permitted by the Copeland Act (Anti-Kickback) Regulations (29 CFR, Part 3)) the full amounts due at time of payment, computed at wage rates not less than those contained in the wage determination decision of the Secretary of Labor which is attached hereto and made a part hereof, regardless of any contractual relationship which may be alleged to exist between the Contractor or subcontractor and such laborers and mechanics; and a copy of the wage determination decision shall be kept posted by the Contractor at the site of the work in a prominent place where it can be easily seen by the workers.

(b) In the event it is found by the Contracting Officer that any laborer or mechanic employed by the Contractor or any subcontractor directly on the site of the work covered by this contract has been or is being paid at a rate of wages less than the rate of wages required by paragraph (a) of this clause, the Contracting Officer may (1) by written notice to the Government Prime Contractor terminate his right to proceed with the work, or such part of the work as to which there has been a failure to pay said required wages, and (2) prosecute the work to completion by contract or otherwise, whereupon such Contractor and his sureties shall be liable to the Government for any excess costs occasioned the Government thereby.

(c) Paragraphs (a) and (b) of this clause shall apply to this contract to the extent that it is (1) a prime contract with the Government subject to the Davis-Bacon Act or (2) a subcontract under such prime contract.

21. EIGHT-HOUR LAWS—OVERTIME COMPENSATION

No laborer or mechanic doing any part of the work contemplated by this contract, in the employ of the Contractor

Standard form 23A (cont.)

or any subcontractor contracting for any part of said work contemplated, shall be required or permitted to work more than eight hours in any one calendar day upon such work, except upon the condition that compensation is paid to such laborer or mechanic in accordance with the provisions of this clause. The wages of every laborer and mechanic employed by the Contractor or any subcontractor engaged in the performance of this contract shall be computed on a basic day rate of eight hours per day and work in excess of eight hours per day is permitted only upon the condition that every such laborer and mechanic shall be compensated for all hours worked in excess of eight hours per day at not less than one and one-half times the basic ra e of pay. For each violation of the requirements of this clause a penalty of five dollars shall be imposed for each laborer or mechanic for every calendar day in which such employee is required or permitted to labor more than eight hours upon said work without receiving compensation computed in accordance with this clause, and all penalties thus imposed shall be withheld for the use and benefit of the Government: *Provided*, That this stipulation shall be subject in all respects to the exceptions and provisions of the Eight-Hour Laws as set forth in 40 U. S. C. 321, 324, 325, 325a, and 326, which relate to hours of labor and compensation for overtime.

22. APPRENTICES

Apprentices will be permitted to work only under a bona fide apprenticeship program registered with a State Apprenticeship Council which is recognized by the Federal Committee on Apprenticeship, U. S. Department of Labor; or if no such recognized Council exists in a State, under a program registered with the Bureau of Apprenticeship, U. S. Department of Labor.

23. PAYROLL RECORDS AND PAYROLLS

(a) Payroll records will be maintained during the course of the work and preserved for a period of three years thereafter for all laborers and mechanics working at the site of the work. Such records will contain the name and address of each such employee, his correct classification, rate of pay, daily and weekly number of hours worked, deductions made and actual wages paid. The Contractor will make his employment records available for inspection by authorized representatives of the Contracting Officer and the U. S. Department of Labor, and will permit such representatives to interview employees during working hours on the job.

(b) A certified copy of all payrolls will be submitted weekly to the Contracting Officer. The Government Prime Contractor will be responsible for the submission of certified copies of the payrolls of all subcontractors. The certification will affirm that the payrolls are correct and complete, that the wage rates contained therein are not less than the applicable rates contained in the wage determination decision of the Secretary of Labor attached to this contract, and that the classifications set forth for each laborer or mechanic conform with the work he performed.

24. COPELAND (ANTI-KICKBACK) ACT—NONREBATE OF WAGES.

The regulations of the Secretary of Labor applicable to Contractors and subcontractors (29 CFR, Part 3), made pursuant to the Copeland Act, as amended (40 U. S. C. 276c) and to aid in the enforcement of the Anti-Kickback Act (18 U. S. C. 874) are made a part of this contract by reference. The Contractor will comply with these regulations and any amendments or modifications thereof and the Government Prime Contractor will be responsible for the submission of affidavits required of subcontractors thereunder. The foregoing shall apply except as the Secretary of Labor may specifically provide for reasonable limitations, variations, tolerances, and exemptions.

25. WITHHOLDING OF FUNDS TO ASSURE WAGE PAYMENT

There may be withheld from the Contractor so much of the accrued payments or advances as may be considered necessary to pay laborers and mechanics employed by the Contractor or any subcontractor the full amount of wages required by this contract. In the event of failure to pay any laborer or mechanic all or part of the wages required by this contract, the Contracting Officer may take such action as may be necessary to cause the suspension, until such violations have ceased, of any further payment, advance, or guarantee of funds to or for the Government Prime Contractor

26. SUBCONTRACTS—TERMINATION

The Contractor agrees to insert Clauses 20 through 26 hereof in all subcontracts and further agrees that a breach of any of the requirements of these clauses may be grounds for termination of this contract. The term "Contractor" as used in such clauses in any subcontract shall be deemed to refer to the subcontractor except in the phrase "Government Prime Contractor."

U.S. GOVERNMENT PRINTING OFFICE : 19––O–815770

Standard form 23A (cont.)

sidered to be within the scope of the contract, did not violate nor breach the contract. The claim for damages for delay as a breach of the contract was not allowed by the court.

In *Continental Illinois Nat. Bank and Trust Co.* v. *United States,* 101 F. Supp. 755, 1952, the contract reserved to the government the right to make changes, and the work was halted when the contracting officer decided to redesign a building. The work was delayed 175 days, though forty calendar days would have been a reasonable time within which to complete the new design. The court allowed the contractor his claim for damages for overhead and administrative expenses for the balance of 135 days. The court said: "We think that the Government's taking 175 days for the redesign of the boiler house was inconsiderate of the harm which was being caused the contractor, and was a breach of the contract. The right reserved in the contract to make changes in the work does not mean that the Government can take as much time as it pleases to consider such changes, regardless of consequences to the other party to the contract. *Grand Investment Co.* v. *United States,* 102 Ct. Cl. U.S. 40."

A contractor who agrees to a change order that specifies the amount of the adjustment and accepts it in writing, without any condition or reservation as to his claim or rights, cannot later complain that the amount of the adjustment is inadequate. The case of *Frazier-David Construction Co.* v. *United States,* 97 Ct. Cl. U.S. 1, 1942, was for additional costs that were claimed because the subsurface conditions of the site were found to be materially different from those shown on the drawings and in the specifications. A change order was issued for an increase in price and an extension of time for completion was granted. The contractor did not protest the change order and, upon completion of the work, accepted the payment. The court dismissed his claim and stated: "We are convinced, however, that under all the circumstances the defendant's (government's) representatives have not only acted generously with the plaintiff, but they have been fair to the defendant's interest."

The rule as to a measure of damages in case of a claim of unreasonable delay due to changes in the plans and specifications is illustrated in *Severin* v. *United States,* 102 Ct. Cl. U.S. 74, 1943. The contract was for the construction of a hospital building within five hundred calendar days after notice to proceed. In January, while the work was under way, the contracting officer stopped all work above the ninth floor of the structure; in June the contractor was ordered to proceed with certain changes. After the contractor completed some of the changed work, the government issued an order canceling the changes, and reverting to the original plans. An increase in the price and a ninety-day extension were granted to the contractor. The court allowed damages to the contractor because:

" (he) was delayed by the nonuse of his equipment, the idleness of his supervisory employees, rental cost of equipment, and extra costs of operating his sand and gravel pit, and the extra costs of the subcontractor. . . .

"Due to defendant's (the government's) procrastination and its inability to decide definitely what it proposed to do with reference to the three floors mentioned, plaintiff's (the contractor's) extra costs were all in addition to the work which was performed under the change orders.

"These change orders were purely and solely for the actual cost of the extra work, and the conduits, pipe sleeves, and stairways, and had nothing to do with the delay in the course of the erection of the building above the tenth floor and to the roof of the fourteenth floor."

An unjustified change order under Clause 3 was discussed in the opinion of the court in the case of *Grand Investment Co.* v. *United States,* 102 Ct. Cl. U.S. 40, 1944, in which the government issued a telegraphic stop order to the contractor, which stopped the work for 109 days. The contractor sued for damages for breach of contract as a result of the delay. The court held that the stop order was not justified and created a breach of the contract. Among the items allowed as damages, the court stated as to equipment on the job site, that: "But when the Government in breach of its contract, in effect condemns a contractor's valuable and useful machines to a period of idleness and uselessness, we think that it should make compensation comparable to what would be required if it took the machines for use for a temporary period, but did not in fact use them."

A contract for the construction of an addition to a Veterans Hospital was before the court in the case of *F. H. McGraw & Co.* v. *United States,* 130 F. Supp. 394, 1955. The original cost of the contract was in excess of more than three million dollars, and the time of completion was within 400 days after notice to proceed. The time to complete the work was actually 1,073 days, but the contracting officer found that the contractor was not responsible for the delay, and necessary extensions of time were granted by the government. When the work was about 65 per cent completed, a partial stop order was issued. Its provisions were ambiguous, however, and no representative of the government who was at the site had authorization to interpret the order, which might have permitted the contractor to proceed with areas of the work that were not affected by the partial stop order. The specifications were revised several times, and, eventually, after 159 days the contracting officer issued a change order, increasing the contract price as an equitable adjustment. The court held that the government had unreasonably delayed the contractor while changes were made in the specifications, which were not so extensive as to consider them as outside the scope of the contract. The government claimed that changes were permitted under Clause 3 of the standard form of the contract, and the court ruled that the issuance of the partial stop order was justified in this case, but held the government responsible for damages, because: "It is settled that the defendant (the government) is allowed under the contract only a reasonable time within which to make permitted changes in the specifications, and that the defendant is liable for breach of its contract if it unreasonably delays or disrupts the contractor's work. *J. A. Ross & Co.* v. *United States,* 126 Ct. Cl. U.S. 323, 332; *Continental Illinois National Bank and Trust Co.* v. *United States,* 101 F. Supp. 755; *James Stewart & Co.* v. *United States,* 105 Ct. Cl. U.S. 284, 328; *Severin* v. *United States,* 102 Ct. Cl. U.S. 74, 85; *Silverblatt & Lasker, Inc.* v. *United States,* 101 Ct. Cl. U.S. 54, 82, and the cases there cited." The court pointed out that so long a delay in deciding on changes was not contemplated by the parties, and "For all delay of over a month we think defendant is liable in damages." The decision also stated: "Here the Contracting Officer has made a finding on the amount to which plaintiff is entitled as an equitable adjustment, and this was done in the way the contract required. In such case, it is binding on us (Court of Claims), unless it is not supported by substantial evidence. *United States* v. *Callahan Walker Construction Co.,* 317 U.S. 56."

In the case of *Silverblatt & Lasker, Inc.* v. *United States,* 101 Ct. Cl. U.S. 54, 1944, the contractor sued for a breach of contract for the construction of a post office in which the government changed the stone specification from bluestone and granite to rubble stone. The court decided the change was within the general scope of the contract, and further, the contractor agreed to make the change. The "changes" clause provided for an equitable adjustment, and the court pointed out: "Since what constitutes an equitable adjustment is a question of fact (*United States* v. *Callahan Walker Construction Co.,* 317 U.S. 56), the 'changes' clause authorized the contracting officer and the head of the department to settle the dispute over the amount to be paid for the change."

The case of *Griffiths* v. *United States,* 74 Ct. Cl. U.S. 245, 1932, involved, among other items, a claim for damages on account of delay caused by changes in the plans. The contract provided that changes could be made within the general scope of the work. The court ruled against the contractor and said:

"The provisions of a contract authorizing changes in the plans and specifications carries with it the reasonable implication that if such changes are made, a delay in the prosecution of the work may result. *Moran Brothers Company* v. *United States,* 61 Ct. Cl. U.S. 73.

"The changes made in this case were authorized in the contract, and the contract price related as much to the privileges of making them as to any other provision in the contract. The right to make the changes was a right the government contracted for and it cannot be held liable in damages beyond the extra cost involved in making the changes themselves. The government cannot be subjected to damages for exercising a privilege bought and paid for by it, in the absence of a showing it has abused such privilege. *McCord* v. *United States,* 9 Ct. Cl. U.S. 155."

Significant criteria are disclosed in certain court decisions concerning the exercise of the right of large numbers of changes. In *Magoba Construction Co. Inc.* v. *United States,* 99 Ct. Cl. U.S. 662, 1943, it was held by the court that the issuance of orders for 62 changes during the progress of the work was not a violation of the "Changes" clause in the contract in the sum of $2,050,000 for the reconstruction of a post office building.

When the provisions of a contract specify a certain period of time within which a contractor may protest or appeal an order by the contracting officer, it is good practice for the contractor to recognize and to comply with the particular provisions in order to safeguard his interests. Occasionally, it is disclosed that the act or omission of the contracting officer constitutes a waiver of the requirements in this respect, as was discussed in *The Arundel Corporation* v. *United States,* 96 Ct. Cl. U.S. 77, 1942. The contract was for the construction of a lock and dam on the Savannah river, and the document provided that any claim for adjustment for a change in the plans and specifications should be asserted within ten days from the date the change was ordered, "unless the contracting officer shall for proper cause extend such time." The contractor did not protest for an adjustment of a change order that required him to install additional concrete forms for chain storage recesses in the dam. However, the contractor gave the claim to the contracting officer who considered it on its merits without any mention of the fact that it had been filed too late. The court decided that the consideration of the claim by the contracting officer indicated that he did not rely upon the contractor's failure to file a protest, and he thereby actually waived the omission of a protest by the contractor as required by the provisions of the contract.

The provisions of Clause 3 state, in part: "The Contracting Officer may at any time, by a written order, and without notice to the sureties, make changes in the drawings and/or specifications of this contract and within the general scope thereof." In some cases the order to the contractor to make changes was oral, not written as required by the related provisions. In such a situation the question arises of an implied contract resulting from the oral order for which the Government is, nevertheless, responsible. For example, in the case of *W. H. Armstrong & Co.* v. *United States,* 98 Ct. Cl. U.S. 519, 1943, the contractor agreed to build officers' quarters, but the common bricks were to be furnished by the government. The contractor inspected the supply of bricks to be furnished in order to estimate the amount of labor and mortar to be used in laying the bricks, but before the contractor completed the brick work, the contracting officer issued an oral order to discontinue the use of the bricks originally provided, and to use bricks that had been salvaged elsewhere and that were irregular in size and shape. When the contractor protested orally to the contracting officer, the latter agreed orally to present the extra costs to the Constructing Quartermaster for audit. The court found in favor of the contractor's claim for additional compensation, and commented that the government had the power under the contract provisions, to modify the terms of the contract, and that the order was either for extra work or change order under the respective clauses of the contract. The court ruled that the order was for extra work requiring extra labor and extra mortar. "When one party writes a contract in its own language, as the government did here, and inserts in it two separate provisions, either of which might apply to a given state of facts, but different legal consequences would result if one, rather than the other, of the two provisions was applied, the other party to the contract is, in the absence of evidence of a contrary intent, entitled to have applied the provision which would be least burdensome to him." The failure of the contracting officer to write his order for the extra work and materials, and the omission of the contractor to insist upon a written order, was considered by the court as follows:

"If the Constructing Quartermaster had, at the time he gave his order, said in writing what he said orally, that would not have been a compliance with Article 5, which requires that the 'price (be) stated in such order.' Strictly, plaintiff (the contractor) could not have protected itself against what has here happened to it except by refusing to obey the order and demanding that the Constructing Quartermaster do, in writing, what was impossible to do at all, viz: fix a price which could not prudently be fixed at that time. Perhaps that would have been correct practice.

"Our question is, however, not what would have been correct practice when the extras were ordered, but what to do with a case in which a responsible representative of the Government has not done, nor has the contractor insisted upon his doing what the contract enjoined him to do at that time; a case in which, moreover, the contractor has, pursuant to an oral rather than a written order, performed just what the Government wanted to be done, and the Government has taken the product of that performance but refuses to pay for it. If we say the position of the Government is legally correct, we thereby refuse to remedy a grave injustice because of the omission of a formality, because the words of the Contracting Officer were spoken and not written and did not say what was then impossible to say."

In deciding that the contractor was entitled to be paid following his full compliance with the oral order, the court pointed out: "We recognize that the decision of the Supreme Court of the United States in *Plumley* v. *United States,* 226 U.S. 545, is contrary to what we have said. We see no distinction, in principle, between that case and the cases of *United States* v. *Andrews,* 207 U.S. 229 and *St. Louis Hay & Grain Co.* v. *United States,* 191 U.S. 159, discussed above. All the authorities in comparable legal situations are in accord with the latter cases, and we think it is our duty to follow this general trend of authority, which leads to what we regard as a just decision, rather than the *Plumley* case, which seems to stand alone."

A comparable oral order was involved in the case of *Long* v. *United States,* 127 F. Supp. 623, 1955, and the court, upon citing the *Armstrong* case, ruled in favor of the contractor on the ground that: "We feel that there then arose an implied contract under which the defendant (the Government) was obligated to pay the value of the services rendered by the plaintiffs (contractor)."

Part 1-7, "Contract Clauses," of the *Federal Procurement Regulations* was officially established on March 10, 1959, and includes the following paragraph in section 1-7.101-3, under the heading of "Extras": "Except as otherwise provided in this contract, no payment for extras shall be made unless such extras and the price therefor have been authorized in writing by the Contracting Officer."

The last sentence in Clause 3, states: "Except as otherwise herein provided, no charge for any extra work or material will be allowed."

Clause 4 – Changed Conditions: Clause 4 of Standard Form 23A, entitled "Changed Conditions," is expressly applicable to subsurface conditions at the site that are materially different from those shown on the plans and specifications and also to conditions that were unforeseen and unknown and, therefore, not shown on the plans and specifications. If the contractor finds subsurface or latent physical conditions at the site that are materially different from those indicated in the contract, he must give written notice of his finding to the contracting officer before the conditions are disturbed. Similar notice to the contracting officer must be given in case the contractor discovers any unforeseen or unknown physical conditions at the site that are not generally recognized as being inherent in the character of the work of the contract. This requirement of notice also applies to conditions that are considered to be unusual or that differ materially from those ordinarily found in the type of work called for by the contract. Upon receiving such notice, the duty of the contracting officer is to investigate promptly the conditions described in the notice from the contractor. If the contracting officer decides that the conditions warrant it, he is authorized to make an equitable adjustment of the increase or decrease in the cost of the work or in the time required to accomplish the desired result, and in appropriate cases, he is expected to issue a written modification of the contract to provide for the revisions. The contractor is warned that only those claims of changed conditions will be allowed that are directed to the attention of the contracting officer, as described here. However, the contracting officer has the right to consider and adjust any claim that is asserted by the contractor before the date of final settlement of the contract. Finally, the clause provides that in case the contractor and the contracting officer cannot agree upon the adjustment to be made for any changed conditions, the dispute is then to be determined by the provisions of Clause 6 – "Disputes."

The court decisions reviewed under this heading are, like the other judicial answers throughout this book, some of the leading, authentic commentaries on the subject of "changed conditions" as expressed in the federal construction contract. At best, the results of the litigation reflect judicial reasoning upon specific facts, and, subject to the advice of legal counsel, the contractor and the government's representative are entitled to rely upon the decisions where they can be applied in a similar state of facts. In line with these observations, it can also be stated that it is preferable to provide in a construction contract for compensation by way of an equitable adjustment rather than by a suit by the contractor for breach of contract by the government. Further, this changed conditions clause is considered as an inducement to contractors to compute a bid upon normal conditions at the site, because there is a provision in the contract for compensation for additional costs if abnormal conditions are discovered when the work operations are in progress.

In the case of *United States* v. *Rice,* 317 U.S. 61, 1942, the contract was for the installation of plumbing, heating, and electrical equipment in a public building, which was to be erected by another contractor who was to finish the work at the same time as the building contractor would complete his contract. When the contractor Rice brought his tools, equipment, and personnel to the site, he found that the building contractor had been stopped by the government because of the discovery of an unsuitable soil condition. The site was changed, the specifications were altered, and, as a consequence, Rice was delayed considerably. An extension of time was granted by the government and it also readjusted the amount due because of the structural changes.

Clause 3 of the contract, "Changes," provided as follows: "The contracting officer may at any time, by a written order, and without notice to the sureties, make changes in the drawings and (or) specifications of this contract and within the general scope thereof. If such changes cause an increase or decrease in the amount due under this contract, or in the time required for its performance, an equitable adjustment shall be made and the contract shall be modified in writing accordingly." This article governed the procedure under which the government could alter the specifications of the contract for general causes.

In the *Rice* case, Clause 4 of the contract, "Changed Conditions," stated: "Should the contractor encounter, or the Government discover, during the progress of the work, subsurface and (or) latent conditions at the site materially differing from those shown on the drawings or indicated in the specifications, the attention of the contracting officer shall be called immediately to such conditions before they are disturbed. The contracting officer shall thereupon promptly investigate the conditions, and if he finds that they materially differ from those shown on the drawings or indicated in the specifications, he shall at once, with the written approval of the head of the department or his representative, make such changes in the drawings and (or) specifications as he may find necessary, and any increase or decrease of cost and (or) difference in time resulting from such changes shall be adjusted as provided in Article 3 of this contract." These provisions directed the procedure under which the government could alter the contract to meet unanticipated physical conditions. Rice sued the government for about $26,000 for delay damages. The questions were whether the delay in commencing the construction was a breach of contract by the government and whether the mechanical contractor was entitled to an equitable adjustment of damages because of the delay. The government contended that the change in the specifications, resulting in a delay, was not a breach of the contract, but was permitted by the contract, as was also the extension of time to cover the period of delay. The court disallowed the claim of the mechanical contractor for damages and stated that there were "two consequences when the Government discovered that the building could not be constructed as originally planned. One was an alteration of the specifications which resulted in a slight cut in respondent's (mechanical contractor's) outlay and his compensation. The other was the delay itself and for this the time necessary to perform the contract was equitably adjusted by extension thereby relieving respondent of liquidated damages which could otherwise have been imposed. Under the terms of the contract, it is entitled to no more." The opinion also stated: "If there are rights to recover damages when the government exercises its reserved power to delay, they must be found in the particular provisions fixing the rights of the parties."

The court also commented on the meaning of clauses 3 and 4 as follows:

> "Both clauses essentially provide that if changes are made affecting an increase or decrease of cost or affecting the length of time of performance, an equitable adjustment shall be made.
>
> "Clearly questions of interpretation in clauses so similar should, if possible, be resolved in the same fashion in each of them. Clause 4 was added to the standard form contract since Clause 3, and we therefore turn first to decisions interpreting the latter clause. The Court of Claims, relying on principles announced in the *Chouteau (95 U.S. 61), Wells Bros. Co. (254 U.S. 83),* and *H. E. Crook Co. (270 U.S. 6)* cases, has uniformly held that the 'increase or decrease of cost' language in Article 3, and in similar clauses, is not broad enough to include damages for delay; that 'It was never contemplated . . . that delays incident to changes would subject the Government to damage beyond that involved in the changes themselves.' "

The decision in the *Rice* case followed the reasoning in *Chouteau* v. *United States,* 95 U.S. 61, 1877, which involved a Navy contract to build a vessel. The contract provided that the Navy Department could make any necessary alterations and additions to the plans and specifications at any time during the progress of the work and should pay any extra expense caused by such changes, at fair and reasonable rates, to be determined when the changes were required. The contractor sued for about $56,000 for extra work due to important changes in the plans and specifications. The court found that there had been delay in completing the vessel due to changes ordered by the Navy Department and also because of a rise in prices of labor and materials. The decision stated that the contract anticipated changes, and provision was also made in the contract to pay for extra work. "For the reasonable cost and expenses of the changes made in the construction, payment was to be made; but for any increase in the cost of the work not changed no provision was made."

The court was called upon to apply the provisions of the changed conditions clause in the case of *The Arundel Corporation* v. *United States,* 103 Ct. Cl. U.S. 688, 1945, to a contractor's claim for an increase in the unit price of dredging of material because a decreased amount of work incurred increased costs. Before the contract was executed, a hurricane caused the current in the canal to scour the area of a large amount of the material. The court held that the action of the hurricane was not a changed condition that would entitle the contractor to an adjustment. The opinion stated: "We think the government did not, by Article 4 – 'Changed Conditions,' assume any obligation to compensate plaintiff through an increase in the contract unit price for any increase in its anticipated dredging cost per cubic yard, or reduction of its anticipated profit not caused by any act or fault of the government, but brought about and caused by a hurricane which neither party expected or could anticipate. *The Arundel Corporation* v. *United States,* 96 Ct. Cl. U.S. 77. The plaintiff (contractor) assumed the risk of the amount of material to be dredged being reduced, as it was by the hurricane, an act of God, just as the government would have had to assume the risk of having to pay for an increase in the material necessary to be dredged for the same reason, as was the case in *Tacoma Dredging Co.* v. *United States,* 52 Ct. Cl. U.S. 447, where a flood caused an increase of 67,000 cubic yards. It is a general principle of law that neither party to a contract is responsible to the other for damages through a loss occasioned as a result of an act of God, unless such an obligation is expressly assumed. Here, the contract was silent in that regard and whatever loss plaintiff may have sustained must be borne by it, and not by the government."

Clause 5 – Termination for Default—Damages for Delay—Time Extensions: In case of delay in the work by the contractor, Clause 5 gives the government the right to terminate the contractor's operations by written notice, either the part of the work that has been delayed or all of the work under the contract, and the contractor and his surety are liable for any excess cost of completing the work and for liquidated damages. If no sum for liquidated damages is specified, then the actual damages caused by the delay are recoverable from the contractor and his surety. The necessary materials, appliances, and plant on the site are subject to possession and use by the government when the right of contract termination is exercised. Even if the government does not exercise its right to terminate the contract, the contractor and his surety are liable for liquidated damages or for actual damages because of delay. However, Clause 5 also provides that the contractor's right to proceed with the work because of delay is not to be terminated and he is not to be assessed for liquidated damages by reason of delays "due to unforeseeable causes beyond the control and without the fault and negligence of the contractor, but not restricted to acts of God or of the public enemy, acts of the government, in either its sovereign or contractual capacity, acts of another contractor in the performance of a contract with the government, fires, floods, epidemics, quarantine restrictions, strikes, freight embargoes, and unusually severe weather, or delays of subcontractors or suppliers due to such causes." The clause also provides that the contractor must within ten days after the delay begins, notify the contracting officer in writing, stating the cause of the delay. When the contracting officer receives the delay notice, he is required to investigate the cause of the delay, and if he determines that the notice of delay is justified, he must extend the contract deadline. His decision is "final and conclusive," but the contractor has the right to appeal by the procedure in Clause 6 – "Disputes." The safe and practical procedure for a contractor to follow when he is delayed in his work is to give the contracting officer the written notice within the prescribed ten-day period. Clause 5 also provides for the granting of an extension of time by the contracting officer before the date of final settlement of the contract without the ten days notice by the contractor.

In the case of *Wells Bros. Co.* v. *United States,* 254 U.S. 83, 1920, the contract was for the construction of a post office and courthouse building. On the day after the contract was signed, the federal government "ordered and directed" the contractor to delay ordering the limestone specified in the contract, because a change was contemplated in the exterior face stonework "Which would require an additional appropriation by Congress." The contractor agreed to wait two weeks only, but he was not allowed to order the stone for about ten months, when an appropriation was made by Congress and a supplemental agreement was made by which marble was substituted for the limestone, increasing the amount of the contract by $210,500. An extension of time for completion was granted to the contractor by the contracting officer. While awaiting the action by Congress, the contractor completed the excavation, foundation, and structural steel work. A six-month delay was ordered by the federal government until Congressional approval of the Parcel Post was completed. The contractor sued for damages because of both delay periods, and the question before the court was whether or not the terms of the contract authorized the federal government to require the delays without liability for damages to the contractor. The contract provisions permitted the United States to suspend all or a part of the work "without expense to the United States," and one day additional to the time stated in the contract would be allowed for each day of such delay. The contract also provided that "no claim shall be made or allowed to the contractor for any damages which may arise out of any delay caused by the United States." The court disallowed the contractor's claim, stating that the terms of the written contract were plain and comprehensive, and the opinion pointed out: "Men who take million-dollar contracts for government buildings are neither unsophisticated nor careless. Inexperience and inattention are more likely to be found in other parties to such contracts than the contractors, and the presumption is obvious and strong that the men signing such a contract as we have here protected themselves against such delays as are complained of by the higher price exacted for the work."

The immunity of the federal government against liability for delay damages to a contractor under a contract to install a heating plant in a public building was reviewed in the case of *Crook Co.* v. *United States,* 270 U.S. 4, 1925. The work was to be done 200 days after the copy of the contract was delivered to the contractor. The government reserved the right to make changes and to interrupt the stipulated continuity of the work. The contract disclosed that the building was under construction by contractors who might not keep up to time, but an approximate date of completion of the construction was stated. The construction contractors were a year behind the schedule for completing their work.

"The government did fix the time (of completion) very strictly for the contractor. It is contemplated that the contractor may be unknown, and he must satisfy the government of his having the capital, experience and ability to do the work. Much care is taken therefore to keep him up to the mark. Liquidated damages are fixed for his delays. But the only reference to delays on the government's side is the agreement that if caused by its acts, they will be regarded as unavoidable, which, though probably inserted primarily for the contractor's benefit as a ground for extension of time, is not without a bearing on what the contract bound the government to do. Delays by the building contractors were unavoidable from the point of view of both parties to the contract in suit. The plaintiff (contractor) agreed to accept in full satisfaction for all work done under the contract, the contract price, reduced by damages deducted for his delays and increased or reduced by the price of changes as fixed by the Chief of the Bureau of Yards and Works. Nothing more is allowed for changes, as to which the government is master. It would be strange if it were bound for more in respect of matters presumably beyond its control. The contract price, it is said in another clause, shall cover all expenses of every nature connected with the work to be done. . . . The plaintiff's (the contractor's) time was extended and it was paid the full contract price. In our opinion it is entitled to nothing more."

In the case of *United States* v. *Howard P. Foley Co.,* 329 U.S. 64, 1946, the lawsuit was for damages for delay in making the site available to an electrical contractor who agreed, for a fixed fee, to furnish and install a field lighting system at a national airport, within 120 days after notice to proceed. The job was not finished until 277 days after the notice was given by the contracting officer. Hydraulic dredging was used to build up the airport site from under water. As parts of the pavement for runways and taxiways were finished, the contractor was to move in and install the necessary fixtures. The dredging required more time than the government engineers had anticipated because some of the dredged soil did not have adequate stability for runways and taxiways and had to be replaced; all of which resulted in delayed completion of the runway sections, which also delayed the work of the electrical contractor. The court commented that the government's representatives were diligent in their efforts and no fault could be attributed to them. The decision stated: "Consequently, the government cannot be held liable unless the contract can be interpreted to imply an unqualified warranty to make the runways promptly available." In finding that no such warranty is expressed in or disclosed by the contract, the decision disallowed the claim of the contractor and it held:

"Here as in the former cases, *H. E. Crook Co.* v. *United States,* 270 U.S. 4, 1926; *United States* v. *Rice,* 317 U.S. 61, 1942, there are several contract provisions which showed that the parties not only anticipated that the government might not finish its work as originally planned, but also provided in advance to protect the contractor from the consequences of such governmental delay, should it occur. The contract reserved a governmental right to make changes in the work which might cause interruption and delay, required respondent (the contractor) to coordinate his work with the other work being done on the site, and clearly contemplated that he would take up his work on the runway sections as they were intermittently completed and paved. Article 9 of the contract, entitled 'Delays—Damages,' set out a procedure to govern both parties in case of respondent's delay in completion, whether such delay were caused by respondent, the government, or other causes. If delay were caused by respondent (the contractor), the government could terminate the contract, take over the work, and hold respondent and its sureties liable. Or, in the alternative, the government could collect liquidated damages. If, on the other hand, delay were due to 'acts of the government' or other specified events, including 'unforeseeable causes,' procedure was outlined for extending the time in which respondent was required to complete its contract, and relieving him from the penalties of contract termination or liquidated damages.

"In the *Crook* and *Rice* cases we held that the government could not be held liable for delay in making its work available to contractors, unless the terms of the contract imposed such liability. Those contracts, practically identical with the one here, were held to impose none." In concluding the opinion, it was stated: "The question on which all these cases turn is, did the government obligate itself to pay damages to a contractor solely because of delay in making the work available? We hold again that it did not for the reasons elaborated in the *Crook* and *Rice* decisions."

The judicial ruling in the *Foley* case was cited as support in *Hagerman* v. *United States,* 180 F. Supp. 181, Wyo., 1960, in which the damages clause in the standard form of United States construction contract was likewise held to relieve the federal government from liability for delay damages on the theory that the clause provided a procedure to govern both of the contracting parties in such a situation.

The government's claim of immunity from liability for delay damages was denied in the case of *Ozark Dam Constructors* v. *United States,* 120 Ct. Cl. U.S. 354, 1955. The contract was undertaken as a joint venture to construct a federal dam on the White River in Arkansas, but the government and not the contractor was to furnish the cement to be delivered f.o.b. at Cotter, Arkansas, where the contractor would unload the material and transport it to the site at his expense. In the specifications for furnishing the cement, it was provided, in part, that "The government will not be liable for any expense or delay caused the contractor by delayed materials except as provided under Article 9 of the contract." Article 9 was the standard "Damages for Delay" article and provided that if the work was delayed for causes beyond the control of the contractor, including strikes, the contract should not be terminated by the government, and the contracting officer should extend the contractor's deadlines. Upon thirty days' notice of the contractor's need for the cement, the government would have the cement shipped by railroad from a mill at Independence, Kansas, to Cotter, Arkansas. Twelve thousand barrels of the material were requested for September deliveries and forty-one thousand barrels for October. A strike on the railroad prevented deliveries in those months, and the lawsuit was to collect damages from the government for the nondelivery. The court pointed out that the government had notice of the impending strike of the railroad employees, and in the months preceding the delivery requirements of the cement, the contractor and the government's representatives discussed the possibility of the strike. The government, however, did nothing to seek out possible alternative ways of making delivery of the material as it was required. Delivery could have been made by another railroad to a point 68 miles to the job or it could have been transported by trucks. The work was delayed 43 days, and damages were claimed by the contractor in the amount of $473,000. During the first month of the delay period the contractor asked the contracting officer for a formal order suspending the work, which was permitted under the contract, but no order was issued by the government.

The contracting officer denied the damage claim of the contractor "on the grounds (1) that under the contract plaintiffs were entitled only to an extension of time; (2) the strike had not resulted from the fault or negligence of defendant, and (3) that suitable hauling equipment in sufficient quantities to transport bulk cement to the job site, was not available during the period of the strike." An appeal was taken by the contractor to the Secretary of the Army, as required by Article 15 of the contract, but the appeal was denied on the basis of the provision in the specifications that: "The Government will not be liable for any expense or delay caused the contractor by delayed deliveries except as provided in Article 9 of the contract," which was intended to entitle the contractor to an extension of time only. The opinion of the court stated: "A contract for immunity from the harmful consequences of one's own negligence always presents a serious question of public policy. That question seems to us to be particularly serious when, as in this case, if the government got such an immunity, it bought it by requiring bidders on a public contract to increase their bids to cover the contingency of damages caused to them by the negligence of the Government's agents. Why the government would want to buy and pay for such an immunity is hard to imagine. If it does, by such a provision in the contract, get the coveted privilege, it will win an occasional battle, but lose the war." The opinion also held with relation to the omission of the government to accomplish delivery of the material by other means, that:

"The possible consequences were so serious, and the actions necessary to prevent those consequences were so slight, that the neglect was almost willful. It showed a complete lack of consideration for the interest of the plaintiffs (the contractor). If the plaintiffs really included in their bid an amount to cover the contingency of such inconsiderate conduct on the part of the Government's representatives, the government was buying and the people were paying for things that were worth less than nothing.

"Our conclusion is that the nonliability provision in the contract, when fairly interpreted in the light of public policy, and of the rational intention of the parties, did not provide for immunity from liability in circumstances such as are recited in the plaintiffs' (contractor's) petition."

The courts have affirmed the rule that the government is liable for delay caused by it, unless expressly exempt by the provisions of the contract. Whereas a standard clause in a government construction contract may provide for an extension of time for completion of the work because of delays from unforeseen causes, including delays caused by the Government, there is an implied provision in every contract that neither party to the contract will interfere with performance of the contract by the other party that will hinder or delay such performance. In the case of *George A. Fuller Company* v. *United States,* 108 Ct. Cl. U.S. 70, 1947, the decision pointed out that:

"It is a necessary corollary to this principle that one who, while not preventing the other party from carrying out the contract, nevertheless hinders or delays him in doing so, breaches the contract, and is liable for the damage which the injured party has sustained thereby. The Supreme Court so held in *United States* v. *Smith,* 94 U.S. 214. In this case it was said: 'Under such circumstances, the law implies that the work should be done within a reasonable time, and that the United States would not unnecessarily interfere to prevent this.' The Supreme Court accordingly affirmed the judgment of this court awarding damages for the unlawful delay.

"It is true in the cases of *Wells Brothers Co.* v. *United States,* 254 U.S. 83, and *Wood, et al.* v. *United States,* 258 U.S. 180, it was held that the defendant was not liable for damages for delay, but this was because of express provisions in these contracts exempting the Government from liability therefor. In the *Wells Brothers Co.* case, the contract provided that 'no claim shall be made or allowed to the contractor for any damages which may arise out of any delay caused by the United States.' There was a similar provision in the contract in the case of *Wood, et al.* v. *United States.* But, in the absence of such an express provision, neither this court nor the Supreme Court has ever exempted the government from liability for damages for delays not contemplated by the contract."

In commenting on the liability of the government for damages caused by its delay, the court stated that this liability exists when the language of the particular clause discloses that it was not intended to deprive the contractor of its right to damages for delays caused by the acts of the government. Since the party drawing the contract so construes it, the courts cannot give it a contrary construction.

"These holdings in no way conflict with the decisions of the Supreme Court in *Crook Co.* v. *United States,* 270 U.S. 4 and *United States* v. *Rice,* 317 U.S. 61. These cases exempted the government from liability for damages caused by delays incident to the making of changes, but this was for the reason that the government had reserved the right to make changes and because it necessarily followed that some delay would result therefrom. Here there was no reservation of a right to delay furnishing the models." There is a difference in the liability of the government between a delay caused by the expressly reserved right to make changes and one caused by the government's failure to furnish materials, which the contractor was to use in performing his contract.

A summary of the general rules of the responsibility of the government for delay, whether because of late delivery of the site or of materials to the contractor, is expressed in the case of *Peter Kiewit Sons' Co.* v. *United States,* 151 F. Supp. 726, 1957, as follows:

"Generally the Government is not liable for delays in making the work or material available to a contractor. *United States* v. *Rice,* 317 U.S. 61; *United States* v. *Howard P. Foley Co.,* 329 U.S. 64. However, where the Government or its authorized representatives are guilty of some act of negligence or willful misconduct which delays the contractor's performance, the Government is liable for the resulting damages. *Chalender* v. *United States,* 127 Ct. Cl. U.S. 557. This is so because there is in every Government contract, as in all contracts, an implied obligation on the part of the Government not to willfully or negligently interfere with the contractor in the performance of his contract. *Challender* v.

United States, 127 Ct. Cl. U.S. 557; *George A. Fuller Co.* v. *United States,* 108 Ct. Cl. U.S. 70. When the contract does not specify particular dates upon which delivery of the material is to be made, the implied obligation just referred to is an obligation not to willfully or negligently fail to furnish the materials in time to be installed in the ordinary and economical course of the performance of the contract. *Walsh* v. *United States,* 121 Ct. Cl. U.S. 546; *Thompson* v. *United States,* 130 Ct. Cl. U.S. 1. If the Government exerts every effort to supply the contractor with the necessary materials on time, it cannot be held that it has willfully or negligently interfered with performance. *Otis Williams Co.* v. *United States,* 120 Ct. Cl. U.S. 249; *W. E. Barling* v. *United States,* 126 Ct. U.S. 34."

Clause 6 – Disputes: Clause 6 is one of the most important clauses in Standard Form 23A. Both the contractor and the contracting officer must be familiar with all of the provisions in the clause, particularly because they govern the express and related rights of the contracting parties under the terms of the three clauses just described.

Clause 6 is generally recognized as a highly effective mechanism at an administrative level, so that adjustments may be made when the contracting parties disagree and so that mistakes can be corrected. As the procedure is exclusive in its nature and governs the method by which claims can be made and adjudicated, the use of the mechanism has a practical value to the contracting parties in that claims can be amicably adjusted and large damage claims are avoided.

Whereas special provisions in certain contracts may be included on disputes, the standard Clause 6 authorizes a decision by the contracting officer of a question of fact upon which agreement cannot be reached by the contracting parties. His decision must be in writing, and the contractor is entitled to a copy. The contractor then has the right to appeal the decision, provided he acts within thirty days by sending a written notice of appeal to the contracting officer, addressed to the head of the department. The procedure of the appeal is intended to give the contractor an opportunity to submit proof of his claim. Pending a final decision, the contractor is required to comply with the order of the contracting officer. The decision of the head of the department or his representatives, who are authorized to conduct the appeal hearings, is final and conclusive, unless the court determines that this decision was "fraudulent, arbitrary, capricious, or so grossly erroneous as necessarily to imply bad faith." If the contractor does not appeal to the head of the department, then the decision of the contracting officer is final and conclusive.

Probably no court decision concerning federal construction contracts in recent years has been more widely discussed and reviewed than the one handed down in the case of *United States* v. *Wunderlich,* 342 U.S. 98, 1951. The United States Court of Claims upheld the claim of the contractor in a dispute over the amount of compensation for the use of equipment for extra work, which was necessary because of unforeseen conditions at the site of the work. The contract was the standard form, which provides that administrative decisions by the government on disputes are final and conclusive. The disputes clause stipulated that all disputes involving question of fact must be decided by the contracting officer, with the contractor retaining the right to appeal to the head of the department, if necessary, "whose decision shall be final and conclusive upon the parties thereto." The contractor was not satisfied with the decision by the appropriate department head, who in this case was the Secretary of the Interior, and brought suit in the Court of Claims. The Court of Claims set aside the decision of the Secretary, and, when the case was heard in the United States Supreme Court, the judgment of the Court of claims was reversed. The opinion of the Supreme Court stated:

"The same Article 15 (disputes clause) was before this Court recently, and we held, after a review of the authorities, that such Article was valid. *United States* v. *Moorman,* 338 U.S. 457. Nor was the *Moorman* Case one of first impression. Contracts, both governmental and private, have been before this Court in several cases in which provisions equivalent to Article 15 have been approved and enforced 'in the absence of fraud or such gross mistake as would necessarily imply bad faith, or a failure to exercise an honest judgment. . . .' " The decision pointed out: "In *Ripley* v. *United States,* 223 U.S. 695, gross mistake implying bad faith is equated to 'fraud.' Despite the fact that other words such as 'negligence,' 'incompetence,' 'capriciousness,' and 'arbitrary,' have been used in the course of the opinion, this Court has consistently upheld the finality of the department head's decision unless it was founded on fraud, alleged and proved. So fraud is in essence the exception. By fraud we mean conscious wrong-doing, an intention to cheat or be dishonest. The decision of the department head, absent of fraudulent conduct, must stand under the plain meaning of the contract. . . .

"Respondents (the contractor) were not compelled or coerced into making the contract. It was a voluntary undertaking on their part. As competent parties they have contracted for the settlement of disputes in an arbitral manner. This, we have said in the *Moorman* case,

Congress has left them free to do. . . . The limitation upon this arbitral process is fraud, placed there by this Court. If the standard of fraud that we adhere to is too limited, that is a matter for Congress."

Legislation to overcome the effect of the Supreme Court decision in the *Wunderlich* case was introduced in Congress in 1954 to "permit review of decisions of Government contracting officers involving questions of fact arising under Government contracts in cases other than those in which fraud is alleged, and for other purposes." (House Report No. 1380, March 22, 1954, on S. 24). The legislation was passed and approved in 1954, and it constitutes Chapter 5 of Title 41 of the *United States Code,* which contains the following significant statutory provisions on judicial review of administrative decisions:

"Sec. 321. Limitation on pleading contract provisions relating to finality; standards of review.

No provision of any contract entered into by the United States, relating to the finality or conclusiveness of any decision of the head of any department or agency or his duly authorized representative or board in a dispute involving a question arising under such contract, shall be pleaded in any suit now filed or to be filed as limiting judicial review of any such decision to cases where fraud by such official or his said representative or board is alleged; *provided,* however, that any such decision shall be final and conclusive unless the same is fraudulent or capricious or arbitrary or so grossly erroneous as necessarily to imply bad faith, or is not supported by substantial evidence.

Sec. 322. Contract provisions making decisions final on questions of law.

No government contract shall contain a provision making final on a question of law the decision of any administrative official, representative, or board."

These statutory provisions are intended to prescribe "fair and uniform standards for the judicial review of such administrative decisions in the light of the reasonable requirements of the various government departments and agencies, of the General Accounting Office, and of government contractors." They will "also prohibit the insertion in government contracts . . . of provisions making the decisions of government officers final on questions of law arising under such contracts." (House Report No. 1380, March 22, 1954, on S. 24.)

The House Report No. 1380 also included the following statements which are pertinent to standards of administrative decisions relative to public contracts: "After extensive hearings it has been concluded that it is neither to the interests of the government nor to the interests of any of the industry groups that are engaged in the performance of government contracts to repose in government officials such unbridled power of finally determining either disputed questions of law or disputed questions of fact arising under government contracts, nor is the situation presently created by the *Wunderlich* decision consonant with tradition that everyone should have his day in court and that contracts should be mutually enforceable. A continuation of this situation will render the performance of government work less attractive to the responsible industries upon whom the government must rely for the performance of such work, and will adversely affect the free and competitive nature of such work. It will discourage the more responsible element of every industry from engaging in government work and will attract more speculative elements whose bids will contain contingent allowances intended to protect them from unconscionable decisions of government officials rendered during the performance of their contracts."

Briefly, the provisions of Chapter 5 of Title 41 of the *United States Code,* as quoted in this section, do not create any new rights that were not available to the contractor before the 1954 statute was enacted. The standards of review in section 321 are intended to provide for substantial evidence for an administrative decision so as to afford a mutual opportunity for rebuttal. Section 322 is aimed at the so-called "all disputes clause," which permits representatitves of the federal government to decide any questions of law as well as fact with finality; the language of the law also prevents the insertion of such a clause in the contract by reference in the drawings, plans, specifications, or other documents.

9.6 The Davis-Bacon Act

The Davis-Bacon Act became a federal law in 1931, and at that time its provisions were applicable to federal contracts for the construction, alteration, or repair of public buildings for which the cost exceeded $5,000. In 1955, the statute was amended to include "public works," reduced the effective cost to $2,000, and also called for a schedule of wage rates with each contract to which the statute was related. The provisions of the law are contained in section 276a of Title 40 of the *United States Code,* and the statute requires that the wages of workmen on a government construction project shall be "not less" than the "minimum wages" specified in the schedule of prevailing wages as determined by the Secretary of Labor for similar work on similar projects in the "city, town, village, or other civil subdivision of the state in which the work is to be performed." The contractor or his subcontractor is required to pay all "mechanics and laborers employed directly upon the site of the work, unconditionally and not less than once a week, . . . computed at wage rates not less than those stated in the advertised specifications, regardless of any contractual relationship which may be alleged to exist between the contractor or subcontractor and such laborers and mechanics." When the contracting officer considers it necessary, he is empowered to withhold from the contractor any accrued payments to pay to the workmen "the difference between" the required rate of wages and that which was actually paid to the laborers or mechanics on the work.

Many contractors apparently compute their bids on federal construction projects by using the wage schedules of prevailing rates as determined by the Secretary of Labor and as shown in the specifications. A question arises as to the rights of the contractor when the wage schedule does not accurately show the prevailing rate when the contract is signed, and also when the Secretary of Labor determines during the progress of the work under the contract that the prevailing rate of wages is higher than when the contract was executed.

A leading case on the construction and effect of the Davis-Bacon Act as to the schedule of wage rates in a federal construction project is *United States* v. *Binghamton Construction Co., Inc.*, 347 U.S. 171, 1954. The contractor sued the government on the grounds that he paid higher wages than that specified in the schedule of the Secretary of Labor, which was included in the government's invitation for proposals and was also incorporated in the contract. It appeared that the contractor's computation of his bid was based upon an already outdated wage rate schedule and one that was inadvertently used in place of a later, accurate determination. The Court of Claims awarded damages against the government because of misrepresentation, and also because, in the judgment of the court, the contractor had the right to rely on the schedule as included in the invitation for bids. The Supreme Court of the United States unanimously reversed the Court of Claims, and ruled that the schedule of wage rates in the specifications was not a warranty as to the local prevailing rates of wages. The statute was designed, the Supreme Court said, for the benefit of the laborers and mechanics as a minimum wage law, and did not assure the successful bidder that the schedule was in fact the prevailing rates of wages in the contract area. The opinion of the Supreme Court pointed out: "The Act (Davis-Bacon Act) itself confers no litigable rights on a bidder for a Government construction contract. The language of the Act and its legislative history plainly show that it was not enacted to benefit contractors, but rather to protect their employees from substandard earnings by fixing a floor under wages on government projects. Congress sought to accomplish this result by directing the Secretary of Labor to determine, on the basis of prevailing rates in the locality, the appropriate minimum wages for each project. The correctness of the Secretary's determination is not open to attack on judicial review."

In the case of *Poirier & McLane Corp.* v. *United States,* 120 F. Supp. 209, 1954, the contract was for erection of a flight hangar. The contract specified, in accordance with the Davis-Bacon Act, that the prevailing rate of wages was 85 cents per hour for unskilled laborers. The contract also provided that the contractor was to make his own investigation to determine the availability of workmen as well as the wage scale for them. If no laborers could be recruited at the schedule of wage rates in the contract, the contractor was to request the Wage Adjustment Board for an increase, and the contractor would absorb the additional cost. The hourly pay-scale was actually $1 for unskilled labor, and the contractor asked the Government to change the contract rate retroactively to $1 and increase the contract price accordingly. The Secretary of Labor determined later that the wage rate was inadvertently set at 85 cents and should have been $1. The contractor had investigated the labor conditions, and the notice from the Secretary of Labor, admitting the inadvertence, showed that the schedule in the specifications misrepresented the prevailing rate of wages. The contractor's claim was allowed by the court on the grounds of misrepresentation by the government of the prevailing rate of wages for unskilled laborers. There was a mutual mistake of fact, and the contract, as executed, did not show the true intent of the contracting parties. In distinguishing the result in this *Poirier & McLane* case from that in the *Binghamton Construction Co.* case, which is discussed in the preceding paragraph, the court said that, as to the former decision, "the contractor did make an adequate and thorough investigation and the provisions of the contract fixed not only the minimum wages, but also the maximum wages which the contractor could pay at the time the contract was made, on the basis of the determination by the Secretary of Labor. The change in that decision, and therefore, in the contract, brought about the increased costs here involved, and the contracting officer required plaintiff (the contractor) to pay the new increased wage rates."

When a contracting officer incorrectly demanded that a contractor pay 80 cents an hour for laborers, the government was held liable for the wrongful act of its representative. In the case of *A. J. Paretta Contracting Co.* v. *United States,* 109 Ct. Cl. U.S. 324, 1947, the specified minimum rate for laborers was 65 cents per hour, but the Federal Wage Adjustment Board allowed a permissive, not a required, increase to 80 cents. The contracting officer then notified the contractor that the latter's failure to pay 80 cents was a disregard of the contract provisions for which the contractor could be penalized. The contractor's claim was allowed by the court, notwithstanding the lack of authority of the contracting officer to order the contractor to pay the higher rate of wages.

A contractor who was engaged in constructing the foundations for a building was forced to pay wages that were higher than those specified in the schedule of prevailing rates, and he sued to recover his additional costs in the case of *United States* v. *Beuttas,* 324 U.S. 768, 1945. The government delayed the start of the work, and while the work was suspended, asked for bids for the erection of the superstructure. In the latter project, the bidders had to pay higher rates for the same classes of workmen that were engaged in the Beuttas' project. The Beuttas' employees learned of this difference and demanded wages equal to those to be paid to the workmen who were to construct the superstructure. The employer had to pay the additional amount in order to settle a strike. The Beuttas firm sought reimbursement from the contracting officer, who disallowed it; the head of the Department also denied the contractor's claim. The Supreme Court of the United States reversed a decision of the Court of Claims which was in favor of the contractor, and decided that the contractor was not entitled to recover his additional costs for labor. The opinion stated: "They (the judges of the Court of Claims) held that by inviting bids for the superstructure at minimum wages higher than those fixed in the respondent's (Beuttas') contract for the foundations, the petitioner (the Government) breached this implied condition (that neither party will hinder the other in the performance of the work or increase the cost of performance). But, of course, this can be true only if the Government caused the increase in wages. And the findings, in our judgment, render a conclusion to that effect impossible. The government did not stipulate that the two groups, those employed on the foundations and those employed on the superstructure, should work at the same time at different wages. On the contrary, it had every reason to suppose that the former would finish their work at the then agreed rate before the latter came on the work. And this is what happened. There is no finding that petitioner's (the government's) officers or agents had reason to expect that the fixing of a wage rate under another contract for a separate portion of the work, which was to be commenced after completion of respondent's contract, would cause a breach of their existing wage agreement by respondent's employees. Moreover, it is found that the petitioner took no part in the wage controversy between the respondents and their employees. It follows that there is no basis for a holding that the government knowingly hindered respondents in the performance of the contract or culpably increased their costs."

According to Part 5 of the Regulations of the Secretary of Labor, a wage determination becomes obsolete after ninety days, when it must be renewed, including all modifications. However, no modification is required if it is received later than five days before a public letting or bid opening, and if an award is made within thirty days of the opening of the bid and within ninety days of the date of the determination.

9.7 Eight Hour Laws, Anti-Kickback Act, and Walsh-Healey Act

Chapter 5 of Title 40 of the *United States Code* is entitled "Hours of Labor on Public Works," and its provisions are included in Clause 21 of Standard Form 23A. The limitation or restriction to "eight hours in any one calendar day" is found in section 321 of Chapter 5, and punishment for a violation by a fine not to exceed $1,000, or imprisonment for not more than six months, or both, is authorized by section 322. Section 324 stipulates a violation penalty against a contractor of $500 "for each laborer or mechanic for every calendar day in which he shall be required or permitted to labor more than eight hours upon said work," but the contractor has the right to appeal to the head of the department having jurisdiction over the contract. Certain contracts are excepted in section 325 from the limitations, and the President may, by executive order, waive the provisions because of war conditions: no penalties can be imposed when violations are due to emergency conditions and are excused by the President. Section 325a declares that the computation of wages of every laborer and mechanic employed by the contractor or subcontractor shall be on a "basic rate of eight hours per day and work in excess of eight hours per day shall be permitted upon compensation for all hours worked in excess of eight hours per day at not less than one and one-half times the base rate." The President is authorized by section 326 in case of national emergency to suspend the provisions of the Eight Hour Law, but the computation of wages as directed by section 325 shall apply in such case.

A judicial interpretation of the statute is found in the case of *Foley Bros. Inc.* v. *Filardo,* 336 U.S. 281, 1949. Filardo, an American citizen, was employed by the contractor on public contracts with the United States in foreign countries, and he sued for overtime pay for working in excess of eight hours per day as provided in the Federal Eight Hour Law. The statute provides, in part, that: "Every contract made to which the United States . . . is a party, . . . shall contain a provision that no laborer or mechanic doing any part of the work contemplated by the contract, in the employ of the contractor or any subcontractor . . . shall be required or permitted to work more than eight hours in any one calendar day upon such work; . . ." *(40 U.S.C. 324)*. Penalties are specified for violations of the statute. In 1941, the contractor undertook, on a cost-plus basis, to perform work on behalf of the United States in the East and Near East and agreed to "obey and abide by all applicable laws, regulations, ordinances, and other rule of the United States of America." The Eight Hour Law was not specified as being included in the contract. The question before the United States Supreme Court was whether or not Congress intended the Eight Hour Law to be applicable to foreign work of public construction for the federal government. In ruling that Congress did not intend to have the law apply to such contracts, the court pointed out:

"First. The canon of construction which teaches that legislation of Congress, unless a contrary intent appears, is meant to apply only within the territorial jurisdiction of the United States. *Blackmer* v. *United States,* 284 U.S. 421, 437, is a valid approach whereby unexpressed congressional intent may be ascertained. It is based on the assumption that Congress is primarily concerned with domestic conditions. We find nothing in the Act itself, as amended, nor in the legislative history, which would lead to the belief that Congress entertained any intention other than the normal one in this case. . . .

"Second. The legislative history of the Eight Hour Law reveals that concern with domestic labor conditions led Congress to limit hours of work. The genesis of the present statute was the Act of June 25, 1868, 15 Stat. 77, c. 72, which was apparently aimed at unemployment resulting from decreased construction in Government navy yards. *Congressional Globe, 40th Cong. 2d sess. Part I, p. 335.* In 1892, when the coverage of this Act was extended to employees of government contractors and when criminal penalties were added, *27 Stat. 340, c 352, 40 U.S.C.Ann. 321,* the considerations before Congress were domestic unemployment, the influx of cheap foreign labor, and the need for improved labor conditions in this country. *H.R. Rep. 1267, 52d Cong. 1st Sess.* The purpose of the new legislation was to remedy the defects in the Act of 1868. *23 Cong. Rec. 5723.*

"The Act was amended in 1912 to include *'every contract'* (italics supplied). The insertion of the word 'every' was designed to remedy a misinterpretation according to which the Act did not apply to work performed on private property by government contractors. *48 Cong. Rec. 381, 385, 394-95.* Nothing in the legislative history supports the conclusion of respondent (Filardo) and the court below that 'every contract' must of necessity, by virtue of the broadness of the language, include contracts for work to be performed in foreign countries. . . .

"The 1940 amendment which permitted work in excess of eight hours per day upon payment of overtime, *40 U.S.C.Ann. 325a,* passed without any discussion indicative of geographical scope. *86 Cong. Rec.* 11216-11217."

The court reviewed the Eight Hour Law *(40 U.S.C.Ann. 321-326),* in the case of *Foley Bros. Inc.* v. *Filardo* from which excerpts are quoted in the preceding paragraphs, as follows: "This Act provides that 'Every contract to which the United States . . . is a party . . . shall contain a provision that no laborer or mechanic doing any part of the work contemplated by the contract, in the employ of the contractor or any subcontractor . . . shall be required or permitted to work more than eight hours in any one calendar day upon such work; June 19, 1912.' Penalties are specified for violations. In 1940, the prohibition against workdays of longer than eight hours was modified as follows: 'Nothwithstanding any other provisions of law, the wages of every laborer and mechanic employed by any contractor or subcontractor engaged in the performance of any contract of the character specified in sections 324 and 325 of this title, shall be computed on a basic day rate of eight hours per day and work in excess of eight hours per day shall be permitted upon compensation for all hours worked in excess of eight hours per day at not less than one and one-half times the basic rate of pay.' . . . *40 U.S.C.Ann. 325a. . . .*"

The Copeland Anti-Kickback Act (*40 U.S.C. 276 b, c*) was placed on the federal statute books in 1934 and provided for a fine of not more than $5,000 or imprisonment of not more than five years, or both, for any contractor who in any manner induces an employee "to give up any part of the compensation to which he is entitled under his contract of employment." The statute applies to any public building work "financed in whole or in part by loans or grants from the United States." The Secretary of Labor is directed to make reasonable regulations for contractors and subcontractors each of whom is required to "furnish weekly a sworn affidavit with respect to the wages paid each employee during the preceding week."

The Walsh-Healey Public Contracts Act of 1936 (*41 U.S.C.Ann. 35*) authorizes the United States Secretary of Labor to establish wages, hours, and working conditions for employees engaged in work on federal contracts "for the manufacture or furnishing of materials, supplies, articles, or equipment in any amount exceeding $10,000." The statute does not apply to construction contracts, but does govern so-called "supply" contracts.

9.8 Renegotiation of contracts

Renegotiation of defense contracts was first incorporated in federal law in 1942 as section 1191 of Title 50 of the *United States Code,* and the statute constituted the legislative response to the public demand to "Take the profit out of war!" That law provided a method whereby the government could validly exercise a statutory concept of refixing contract prices. The 1942 statute was a wartime measure and has been supplemented and amended from time to time, particularly in 1943, 1948, and 1951. The probability of excessive profits by government contractors who are engaged in defense projects, and the technique of control of such profits, seem to be permanent considerations, as long as the federal government continues to spend vast sums of money for defense procurement.

While Congress debated the subject of excessive profits on defense contracts, the United States Supreme Court handed down a decision in the case of *United States* v. *Bethlehem Steel Corp.,* 315 U.S. 289, 1942, which pointed out the antiquity of the problem of war profits and also produced a judicial challenge that hastened Congressional recognition of an intolerable situation. The parties to the lawsuit were involved in a dispute on the amount of profits that were claimed by Bethlehem Steel under thirteen wartime contracts for building ships. The War Powers Act of June 15, 1917, empowered the President: (1) to commandeer shipbuilding plants, (2) to purchase ships at a reasonable price with a provision for subsequent revision by the courts in case the seller regarded the price as unfair, and (3) to buy or to contract for building ships at prices to be established by negotiation. The United States Shipping Board Emergency Fleet Corporation, upon receiving power delegated by the President, decided upon the third method to make purchases through ordinary business bargaining. Bethlehem Steel Corporation claimed about 109 million dollars based upon a formula to determine actual costs which were used by the Fleet Corporation in other shipbuilding contracts. The total profits claimed by Bethlehem Steel were about 24 million dollars or a little more than 22 per cent of the computed cost, but this figure did not include any profits of the parent steel company, which sold the steel used in the ships at the maximum prices established by the War Industries Board.

In considering the 22 per cent profit, the opinion of the court stated that in 1917, average profits on the sales by salmon canneries to the government were about 52 per cent of cost, and, in peacetime, the profits on building cruisers were from 25 per cent to 37 per cent in 1927 and about 22 per cent in 1929. A large American steel producer made 49, 58, and 46 per cent in 1916, 1917, and 1918, while other steel producers earned from 30 per cent to 32 per cent. Lumber producers received as much as 121 per cent on their investment in 1918, and petroleum products returned as high as 122 per cent profit. Sulphur was sold in 1917 at more than 200 per cent to 236 per cent above cost.

The court ruled: "The problem of war profits is not new. In this country, every war we have engaged in has provided opportunities for profiteering and they have been too often scandalously seized. . . . To meet this recurrent evil, Congress has at times taken various measures. It has authorized price fixing. It has placed a fixed limit on profits, or has recaptured high profits through taxation. It has expressly reserved for the government the right to cancel contracts after they have been made. Pursuant to Congressional authority, the government has requisitioned existing production facilities or itself built and operated new ones to provide needed war materials. It may be that one or some or all of these measures should be utilized more comprehensively, or that still other measures must be devised. But if the Executive is in need of additional laws by which to protect the nation against war profiteering, the Constitution has given to Congress, not to this Court, the power to make them."

There was no finding of fraud against Bethlehem Steel Corporation, and the court also held that there was neither duress applied against the government by the corporation during a national emergency, nor were the profits grossly in excess of customary standards.

The constitutionality of the Renegotiation Act of 1942 was disputed in the case of *Lichter* v. *United States,* 334 U.S. 742, 1948. The building construction contractor was charged by the Undersecretary of War with excessive profits of $70,000 for subcontracts that had a total price of $710,000. Some of the subcontracts were made before the Act was passed, but none of the subcontracts contained clauses permitting or requiring their renegotiation. The court said: "The Renegotiation Act was developed as a wartime policy of Congress comparable to that of the Selective Service Act. The authority of Congress to authorize each of them sprang from its war powers."

Although Congress has the power, in time of war, to draft property and workmen: "The plan for renegotiation of profits which was chosen in its place by Congress appears in its true light as the very symbol of a free people united in reaching unequalled productive capacity and yet retaining the maximum of individual freedom consistent with a general mobilization of effort.

"Contracts were awarded (for wartime production) by negotiation whenever competitive bidding no longer was practicable. Contracts were let at cost-plus-a-fixed-fee. Escalator clauses were inserted. Price ceilings were established. A flat percentage limit on the profits in certain lines of production were tried. Excess profits taxes were imposed. Appeals were made for voluntary refunds of excessive profits. However, experience with these alternatives convinced the Government that contracts at fixed initial prices still provided the best incentive to production."

The decision of the United States Supreme Court in the *Lichter* case pointed out with reference to the validity and constitutionality of the Renegotiation Act, that: "The constitutional argument is based upon the claim that the delegation of authority (to determine excessive profits) contained in the Act carried with it too slight a definition of legislative policy and standards. Accordingly, it is contended that the resulting determination of excessive profits which were claimed by the United States amounted to an unconstitutional exercise of legislative power by an administrative official instead of a mere exercise of administrative discretion under valid legislative authority. We hold that the authorization was constitutional."

9.9 Renegotiation Act of 1951

The Renegotiation Act of 1951 is in sections 1211-1224 of Title 50, Appendix, of the *United States Code,* and the "Congressional declaration of policy" (section 1211) is expressed as follows: "It is hereby recognized and declared that the Congress has made available for the execution of the national defense program extensive funds, by appropriation and otherwise, for the procurement of property, processes, and services, and the construction of facilities necessary for the national defense; that sound execution for the national defense program requires the elimination of excessive profits from contracts made with the United States, and from related subcontracts, in the course of said program, and that the considered policy of the Congress, in the interest of the national defense and the general welfare of the nation, requires that such excessive profits be eliminated as provided in this title. . . ."

The contracts that are subject to renegotiation under the 1951 Act are stated in subdivision (a) to be those with the Department of Defense, the Department of the Army, the Department of the Navy, the Department of the Air Force, the Maritime Administration, the Federal Maritime Board, the General Services Administration, the National Aeronautics and Space Administration, and the Atomic Energy Commission, as well as "any other agency of the Government exercising functions having a direct and immediate connection with the national defense which is designated by the President during a national emergency proclaimed by the President, or declared by the Congress after the date of the enactment of the Renegotiation Amendments Act of 1956" (section 1213) "to the extent of the amounts received or accrued by a contractor or subcontractor on or after the first day of January 1951, whether such contracts or subcontracts were made on, before, or after such first day . . ." (section 1212).

The terms "renegotiate" and "renegotiation" include a "determination by agreement or order under this title . . . of the amount of any excessive profits" (section 1213).

Subdivision (e) of section 1213 defines the term "excessive profits" as follows:

"The portion of the profits derived from contracts with the Departments and subcontracts which is determined in accordance with this title . . . to be excessive. In determining excessive profits favorable recognition must be given to the efficiency of the contractor or subcontractor, with particular regard to attainment of quantity and quality production, reduction of costs, and economy in the use of materials, facilities and manpower; and in addition there shall be taken into consideration the following factors:

1. Reasonableness of costs and profits, with particular regard to volume of production, normal earnings, and comparison of war and peace-time products;

2. The net worth, with particular regard to the amount and source of public and private capital employed;

3. Extent of risk assumed, including the risk incident to reasonable pricing policies;

4. Nature and extent of contribution to the defense effort including inventive and development contribution and cooperation with the Government and other contractors in supplying technical assistance;

5. Character of business, including source and nature of materials, complexity of manufacturing technique, character and extent of subcontracting, and rate of turn-over;

6. Such other factors the consideration of which the public interest and fair and equitable dealing may require, which factors shall be published in the regulations of the Board from time to time as adopted."

Provision is made in subdivision (m) of section 1213 for allowance of renegotiation loss carry-forward, and by section 1214, the renegotiation clause must be inserted in each contract made by the affected departments and must provide, among other things, for a withholding by the government of any excessive profits.

Section 1215 is entitled "Renegotiation proceedings," and in subdivision (a) the procedural steps before the Renegotiation Board in compliance with the provisions in the contract on this subject are contained. The renegotiation proceedings are commenced by a notice sent by registered mail to the contractor. The Board is directed by section 1215 to seek an agreement with the contractor or subcontractor "with respect to the elimination of excessive profits received or accrued, and with respect to such other matters relating thereto as the Board deems advisable." In case no agreement is made by the Board, an order must be issued by it "determining the amount, if any, of such excessive profits, and forthwith give notice thereof by registered mail to the contractor or subcontractor." Unless the contractor who objects to and seeks a review of the determination of the Board files a petition with the Tax Court of the United States within the ninety-day period allowed by section 1218, "such order shall be final and conclusive and shall not be subject to review or redetermination by any court or other agency." The statute also permits the Board to consider the aggregate or over-all situation of the contractor or subcontractor and is not restricted to the consideration of each separate contract held by the same party. When the Board makes a determination, the contractor may request a statement of the "facts used as a basis therefor, and of its reasons for such determination. Such statement shall not be used in The Tax Court of the United States as proof of the facts or conclusions stated therein."

Subdivision (b) of section 1215 prescribes a reduction in payments to be made to the contractor by withholding from amounts due the contractor in order to eliminate excessive profits, or the government may sue for a recovery of the money. The surety on a contract or subcontract is not liable for repayment to the government of any excessive profits under the contract or subcontract. The Board has the right to audit the books of the contractor or subcontractor, and may also seek auditing assistance of the Bureau of Internal Revenue.

Subdivision (f) of section 1215 is entitled "Minimum amounts subject to renegotiation." It provides generally for renegotiation of aggregate amounts received of not more than $250,000 when the fiscal year has ended before June 30, 1953, or $500,000 when the fiscal year ended on or after June 30, 1953, or $1,000,000 for a fiscal year ending after June 30, 1956. When the aggregate amounts received or accrued in any of the fiscal periods mentioned exceeds the respective sums stated, "no determination of excessive profits to be eliminated for such year with respect to such contracts and subcontracts shall be in an amount greater than the amount by which the aggregate exceeds $250,000, in the case of a fiscal year ended before June 30, 1953, or $500,000, in the case of a fiscal year ended on or after June 30, 1953, or $1,000,000, in case of a fiscal year ending after June 30, 1956."

Section 1216 grants exceptions of certain contracts from the Renegotiation Act—for example, any contract with a state or a foreign government, also for certain commodities, or a contract which in the judgment of the Board "does not have a direct and immediate connection with the national defense." A mandatory exemption is directed in subdivision 7 of section 1216 for "any contract, awarded as a result of competitive bidding, for the construction of any building, structure, improvement, or facility, other than a contract for the construction of housing financed with a mortgage or mortgages insured under the provisions of title VIII of the National Housing Act, as now or hereafter amended (sections 1748-1748h of Title 12)."

A Renegotiation Board is created in the executive branch of the government by section 1217, consisting of five members to be appointed by the President, by and with the advice and consent of the Senate. The members of the Board may delegate "in whole or in part any function, power, or duty (other than its power to promulgate regulations and rules and other than its power to grant permissive exemptions under section 106 (d) – section 1216 (d) of this Appendix) to any agency of the government. . . ."

The Renegotiation Act in section 1191 of Title 50, Appendix is to remain in full force and effect, except as amended and modified by sections 1211-1233.

Appendix A

Summary of state requirements

This summary of state requirements for prequalification, postqualification, licenses, and fees is reproduced with the permission of the Bureau of Contract Information, Inc., Washington, D.C.

For a discussion of its use please see Section 49 of this book.

ALABAMA

PREQUALIFICATION *required* in order to bid on highway work.

CONTRACTOR'S LICENSE, *having the effect of Prequalification, required* before undertaking any projects, public or private, costing $20,000 or more.

PREQUALIFICATION

Prospective bidders on highway work must prequalify with State Highway Dept., Montgomery, Alabama. They must furnish sworn statements, on forms prescribed, giving detailed information with regard to financial resources, equipment, past record, personnel, experience, and all other pertinent data. Financial statements must be prepared either by CPA or qualified licensed public accountant approved by State Society of CPA's. Foreign corporations must present certification of Secretary of State showing authorization to do business in State.

Within 30 days after submission, if applicant is found to possess required qualifications, certificate of qualification will be issued, valid for not more than one year. Certificate will set forth amount of work applicant may be allowed to have under contract at any one time, and may specify types of work upon which applicant will be permitted to bid.

All bids must be accompanied by certified check or bid bond equal to not less than 5% of bid price or estimated cost.

Once prequalification has been accomplished, the Highway Department can accept bids and make awards on work up to $20,000 without a license being issued.

LICENSING

Contractors on both public and private work must obtain licenses before undertaking projects costing $20,000 or more.

Licensing supervised by:
State Licensing Board for General Contractors,
Room 604,
State Administration Building,
Montgomery, Alabama

Work performed for Federal government on government-owned land exempt. Also, exempt, construction of one-family residence or one-family private dwelling.

Application for license, on form prescribed by the Board, must be filed not less than 30 days prior to any regular or special meeting of Board and must be accompanied by a certified check for $50. Renewal fee is $35 per year, after qualification.

If application is satisfactory to the Board, applicant may be required to take examination to determine his qualifications. Examination may

be oral or written or both and may cover, in addition to financial responsibility and past record, the qualifications of the applicant in reading plans and specifications, estimating costs, construction, ethics and other matters.

License, when and if issued, will stipulate type or types of work applicant is found qualified to perform, also the maximum amount which may be under contract at any one time.

STATE AND COUNTY PRIVILEGE LICENSE: In addition to the above license requirements, a State and County license must be obtained from the Probate Judge of the county in which the contractor maintains his principal office, or in which the contract is to be performed. Payment in one county of the State shall be sufficient. Cost, based on gross amount of contracts, ranges from $15.50 to $375.50.

TAXES AND FEES

ADMISSION OF FOREIGN CORPORATIONS: Filing Fee of $10. (Secretary of State). Admission tax based on capital to be employed in the State. (Department of Revenue).

ANNUAL CORPORATION FRANCHISE TAX: $2.50 per $1,000 on capital employed in State. If corporations qualify after July 1st of any year, a franchise tax for one-half year is required, or at rate of $1.25 per $1,000 on capital employed in State. (Department of Revenue).

ANNUAL CORPORATION PERMIT FEE: Foreign corporation permit fee based on capital employed, on graduated scale. Schedule of payments on bottom of permit application form. (Department of Revenue).

INCOME TAX: 3% on taxable net income from business within Alabama. Annual return required of every corporation qualified and doing business in Alabama. Non-resident individuals, graduated scale, 1½%—5%, on net income in excess of allowable personal exemption. (Department of Revenue).

SALES TAX: 3% on retail sales of tangible personal property. (Department of Revenue).

USE TAX: 3% on purchase price of tangible personal property bought outside State and brought into State for use, consumption or storage, to be collected by non-resident vendors, or, if not collected, to be paid direct by user to Department of Revenue. No exemption for contractors on jobs for governmental agencies. However, credit may be allowed, in the Alabama sales and use tax levy, on "used" property brought into State where, in State of original purchase, sales tax has been paid at time of purchase. Property must have been "used," however, before being brought into Alabama, to benefit by this provision.

Tax rate on automotive vehicles, truck trailers and semi-trailers, 1½% instead of general rate of 3%. (Department of Revenue).

FUEL TAX: Gasoline, 7c per gallon, regardless of how used. Diesel fuel, 7c per gallon. Tax applies only when used to operate motor vehicles over public highways. (Department of Revenue).

ALASKA

PREQUALIFICATION *required* in order to bid on all public works projects except small jobs which may be declared exempt.

CONTRACTOR'S LICENSE *not* required.

PREQUALIFICATION

Effective March 9, 1960, prospective bidders on public works projects, except small jobs which may be declared exempt, must prequalify with the Department of Public Works, Box 1361, Juneau, Alaska. Submissions must include, on forms to be provided, Experience Questionnaire and Financial Statement, latter to be sworn to, dated as of close of applicant's business year and prepared by and certified to by a certified or licensed public accountant. New statements required annually within 90 days of close of applicant's business year. Commissioner will rate approved applicants as to type of work they qualify to perform and maximum dollar value upon which they may bid.

LICENSING

Although contractors, as such, need not be licensed, all corporations, partnerships, associations and individuals, before transacting business in the State, must obtain a Business License, which also involves payment of a tax on gross receipts. (See below).

TAXES AND FEES

ADMISSION OF FOREIGN CORPORATIONS: Foreign corporations must obtain Certificate of Authority. Application must show total capitalization and proportion allocable to Alaska, estimated gross income and proportion allocable to Alaska, and other pertinent data. Fee schedule, on capital up to and including $1,000,000, $25, plus ten cents for each $1,000 over $100,000 or fraction thereof; over $1,000,000, an additional $10 for each $1,000,000, or fraction thereof. (Commissioner of Revenue).

ANNUAL CORPORATION FRANCHISE TAX: Tax $15, payable on or before January 1st. Annual report filing fee, foreign corporations, $5, domestic, $2.50. (Commissioner of Revenue).

BUSINESS LICENSE AND GROSS RECEIPTS TAX: Initial tax, $25. Report of gross receipts due Feb. 28th following, and annually thereafter. Tax based on volume of receipts. Less than $20,000 gross receipts, $25 (which is inclusive of initial tax). Gross receipts $20,000 to $100,000, ½ of 1% of excess over $20,000. Over $100,000, $400, plus ¼ of 1% of excess over $100,000. (Commissioner of Revenue).

NET INCOME TAX: Corporations, 18% of the amount of the federal tax due from income derived within State; Individuals, 14%, on same base. (Commissioner of Revenue).

ANNUAL TAX ASSURANCE STATEMENT AND BOND: All non-resident corporations and individuals must file sworn statements, on or before June 1st of each year, setting forth, among other things, their estimated gross receipts for current tax year, estimated payroll for same period, estimated total amount of taxes and fees due State, description and estimated market value, if any, of all real property owned in Alaska, upon which unpaid taxes and fees could become first lien and description and estimated fair market value of any resources severed or taken from Alaska during period. Where property subject to tax lien is less than twice estimated amount of taxes and fees to be due, bond must be furnished in amount equal to twice taxes and such fees, minimum bond $1,000. Non-residents must also notarize Commissioner of Administration as authorized to accept service on behalf of taxpayer in suits for collection of taxes and fees. (Commissioner of Revenue).

FUEL TAX: 5c per gallon on gasoline and diesel fuel, with 3c refundable for non-highway use. (Commissioner of Revenue).

ARIZONA

PREQUALIFICATION *required* in order to bid on highway work.

LICENSE *required* in order to bid on any work.

PREQUALIFICATION

Under regulations of the State Highway Commission, contractors desiring to bid on State highway work must prequalify.

Applicants must submit a statement of experience and financial condition on forms supplied by the Arizona Highway Commission. The forms must be fully completed, and the Financial Statement made by a certified or licensed public accountant, in sufficient time for investigation prior to opening bids.

This information is good for fifteen months after its date, but right is reserved by the State to request additional information on current assets and liabilities. The contractor also may be required to show the amount contracted subsequent to the date of his Financial Statement, and to furnish a list of available equipment.

Prequalification is complete when this information has been approved by the Commission and the contractor holds a valid license.

Communications regarding prequalification should be addressed to:
State Highway Engineer
Attention: Division of Contracts and Specifications

Arizona Highway Department
1739 W. Jackson St.
Phoenix, Arizona

LICENSING

Title 32, Chapter 10, Arizona Revised Statutes, requires all contractors to be licensed before doing work or submitting bids on work in the State. Fees range from $50 to $100.

To obtain a license under this Act the applicant must remit the required fee and submit, upon such forms as the Registrar of Contractors shall prescribe, a duly verified application including (1) a complete statement of the general nature of his contracting business, giving the names of principals or officers of the organization (2) if a foreign corporation, a statement showing that the corporation is qualified to do business in the state and in each county in which the contract is to be performed (3) the certification of two reputable citizens of the county in which the applicant resides, that he is of good reputation, recommending that license be granted, and containing the statement that the applicant desires the issuance of a license under the terms of the Act and (4) a contractor's license bond.

An applicant must be of good reputation and must show, by written examination, experience in the kind of work he proposes to contract and a general knowledge of the building, safety, health and lien laws of the State and of the rudimentary administrative principles of the contracting business and of the rules and regulations promulgated by the Registrar of Contractors pursuant to this Act.

A foreign corporation, before applying for a license, must have maintained both an office and a residence for one or more of its principals in the State for at least 90 days prior to date of application, except that office and residency requirements may both be waived to correspond with any such requirements, imposed by applicant's home State upon Arizona residents seeking similiar licenses in such State.

The Examinee will be the individual applicant, member of a co-partnership, corporate officer or responsible managing employee of the applicant and must have had four years experience in the classification for which the applicant applies for license. This experience must have been as a journeyman, foreman, supervising employee or contractor and must have been within the ten years immediately preceding the date of filing the application.

The application must be complete in every detail before filing. When a satisfactorily completed application has been filed the applicant will be notified of the time and place of the written examination. Following a successful examination the Registrar will contact the references given by the examinee in accordance with the law. The license will either be issued or refused within 60 days from the date of filing application.

All contractors must have licenses issued in

names under which they do business. Licensees are required by law to notify the Registrar within 30 days of a change of address. Applicant must secure privilege sales tax license and the number of the license issued must appear on the application.

All licenses expire June 30th of each year but are renewable on or before July 30th of the same year. Unless renewed by that date licenses will be suspended. Reinstatement will involve payment of double renewal fee.

Application forms and detailed information available from:

Registrar of Contractors
1818 W. Adams St.
Phoenix, Arizona

TAXES AND FEES

ADMISSION OF FOREIGN CORPORATIONS: Total fees for filing (including one agent) $65. Non-profit corporations, fee $25. (State Corporation Commission).

ANNUAL CORPORATION REPORT FEE: Report fee $25. (State Corporation Commission).

INCOME TAX: Foreign and domestic corporations, graduated scale, from 1% on first $1,000 to 5% on income in excess of $6,000. Individuals, graduated scale, from 1% on first $1,000 to 4½% on income in excess of $7,000. (State Tax Commission).

GROSS RECEIPTS TAX (SALES TAX): General contractors pay 1½% on gross receipts, less amounts paid for labor and subcontractors. Subcontractors pay 1½% on gross receipts, less amount paid for labor. Labor costs not inclusive of fees for architects, engineers or other professional employees, or of administrative costs. (State Tax Commission).

FUEL TAX: 5c per gallon. Refund of tax is obtainable, upon proper application, for use other than in motor vehicles. (Superintendent, Division of Motor Vehicles).

ARKANSAS

PREQUALIFICATION *required* in order to bid on highway work.

LICENSE *required* in order to bid on all projects of $20,000 or more, public or private.

PREQUALIFICATION

State Highway Dept., Highway Building, Little Rock, Arkansas, prequalifies contractors for all highway work.

Each prospective bidder will be required to file a Prequalification Questionnaire consisting of a Financial Statement, Experience Record, and Equipment Schedule on a form furnished by the

State Highway Commission. A Questionnaire reflecting the status of the prospective bidder as of any December 31, or any subsequent date within the ensuing calendar year, will become the criterion for proposals issued and work awarded during that calendar year and during a grace period extending through March 31 of the succeeding year. Bidders may file the required questionnaire at any time during the annual period, but no proposal will be released to a bidder unless a Questionnaire has been filed with the Commission and a rating extended at least seven days before the date set for receipt of bids.

It will be required that but one Questionnaire be submitted per calendar year regardless of the number of proposals tendered. *Questionnaires must be transmitted through the U. S. Mail.*

Contractors will be rated in accordance with the maximum amount of work that it is deemed they can satisfactorily prosecute through any given period, generally, such amount not to exceed twenty times the net quick assets reflected in the current financial statement. The maximum amount of work considered will include the unfinished value of going contracts. In the event a contractor submits a low bid on more than one project, and the aggregate amount is greater than the amount the contractor may be allowed to undertake the Commission will exercise its discretion with regard to the particular project or projects to be awarded.

In addition to the annual questionnaire required, the contractor will file an affidavit at each period proposal forms are issued citing active contracts in force and the unfinished value of such work. Proposal forms will not be issued until such affidavit is on file and the amount of work the contractor may be allowed to undertake so determined.

Proposal forms will be issued only upon the written request of the contractor, to the contractor in person or to a member of the contractor's organization.

Proposals must be accompanied by either a certified or cashier's check drawn on a solvent bank or trust company or a bidder's bond executed by an approved surety company. The proposal guarantee shall be made payable to the Arkansas State Highway Commission and in the amount shown in the proposal form. A lesser amount will not be accepted. Individual checks or bidder's bond must be furnished with each proposal submitted.

Checks drawn on banks or trust companies not located in the State of Arkansas shall not exceed twenty-five (25%) per cent of the combined capital and surplus of the banking institution on which drawn, and a statement to this effect, signed by either the president or cashier of such banking institution, must accompany the check.

LICENSING

Contractors on both State and private work must obtain license *before bids are submitted,* if **individual projects amount to $20,000 or more.**

Licensing supervised by:
Contractors Licensing Board
P. O. Box 2421
Little Rock, Arkansas

Application for license, on forms prescribed by the Board, must be accompanied by a certified check for $50 and be submitted at least 30 days prior to any regular or special meeting of the Board. License expires on June 30th of each year. Renewal is by application, with renewal fee, maximum $50 to be determined by the Board.

In determination by the Board of the applicant's qualifications, the following are considered: "(a) experience, (b) ability, (c) character, (d) the manner of performance of previous contracts, (e) financial condition, (f) equipment, (g) any other fact tending to show ability and willingness to conserve the public health and safety, and (h) default in complying with the provisions of this Act, or any other law of the State."

Recipients of certificates of license must record same with the Secretary of State with the date of recording to be shown thereon. Until such recording holders of license certificates shall not exercise any of the rights or privileges conferred thereby. Failure to record within 60 days of date of issuance shall invalidate the certificate of license. Fee $1.00.

License NOT required where contractor bids on or performs contracts for Federal Government, on Government land. Sub-contractors, however, must be licensed.

TAXES AND FEES

ADMISSION OF FOREIGN CORPORATIONS: Filing required of copy of Articles of Incorporation and statement, under oath of president or secretary, showing (1) number and par value of authorized capital shares and (2) value of property owned and used in Arkansas, and everywhere and (3) proportion of issued and outstanding stock represented by business transacted in and out of Arkansas. Filing fee based on capital stock represented by property and business in the State: $10 for such stock up to $100,000; $1 additional for each $10,000 of such stock from $100,001 to $1,000,000; $1 additional for each $20,000 of such stock from $1,000,001 to $10,000,000 or over. Minimum, $12.50. Foreign corporations must maintain offices in State and appoint authorized agents upon whom process may be served. (Secretary of State).

ANNUAL CORPORATION FRANCHISE TAX: Out-of-state corporations must file Franchise Tax Report on or before April 1st of each year showing condition as of January 1st next preceding, or, if newly organized or qualified, as of date of organization or qualification. Tax rate 11/000 of 1% on subscribed capital stock represented by property used in State. Payable July 10-Aug. 10 of each year. (Commissioner of Revenues).

GROSS RECEIPTS TAX (SALES TAX): 3% of gross receipts from taxable sales. (Commissioner of Revenues).

USE TAX: Use tax of 3%. (Commissioner of Revenues).

INCOME TAX: Foreign corporations and non-resident individuals, graduated scale from 1% on first $3,000 to 5% on taxable net income over $25,000 earned within the State. (Commissioner of Revenues).

FUEL TAX: Gasoline, 6½c per gallon, with refund of 4½c per gallon, for agricultural purposes, to those who qualify and secure permit. Diesel fuel, tractor fuel, kerosene, butane, etc., are non-taxable products, with execption of the 3% sales tax, when used off the highways. These fuels become taxable at 6½c per gallon only when used to propel motor vehicles over the highways. (Department of Revenues, Motor Fuel Tax Division).

CALIFORNIA

PREQUALIFICATION *required* in order to bid on State work in excess of $15,000.

LICENSES *required* of general and sub-contractors bidding on projects of $100 or more, public or private, with exceptions noted below.

PREQUALIFICATION

All contractors bidding on State work costing over $15,000 must prequalify with the Department of Public Works, in accordance with provisions of Article 4 of Chapter 3, Part 5 of Division 3, Title 2 of the Government Code. Smaller jobs, usually placed through District Offices, exempt.

The standard form, "Contractor's Statement of Experience and Financial Condition," may be obtained from the State Department of Public Works, Sacramento 7, California.

LICENSING

Contractors on both State and private work must obtain license, with exceptions noted, *before* bids are submitted.

Chapter 9 of Division III of the Business and Professions Code requires all contractors, including both general and sub-contractors, to be licensed.

"The registrar, with the approval of the Board, may adopt rules and regulations necessary to effect the classification of contractors in a manner consistent with established usage and procedure as found in the construction business and may limit the scope of the operations of a licensed contractor to those in which he is classified and qualified to engage.

"Under rules and regulations adopted by the Board and approved by the director, the registrar may investigate, classify and qualify applicants for contractors' licenses by written examination."

Application for license is made on forms prescribed by the Board. Application fee for original license is $20. The renewal fee is from $5-$20, determined annually by the Board.

License is NOT required for any construction, alteration, improvement or repair carried on within the limits and boundaries of any site or reservation, the title of which rests in the Federal Government, nor to bid on or receive contract award on any job involving Federal funds. However, before initial payment for work done on such contracts can be obtained there must be certification by State License Board that license has been issued.

Licensing supervised by:

Registrar of Contractors (Appointed by Contractors State License Bd.)
Sacramento, California.

TAXES AND FEES

ADMISSION OF FOREIGN CORPORATIONS: Filing fee of $357, including incidental fees. Prepayment of first annual minimum franchise tax of $100 required before qualification papers may be filed. (Secretary of State).

ANNUAL CORPORATION FRANCHISE TAX (BANK AND CORPORATION TAX LAW): 5½% on "measure of tax." "Measure" defined as income considered derived from California sources. (Franchise Tax Board).

SALES AND USE TAX: 3% on selling price of tangible personal property bought within or brought into the State for use or consumption. Board also acts as collection agency for county sales tax of 1%, imposed by all but 4 of State's 58 counties, thus making effective sales tax rate of 4%. (State Board of Equalization).

INCOME TAX: Corporations, 5½% on income considered derived from California sources. Non-resident individuals, graduated scale, 1% to 7%, on income earned in State. (Franchise Tax Board).

FUEL TAX (Gasoline and Use Fuel Tax): Gasoline, 6c per gallon. Tax collected by State from oil companies, which add to price of fuel. Tax, less 4% for sales tax on cost, refundable on non-highway use upon application, within thirteen months, to State Controller. Use fuel (diesel) tax, 7c per gallon, on highway use only, payable to State directly by user who must obtain permit from State Board of Equalization.

Liquefied petroleum gas, 6c per gallon. Permit required for highway use. For non-highway use tax refundable. When fuel is delivered into fuel tank of user's vehicle tax is collected by vendor who gives user receipt and remits to State. Vendor must also obtain permit from State. (State Controller and State Board of Equalization).

COLORADO

PREQUALIFICATION *required* in order to bid on highway work.

CONTRACTOR'S LICENSE *not* required.

PREQUALIFICATION

Before submitting bids on highway work contractors must prequalify with Department of Highways.

Application for prequalification is made by filing the standard form, "Experience, Equipment and Financial Statement," with the Chief Engineer, Highway Department Building, 4201 E. Arkansas Ave., Denver 22, Colo.

The certificate of a certified public accountant will be required in all cases where the contractor desires to qualify for work in excess of $50,000. Bonds required on all public works contracts over $1,000. See CH. 86-7-4 through 86-7-7, CRS 1953. Also see STANDARD SPECIFICATIONS, Dept. of Highways, $5.00 per copy.

LICENSING

There is no State law requiring contractors on either State or private work to be licensed.

TAXES AND FEES

ADMISSION OF FOREIGN CORPORATIONS: Application for Certificate of Authority, made in duplicate, must show Articles of Incorporation, home address of corporation, name and address of registered agent in Colorado, names and addresses of officers and directors, full data on capital structure including "stated capital" expressed in dollars, and other pertinent data. Filing fee, $60. (Secretary of State).

ANNUAL CORPORATION FRANCHISE TAX (CORPORATION LICENSE TAX): Sliding scale, ranging from minimum of $10 for less than $50,000 authorized capital stock allocated to Colorado to $250 for $1,000,000 or more of such stock. Annual report required by May 1st. Filing fee $5.00. (Secretary of State).

SALES AND USE TAX: 2% on all sales to "users-consumers" of tangible personal property and on such property imported for use within State. General and sub-contractors deemed "users-consumers" and subject to either sales or use levy. (Department of Revenue).

INCOME TAX: Corporations, 5% on net income earned in State. Non-resident individuals, graduated scale on income earned in State. Allocation may be determined by formula adopted by 1958 Legislature. (Department of Revenue).

FUEL TAX: 6c per gallon. No refund for non-highway use in construction equipment other than stationary engines and motor vehicles not licensed to operate on streets or highways. (Department of Revenue).

CONNECTICUT

PREQUALIFICATION *required* in order to bid on highway work.

CONTRACTOR'S LICENSE *not* required.

PREQUALIFICATION

By ruling of the Highway Department, contractors desiring to bid on State Highway work must prequalify. Application for prequalification is by standard form No. CON. 16. Information required includes statement of type of organization and officers or principals, financial statement, performance record of the organization and employment record of the principal individuals in the concern.

Forms are obtained from State Highway Department, Wethersfield, Connecticut.

LICENSING

There is no State law requiring contractors on either public or private work to be licensed.

TAXES AND FEES

ADMISSION OF FOREIGN CORPORATIONS: Entrance fee of $100. (Secretary of State).

ANNUAL CORPORATION FRANCHISE TAX: (Permit To Foreign Corporations to do Business in Connecticut): $100 per year. (Secretary of State).

INCOME TAX: Corporations, 3¾% on net income earned in State; or 1.9 mills per dollar on the amount of capital invested in Connecticut business; or $20, whichever is greater. Unincorporated businesses must pay tax based on proportion of gross receipts allocable to Connecticut—$1 per $1,000 up to and including the first $60,000, $2 per $1,000 above $60,000, retail; 25c per $1,000 up to and including the first $60,000, 50c per $1,000 above $60,000, wholesale. (Tax Commissioner).

SALES AND USE TAX: 3% on sales of materials, supplies and motor vehicles, whether used in the performance of a contract (Use Tax) or sold at retail (Sales Tax).

Use Tax also applies to materials, supplies and motor vehicles or other equipment bought outside Connecticut for use within the State. Materials and supplies are exempt only if they are physically incorporated in and become a permanent part of the projects being performed under a contract with an exempt agency, such as Federal, State, County and Municipal governments or charitable and religious organizations. (Tax Commissioner).

FUEL TAX: 6c per gallon, with refund for non-highway use, provided proper application is made within six months of date of purchase. (Tax Commissioner).

DELAWARE

PREQUALIFICATION *required* on highway work.

LICENSE *required* of all contractors doing business in the State.

PREQUALIFICATION

Prospective bidders on highway work must prequalify annually, in January, or at least prior to being awarded any contract, by submitting sworn statements, on forms to be provided, relating to their experience, equipment and finances.

LICENSING

Licensing of contractors is for revenue only. No examination for license is required.

Any individual, co-partnership, firm or corporation, or other association of persons resident of the State of Delaware, acting as a unit desiring to engage in, prosecute, follow or carry on the business of Contracting as herein defined shall obtain a license from the State Tax Department and pay a license fee.

Any individual, co-partnership, firm or corporation or other association of persons not residents of the State of Delaware, acting as a unit desiring to engage in, prosecute, follow or carry on the business of contracting as herein defined shall obtain an original license from the State Tax Department for each contract or contracts on hand before engaging in business and shall pay a license fee for the execution of such contracts.

Every architect and/or mechanical engineer and/or general contractor engaging in the practice of such profession shall furnish within ten (10) days after any contract or contracts in the preparations or plans for which they were engaged are entered into with a contractor or sub-contractor not a resident of this State, a statement of the total value of such contract or contracts together with the names and addresses of the contracting parties. Failure to furnish each such statement shall subject each architect and/or mechanical engineer and/or general contractor to a penalty of Twenty-Five Dollars ($25) which shall be collected and paid in the same manner as provided for the collection of delinquent licenses as provided in this Article.

Every architect and/or mechanical engineer

and/or general contractor, engaging in the practice of such profession, before the payment of any award or amount payable to any contractor or sub-contractor not a resident of this State, shall ascertain from said non-resident contractor or sub-contractor and/or the State Tax Department, whether he has obtained a license and satisfied his liability to the State under this Section, and if said license has not been obtained and the license liability paid by said non-resident contractor or sub-contractor, the architect and/or mechanical engineer and/or general contractor shall deduct from the award or amount payable to said non-resident contractor or sub-contractor, the amount of said license liability and shall pay same to the State Tax Department within ten days after final payment and settlement with the non-resident contractor or sub-contractor. Failure to ascertain the payment of license liability of any contractor or sub-contractor not a resident of this State, by any architect and/or mechanical engineer and/or general contractor, in accordance with this section, shall render the architect and/or mechanical engineer and/or general contractor personally liable for the license liability of the non-resident contractor or sub-contractor.

Licensing is under the supervision of:

State Tax Department
Wilmington, Delaware

License is NOT required where contractor performs contracts for Federal Government, on Government land.

TAXES AND FEES

ADMISSION OF FOREIGN CORPORATIONS: Entrance fee of $25, plus incidental fees of approximately $13. (Secretary of State).

FUEL TAX: 5c per gallon. Refund of tax on fuels used off the highway is obtainable upon proper application and proof of such use. (State Highway Department).

DISTRICT OF COLUMBIA

PREQUALIFICATION *not* required in order to bid.

CONTRACTOR'S LICENSE *not* required.

PREQUALIFICATION

It is not necessary to prequalify in order to bid. Low bidders on public work may be required to demonstrate capacity to handle work before contracts are awarded.

LICENSING

Contractors, as such, need not be licensed. But any corporation or unincorporated business must obtain an annual license before doing business in the District; fee, $10.

TAXES AND FEES

ADMISSION OF FOREIGN CORPORATIONS: Foreign corporations must obtain certificate of authority. Application must be on form provided for purpose and accompanied by recent certified copy of charter or Articles of Incorporation, with amendments duly certified by proper authority in home state, and designation of local resident, individual or corporation, who can be sued. Registered office of corporation in D.C. must be business address of designated agent. Filing fee, $22. (Superintendent of Corporations).

FRANCHISE TAX: Corporations, 5% of net income, no exemptions. Unincorporated businesses, 5% of net income above $5,000. (Finance Office, Revenue Division. Payments to D. C. Treasurer).

INCOME TAX: Individuals, 2½%—5%, after personal exemptions. Withholding by employers required.

(Finance Office, Revenue Division.
Payments to D. C. Treasurer).

SALES AND USE TAX: 2% on the sales price or rental value of tangible personal property. The contractor may be deemed to be either a vendor or purchaser (consumer), depending on the type of contract and the use of the material.

On contracts with any semi-public institution holding a valid exemption certificate, or with the U. S. Government or District Government, or instrumentalities thereof, the contractor may purchase such materials and supplies as are to be physically incorporated in and become real property without payment of the tax, and shall not charge any such institution, government or instrumentality of such government, reimbursement for any sales or use taxes thereon. Any other type of materials taxable. Contractors, on exempt purchases, must furnish their suppliers with a contractor's Exempt Purchases Certificate. Products such as lumber for forms and materials used in performance of repair or service contracts not exempt. (Sales, Use and Excise Tax Section. Finance Office. Payments to D. C. Treasurer).

FUEL TAX: 6c per gallon, with refund, upon proper application, for non-highway use. (Sales, Use and Excise Tax Section, Finance Office. Payments to D. C. Treasurer).

FLORIDA

PREQUALIFICATION *required* in order to bid on State highway work.

CONTRACTOR'S OCCUPATIONAL LICENSE *required.*

PREQUALIFICATION

Prequalification for bidding on highway work is by application to State Road Department of Florida, Tallahassee, Florida, on a standard form, "Confidential Financial Statement, and Experience Questionnaire."

Each contractor is given a specified classification and a maximum rating which is the total amount of uncompleted work he will be permitted to have under contract at any one time.

LICENSING

State Occupational License Tax of $3, plus $1 for each person employed. Maximum, $250. Also, license must be obtained in each county in which the contractor establishes headquarters or a separate place of business. County tax is 50% of the State tax. Both licenses obtainable from Tax Collector of any county. Certain municipalities also are permitted to levy a tax equal to 50% of the State tax. (State Comptroller).

License IS required where contractor performs contracts for Federal Government, on Government land.

TAXES AND FEES

ADMISSION OF FOREIGN CORPORATIONS: Entrance fee based on amount of authorized capital stock represented by capital to be employed in the State. Ranges from $10 minimum to $937.50 plus 10c per $1,000 for authorized capital stock in excess of $2,000,000. No par value stock, graduated scale, from 20c per share, on 1 to 1250 shares, to 1/10 of 1c per share above 20,000 shares.

In addition there are filing fees ranging from $1 to $10 for obtaining such documents as certificate of incorporation, of amendment, of increase in capital stock, of dissolution and other revisions or changes in corporate status. (Secre- of State).

ANNUAL CORPORATION FRANCHISE TAX (ANNUAL FEE ON CORPORATIONS): Based on capital employed in State, ranges from $10 on capital of $10,000, or less, to $1,000 on capital of over $2,000,000. (Secretary of State).

SALES TAX: 3% on retail sales of tangible personal property. NO exemptions for contractors irrespective of for whom work is being performed. (State Comptroller).

USE TAX: 3% on tangible personal property, including contractors' materials and equipment, purchased and used outside of Florida and brought into Florida for use, consumption or storage. Motor vehicles 1%. However, where such property has been subjected to sales tax in state of purchase, at a rate equal to or greater than the rate effective in Florida, no use tax is imposed *provided* such state extends similar consideration to Florida contractors operating in that state. Currently only Georgia is in this position. (State Comptroller).

FUEL TAX: 7c per gallon on gasoline, whether

for highway or non-highway use. Other fuels taxed at 7c only when used on highways. (State Comptroller).

GEORGIA

PREQUALIFICATION *not* required.

CONTRACTOR'S LICENSE *not* required.

PREQUALIFICATION

It is not necessary to prequalify in order to bid on any State work, but State Highway Department, 2 Capitol Square, Atlanta, Georgia, requires contractors bidding on highway work to submit confidential financial statement.

LICENSING

There is no State law requiring contractors on either public or private work to be licensed.

TAXES AND FEES

ADMISSION OF FOREIGN CORPORATIONS: $1 registration fee and $10 filing fee. Copy of charter and all amendments thereto required. Both fees payable in advance. Filing fee for subsequent amendments, $10. Annual recording fee, $1. (Secretary of State).

ANNUAL FOREIGN CORPORATION FRANCHISE TAX (LICENSE TAX): Based on property and business allocable to State. Graduated scale ranging from $10 on capital and surplus totalling $10,000 or less to $5,000 on $22,000,000 or more of taxable capital and surplus. (State Revenue Commissioner).

SALES TAX: 3% of retail sales, including materials bought in State and those brought into State on which tax equal to or greater than Georgia tax has not been paid. However, no credit will be granted for taxes paid in another state if that state does not grant like credit for taxes paid in Georgia. Prime contractors must withhold 2% from sums paid to subcontractors until subcontractors have filed reports showing sales taxes due State as paid, unless bonds have been posted to assure payment. Prime contractors must notify Dept. of Revenue, on form provided for purpose, within 10 days of .executing any contract with subcontractor. Contractor liable for 3% sales tax on tangible personal property furnished him by one for whom contract is being performed, if sales tax has not been paid on such property by that person. (Department of Revenue, Sales and Use Tax Division).

USE TAX: 3% of the cost price or rental value of tangible personal property used or consumed in the State, including contractors machinery and equipment. (Department of Revenue, Sales and Use Tax Division).

INCOME TAX: Corporations—4% on net income from property owned or business done in State; see Section 92-3102. State and federal income taxes paid not deductible from gross income. (Income Tax Unit, Department of Revenue).

Individuals—on a sliding scale, 1%—6%, applied to taxable net income from business or employment within the State. (Income Tax Unit, Department of Revenue).

FUEL TAX: Gasoline, 6½c per gallon, kerosene, 1c per gallon, for all purposes except farm use; diesel fuel, 6½c per gallon only when used to operate internal combustion motors upon the highways. The 3% sales tax also applies to gasoline and diesel fuel. (State Revenue Commissioner).

HAWAII

PREQUALIFICATION *may be required* to bid on all public work.

LICENSE *required* to bid on all work except Federal-aid highways.

PREQUALIFICATION

State Highway Dept. regulations require that prospective bidders file written notice of intention to bid at least six days before date of opening. Engineer in charge may then require submission of Standard Questionnaire and Financial Statement on forms to be provided. Complete and notarized answers must be made at least 48 hours prior to date set for opening of bids. If deemed satisfactory bids will be received.

LICENSING

All prospective bidders on public or private work, with the exception of Federal-aid highways where federal regulations applicable, must be licensed. Application forms obtainable from Contractors' License Board.

Licenses issuable in three classes—(1) General Engineering (2) General Building (3) Specialty.

License fees—Classes (1) and (2) $100, Class (3) $25.

Licensing year ends June 30th, prior to which date renewal required. Renewal fees same as original license fees.

Licensing supervised by
Contractors' License Board
Room 202, Honolulu Armory
Honolulu 13, Hawaii

TAXES AND FEES

ADMISSION OF FOREIGN CORPORATIONS: All foreign corporations must be domiciled. For requirements contact Treasury Department, Box 40, Honolulu, Hawaii.

GENERAL EXCISE (GROSS INCOME) TAX: Rate, 3½% on gross income of contractors and most other service and manufacturing industries. Licenses must be secured, annual fee $2.50. Monthly returns due, and annual summary within 80 days of close of taxpayers' fiscal year. (Tax Commissioner).

CONSUMERS' TAX: Rate, 3½% on fair value of tangible personal property imported for use, consumption or storage, if purchased from source not subject to General Excise Tax law. (Tax Commissioner).

NET INCOME TAX: Corporations, 5% on net income up to $25,000, 5½% above. Individuals, graduated scale from 3% on first $500 of net income, after personal exemptions, up to 9% on net income over $30,000. (Tax Commissioner).

FUEL TAX: 8½ cts. to 11 cts. per gallon on gasoline, except aviation gasoline, and on diesel oil for highway use. 1 ct. per gallon on diesel oil for non-highway use. (Tax Commissioner).

IDAHO

PREQUALIFICATION is effected through operation of Public Works Contractors License Act.

LICENSE *required* of contractors on public works.

PREQUALIFICATION

Contractors desiring to bid on public work must prequalify by obtaining a license.

LICENSING

The Act provides that it shall be unlawful for any person to engage in the business or act in the capacity of a public works contractor within this State without first obtaining and having a license therefor . . . unless such person is particularly exempted as provided in this Act, or for any public works contractor to subcontract in excess of eighty per cent of the work under any contract to be performed by him as such public works contractor according to the contract prices therein set forth, unless otherwise provided in the specifications of such contracts or to sublet any part of any contract for specialty construction, as defined in the Act, to a specialty contractor who, at or before the time of the original bid opening, was not licensed in accordance with this Act; provided, however, that no contractor shall be required to have a license under this Act in order to submit a bid or proposal for contracts for public works financed in whole or in part by federal aid funds, but at or prior to the award and execution of any such contract by the State of Idaho, or any other contracting authority mentioned in the Act, the successful bidder shall secure a license as provided in this Act.

In the above "person" means any individual, firm, co-partnership, corporation, association or other organization or combination thereof acting as a unit.

Application for license shall be on form prescribed by the Public Works Contractors State License Board, shall include complete statement with regard to applicant's contracting business experience, qualifications, description of work completed during preceding 3-year period, description of machinery and equipment owned, statement of financial condition, certified to by a Certified Public Accountant, and other pertinent data including evidence of good character and reputation.

Applications shall be sworn to and filed at least 30 days in advance of consideration by Board, whose meetings shall be on second Monday of January, April, July and October and at such other times as Board may designate. Applications shall specify class of license desired, A, B, or C, and be accompanied by appropriate fee.

CLASS A—covering contracts for public works with estimated cost of over $50,000. Fee, $100; Renewal, $50.

CLASS B—covering contracts in excess of $25,000 but not more than $50,000. Fee, $50; Renewal, $25.

CLASS C—covering contracts of $25,000 or less. Fee, $10; Renewal, $5.

Licensing year ends June 30th.

Qualifications to be demonstrated by applicant for a license are as follows:

"Such degree of experience, and such general knowledge of the building, safety, health and lien laws of the State, and of the rudimentary administrative principles of the contracting business, as may be deemed necessary by the Board for the safety and protection of the public. The applicant, if an individual, may qualify as to the aforementioned experience and knowledge by personal appearance or by the appearance of his responsible managing employee, and if a co-partnership or corporation, and any other combination or organization, by the appearance of the responsible managing officer or member of the personnel of such applicant. If the person qualifying by examination as to experience and knowledge shall, for any reason whatsoever, cease to be connected with the licensee to whom the license is issued, such licensee shall so notify the Board in writing within ten (10) days from such cessation. If such notice is given, the license shall remain in force for a reasonable length of time, to be determined by rules of the Board, provided, however, that if such licensee fails so to notify the Board within said ten (10) day period, then at the end of such ten (10) day period, the license of such licensee shall be automatically suspended. A suspended license shall be reinstated upon the filing with the Board of an affidavit executed by the licensee or a member of the suspended firm, to the effect that the individual originally examined for the firm has been replaced by another individual who has been qualified by examination as herein provided, and who shall not have had a license suspended or revoked for reasons that should preclude him from personally qualifying as to good character as herein required of an applicant."

Licensing is supervised by:
Public Works Contractors State License Board
1210 Grove St.
Boise, Idaho

License is NOT required where contractor bids or performs contracts for Federal Government, on Government land.

TAXES AND FEES

ADMISSION OF FOREIGN CORPORATIONS: Entrance fee on graduated scale, from $20 on authorized capital stock of $25,000 or less to $200 on authorized capital stock over $1,000,000. (Secretary of State).

ANNUAL CORPORATION FRANCHISE TAX (CORPORATE LICENSE FEE): Varies from $20 to $300, according to amount of authorized capital stock. Fee payable in advance. (Secretary of State).

INCOME TAX: Corporations, 9½% on all net taxable income earned in state. Individuals, graduated scale, 3%-9½%, on net taxable income earned in state. Under opinion of Attorney General, March 17, 1950, it is declared a responsibility of contracting units, under Sections 63-1503 and 63-1504, Idaho Code, where public works are concerned, to require out-of-State contractors, unless tax bond has been posted, to provide evidence that the requirements of the Income Tax Law have been complied with before final settlement is approved. Withholding requirements also in effect. (Office of Tax Collector, Income Tax Division).

FUEL TAX: 6c per gallon on all motor fuels. Refund for non-highway use available, in contracts with private individuals, upon proper application within 270 days of use. No refunds on state highway contracts. On other state contracts terms of contract will determine whether or not refund available. (Motor Fuels Division).

ILLINOIS

PREQUALIFICATION *required* in order to bid on highway work.
CONTRACTOR'S LICENSE *not* required.

PREQUALIFICATION

State of Illinois, Department of Public Works and Buildings, Division of Highways, requires contractors desiring to bid on highway work to submit an "Experience Questionnaire" and "Contractor's Financial Statement," from which the contractor is rated and prequalified.

LICENSING

There is no state law requiring contractors to be licensed.

TAXES AND FEES

ADMISSION ON FOREIGN CORPORATIONS: Filing fee of $20, plus a license fee of 1/20 of 1% of the stated capital and paid-in surplus allocated to this State on the basis of business and property within the State, as compared to the business and property everywhere (minimum, $1,000 stated capital and paid-in surplus); and a franchise tax of 1/20 of 1% of the stated capital and paid-in surplus allocated to this State on the basis of business and property within the State, as compared to the business and property everywhere, with a minimum annual fee of $10, which varies monthly from 18/12 of the amount of tax or minimum in January to 7/12 thereof in December. (Secretary of State).

SALES TAX (RETAILERS' OCCUPATION TAX): Rate, until July 1, 1961, 3% on gross receipts from sales at retail of tangible personal property purchased in state, for use or consumption, and on fair depreciated value of same where purchased outside and imported into state. For contractors, applies on all equipment and materials *not* permanently installed or *not* made an integral part of a structure, as that term is defined in Rule 6, issued July, 1956, irrespective of whether or not sale or provision occurred in connection with carrying out terms of a construction contract. Examples of these items are stoves, refrigerators, furniture, curtains, drapes, trade fixtures and machinery. Exempt items would include screens, storm doors, weather stripping, venetian blinds and permanently installed plumbing and electrical fixtures such as pumps, water heaters, sinks and bathtubs, *provided* sale and installation are effected in connection with fulfilling requirements of a construction contract. Also exempt are sales to State, its political subdivisions and to organizations exclusively of a charitable, religious or educational nature. (Department of Revenue).

FUEL TAX: 5c per gallon, subject to refund, upon proper application, when fuel is consumed in non-highway use. (Department of Revenue).

INDIANA

PREQUALIFICATION *required* on all public work costing over $5,000.
CONTRACTOR'S LICENSE *not* required.

PREQUALIFICATION

Prospective bidders on highway work costing

over $5,000 must prequalify with State Highway Dept., State House Annex, Indianapolis, Indiana. Information required, on forms to be provided, includes detailed financial statement and outline of experience, available equipment and proposed plan for performing work. Financial statement must be prepared by CPA and all material submitted sworn to. Contractors will be rated according to types and amounts of work they are deemed qualified to perform.

On public work other than highways, costing over $5,000, same type of information must be presented to awarding authority, along with and as part of bid.

LICENSING

There is no State law requiring contractors on either public or private work to be licensed.

TAXES AND FEES

ADMISSION OF FOREIGN CORPORATIONS: Fee of 2c per share on capital stock allocated to business in the State. Minimum, $26. (Secretary of State).

GROSS INCOME TAX: 1½% on Gross Income earned in State. However, some contractors may be able to qualify as "retail merchants" and pay retail sales rate of 3/8 of 1%. Amended law states that "a contractor who, operating from a fixed and established place of business, contracts to provide material, labor and other elements of cost to perform a job for another contractor or final user" may so qualify. Both prime and subcontractors, in order to establish this eligibility must, in their accounting and reporting, segregate income claimed to have been received while functioning in capacity of "retail merchants" from other income. Failure to show such segregation will subject taxpayer to full 1½% rate on all gross income.

Law provides for withholding 1½% of all amounts over $1,000 per calendar year paid to non-resident contractors. The Gross Income Tax Division has by memorandum waived this requirement as respects non-resident corporate contractors who are duly licensed and qualified and engaged in business in Indiana. Money received for the performance of contracts with a State or the Federal Government or any department or subdivision of either is *not* exempt under the Gross Income Tax Act.

Employers of four or more non-residents subject to Indiana Gross Income Tax must withhold the 1½% tax on all payments made them in excess of $1,000, in any one year, for personal services performed. Returns of such withholdings due annually. Withholding tables available from Income Tax Division, Department of State Revenue.) Employers of Indiana residents must submit Information Returns showing payments made them in excess of $100 for personal services performed in state, unless tax has been withheld. (Gross Income Tax Division).

FUEL TAX: 6c per gallon on all fuels, with refund for non-highway use available upon proper application. (Department of State Revenue).

IOWA

PREQUALIFICATION *required* in order to bid on highway work.

CONTRACTOR'S LICENSE *not* required.

PREQUALIFICATION

By regulation of the State Highway Commission contractors desiring to bid on State highway work must prequalify. Application for prequalification must be submitted on a standard form entitled "Iowa State Highway Commission Contractor's Financial-Experience-Equipment Statement," available from Iowa State Highway Commission, Ames, Iowa.

LICENSING

There is no State law requiring contractors on either public or private work to be licensed.

TAXES AND FEES

ADMISSION OF FOREIGN CORPORATIONS: Certificate of Authority required to do business in state. Fee, $20. (Secretary of State).

ANNUAL FRANCHISE TAX: Annual reports must be filed on or before March 1st. Fee based upon money and property in use in Iowa, *or* on amount of stated capital in home state, as preferred. Minimum, $5. (Secretary of State).

SALES TAX: 2% on retail sales. (State Tax Commission).

USE TAX: 2% on purchases brought into State. (State Tax Commission).

INCOME TAX: Corporations, 3% of taxable net income earned in State; non-resident individuals, graduated scale on taxable income derived within State. (State Tax Commission, Income Tax Division).

FUEL TAX: 6c per gallon on all motor fuel, with refund available, upon proper application, for non-highway use. 7c per gallon on fuel oil used or sold for purpose of propelling motor vehicles on highways of State. Purchases by federal government and State of Iowa exempt. (State Treasurer).

KANSAS

PREQUALIFICATION *required* in order to bid on highway work.

CONTRACTOR'S LICENSE *not* required, but certain contractors may be required to register and furnish bond to assure payment of taxes.

PREQUALIFICATION

Prequalification on highway work is by the State Highway Commission, Topeka, Kansas.

Prospective bidder must submit to the Commission, once a year, and at such other times as Commission may designate, but in any event at least seven days prior to date set for opening of bids, "a complete statement of his financial condition, equipment, experience and organization, on forms provided for that purpose." The contractor's financial statement must show his net worth and must be certified to by a certified public accountant holding an unrevoked certified public accountant's certificate in Kansas or in any state which has a reciprocity agreement with the State of Kansas. The certification by a certified public accountant may be waived where the contractor's net worth is less than $5,000 or where the contractor does not desire a qualification of over $15,000.

Contractors are classified according to the type of work they are qualified to undertake and rated on the amount of work in dollars allowable in any one or more classifications of work which the bidder may have under contract at any one time.

LICENSING

There is no law requiring contractors bidding on or performing work in the State to be licensed.

However, Chapter 515, Kansas Session Laws, 1957, requires that non-resident contractors on jobs of more than $3,000, *except foreign corporations authorized to do business in the state,* must register, with respect to each such contract, with the Director of Revenue. Registration fee $10. Such non-resident contractors must also designate Secretary of State as their resident agent, who can can be sued, and must be bonded to assure payment of taxes, including employment security taxes, with respect to each contract in excess of $10,000. Director of Revenue *may* require that bond be increased, if deemed necessary to assure payment of taxes. Minimum bond, greater of 10% of amount of contract, or $5,000. Release of bond contingent upon performance of contract and upon statement by Director of Employment Security that all payments due under that law have been made. (Director of Revenue).

TAXES AND FEES

ADMISSION OF FOREIGN CORPORATIONS: Application fee of $25 and filing fee of $2.50. Also fee of 1/10 of 1% on authorized capital stock of less than $100,000; $100 plus 1/20 of 1% on authorized capital stock of $100,000 and over. (Secretary of State).

ANNUAL CORPORATION FRANCHISE TAX (Annual License Fee): Fee, based on portion of issued capital stock allocated to State, ranges from $10 on first $10,000 to $2,500 on amounts in excess of $5,000,000. (Secretary of State).

SALES TAX: 2½% on retail sales. (Director of Revenue).

USE TAX: 2½% on purchases brought into State. (Director of Revenue).

INCOME TAX: Corporations, 3½% on net income from business in State; non-resident individuals, graduated scale on income derived from sources within State; resident individuals, graduated scale on all income. (Director of Revenue).

FUEL TAX: Gasoline, propane and butane, 5c per gallon when used on the highways. Refund permits allow non-highway users of gasoline, who purchase 40 gallons or more from licensed distributors, to obtain a refund of the tax, permits to be obtained from the county clerk of the county in which the applicant resides or maintains a place of business. Special fuels, including Diesel, taxed at 7c per gallon when used on the highways except upon delivery into fuel tanks, or bought for the used of vehicles especially licensed as farm trucks, truck tractors or urban transit buses, which pay 5c per gallon. Refund of Diesel tax allowed non-highway users of 50 gallons or more. (Director of Revenue).

KENTUCKY

PREQUALIFICATION *reqiured* in order to bid on highway work.

LICENSING, as such, *not required,* but effected through prequalification.

PREQUALIFICATION

Highway Department Order No. 432, dated Feb. 1, 1958, issued pursuant to KRS 176-140, sets up following procedures for prospective bidders on highway work. Contractors must apply for Certificate of Eligibility, on form to be provided. Application must be accompanied by (1) Financial statement, certified to by independent CPA, authorized to do business in Kentucky or in home state showing condition at end of fiscal year or other approved date (2) evidence of experience and competency of personnel and adequacy of equipment (3) certified copy of Articles of Incorporation or Partnership Agreement (4) certified or cashier's check or bank draft for $50 and (5) if foreign corporation, certificate from Kentucky Secretary of State that applicant is authorized to do business in state. Applicant must obtain copy of Department's Standard Specifications.

If applicant meets Department's standards Certificate will be issued within not less than 30 days from filing date, except that, if applicant states in writing that he intends to bid on U.S. Bureau of Public Roads projects, issuance may be made within 15 days.

Certificate may carry rating stating type or

types and amount or amounts of work upon which applicant is qualified to bid and have under contracts at any one time.

Certificates must be renewed annually by February 1st. Temporary Certificates will then be issued good for four months. At least 30 days prior to expiration of this period applicant must file new statement of Experience and Financial condition, which, if approved, will be followed by issuance of Certificate good through the following January. Renewal fee, $10. Applications to and fees paid to Department of Highways, Frankfort, Ky.

LICENSING

There is no State law requiring contractors on either public or private work to be licensed.

TAXES AND FEES

ADMISSION OF FOREIGN CORPORATIONS: Filing fee $25, recording fee of $10 and $5 fee for process agent statement. Annual statement of existence, $1. Annual Verification Report, $1. (Secretary of State).

ANNUAL CORPORATION FRANCHISE TAX (LICENSE TAX): 70c on each $1,000 of capital stock allocated to State. Minimum, $10. (Department of Revenue).

INCOME TAX: Corporate, 5% on first $25,000 net income, 7% on balance; Individual, 2% on first $3,000 net income graduated to 6% on excess over $8,000. Surtax imposed equivalent to 10% of first $25 of normal tax, 20% of next $75 of normal tax and 30% of all over $100 normal tax. $13 tax credit for each exemption. (Department of Revenue).

PROPERTY TAXES: State rate of 50c per $100 on tangible personal property, and 25c per $100 on intangible personal property located in Kentucky as of January 1 of each year.

Tangible personal property subject to local tax rate. Real estate subject to state rate of 5c per $100 and full local rate. State-wide basis of assessment approximately 33% of market value. (Department of Revenue).

FUEL TAX: 7c per gallon on all motor fuels with additional 2c where used in equipment having three or more axles. (Department of Revenue).

Kentucky Sales Tax approved by voters in 1959 but being challenged in courts, hence not yet effective.

LOUISIANA

PREQUALIFICATION *not* required in order to bid but must present statement with bids on highway work.

CONTRACTOR'S LICENSE required to bid on most public and private work in excess of $30,000.

PREQUALIFICATION

Contractors are not required to prequalify in order to submit bids to the Department of Highways, but must submit with their bids or prior to bidding a statement of experience and financial condition on a form provided by the Department. Statements to Dept. of Highways, P.O. Box 4245, Capitol Station, Baton Rouge, La.

LICENSING

Act 233, 1956, requires contractors on both public and private work in amounts exceeding $30,000, with certain exceptions, to be licensed. Exceptions include private residential work, federal aid projects and projects for public utilities subject to regulation by State Public Utilities Commission. However, successful bidders on last named shall, before start of work, secure license and pay fee. They must also comply with terms and provisions of Act and with Rules and Regulations of State Licensing Board created by Act.

Application for license shall be accompanied by statement outlining financial condition, experience and all pertinent facts bearing on applicant's responsibility. Board will rate successful applicants on basis of types of work upon which they may be permitted to bid.

License fee, $150. License year ends Dec. 31st. Renewal required on or before first Tuesday of following February.

Licensing supervised by:
State Licensing Board for Contractors
Box 4242, State Capitol Building
Baton Rouge, La.

TAXES AND FEES

ADMISSION OF FOREIGN CORPORATIONS: Certified copy of charter and any amendments thereto must be filed together with statement showing amount of capital employed in state. State resident must be designated as empowered to accept service. Filing charter fee, 25 cents per 100 words, filing power of attorney designating local resident, $3.50.

Fee of 1/20 of 1% on capital stock employed in Louisiana. Minimum $10; maximum $2,500. (Secretary of State).

ANNUAL CORPORATION FRANCHISE TAX: $1.50 per $1,000 of capital stock, surplus, undivided profits and borrowed capital allocable to State. Allocation formula based on arithmetical average of (1) ratio net sales attributable to Louisiana to total and (2) ratio value property and assets situated or used in Louisiana to value everywhere. Taxable base shall not be less than assessed value of real and personal property located in Louisiana. (See Sec.

606 of Title 47, Chap. 5, Revised Stats. of 1950, as amended). Minimum, $10. (Collector of Revenue).

SALES AND USE TAX: 2% on retail sales and on purchases brought into State. (Collector of Revenue).

INCOME TAX: Foreign corporations, 4% on net income derived within State after exemption; non-resident individuals, graduated scale on income from sources in State. (Collector of Revenue).

OCCUPATIONAL LICENSE TAX: Levied on all contractors doing business in Louisiana. Fee based on gross annual receipts in accordance with graduated rate schedules. Contractors on cost-plus basis are covered by Section 378, with rate schedule in Section 349, of Title 47 of the Louisiana Statutes of 1950, Revised; those on lump sum basis under the rate schedule of Section 377. Copy of the Act and application forms available from: Collector of Revenue, Occupational License Tax Division, Baton Rouge 1, Louisiana. (Collector of Revenue).

LOCAL TAXATION: Several Louisiana cities, including New Orleans and Baton Rouge, may levy sales or Occupational License taxes about which contractors should seek information locally.

FUEL TAX: 7c per gallon on all motor fuels. No refunds for non-highway use by contractors. (Collector of Revenue).

MAINE

PREQUALIFICATION *not* required in order to bid.

CONTRACTOR'S LICENSE *not* required.

PREQUALIFICATION

There is no law or regulation requiring prequalification for bidding on any work in the State. On State highway work, the State Highway Commission makes whatever investigation of contractors it deems necessary and may request statements of financial condition, equipment and experience before contracts are awarded.

LICENSING

There is no State law requiring contractors on either public or private work to be licensed.

TAXES AND FEES

ADMISSION OF FOREIGN CORPORATIONS: Entrance fee of $20. (Secretary of State).

ANNUAL CORPORATION LICENSE FEE: $10. (Secretary of State).

SALES AND USE TAX: 3% of sale or purchase price of tangible personal property sold in the State or purchased outside the State for use or consumption within the State. Materials and supplies are exempt when consumed in performance of a contract with an exempt agency, such as Federal, State, County and Municipal governments, hospitals, schools, or churches, only when such materials or supplies are to be physically incorporated in and become a permanent part of the real estate of the exempt agency. (State Tax Assessor).

FUEL TAX: 7c per gallon on gasoline, with refund of 6c per gallon available on such fuel consumed in off-the-highway use. For use of diesel fuel, a license is necessary and the 7c per gallon tax is applied only upon such fuel used on the highway. (State Tax Assessor).

MARYLAND

PREQUALIFICATION *required* in order to bid on highway work, by both general and subcontractors.

LICENSE *required* of all contractors obtaining work exceeding $5,000 per annum, public or private.

PREQUALIFICATION

Prospective bidders on highway work, both general and subcontractors, must prequalify with State Roads Comm., 300 W. Preston St., Baltimore 1, Md.

Statements must be on forms, to be supplied, submitted 15 days prior to any contract letting upon which it is desired to bid, or, in any event at least once during the period July 1-June 30. Information submitted must include sworn financial statement, outline of experience and equipment owned, and proposed plan for carrying on the work. Prime contractors must not sublet more than 40% of value of contract.

LICENSING

No license is required to bid, but under Chapter 704, Section 184, Acts of 1916, any person or corporation accepting orders or contracts for doing work in a gross yearly volume exceeding $5,000, must be licensed. This applies to all work, public, or private, performed within the State, with the exception of Federal work. The annual license fee is $15, plus $1 clerk fee.

Only one license is necessary, which is valid in any part of Maryland. License year begins May 1st. Fees prorated quarterly.

Licensing administered by:
State License Bureau
Treasury Department
301 W. Preston St., Room 305
Baltimore 1, Maryland

In Baltimore City, the license is secured from the Clerk of the Court of Common Pleas, and in the counties from the Clerks of the Circuit Court. If it is not convenient for the contractor to apply for license personally in county in which work is located the State License Bureau will secure the license for him.

License issued upon application and payment of the fee. No examination is required.

License not required on contracts for Federal Government on government land.

TAXES AND FEES

ADMISSION OF FOREIGN CORPORATIONS: Entrance fee of $25 for corporations doing intrastate business. This fee must accompany an application for qualification under the foreign corporation laws, also a certified copy of the corporation's Charter and all amendments. For interstate business, no fee. (State Dept. of Assessments and Taxation, Baltimore, Md.).

ANNUAL CORPORATION TAX (Foreign Corporation Filing Fee): $25, with annual report. (State Dept. of Assessments and Taxation, Baltimore, Md.).

SALES TAX: 3% on all purchases of equipment, materials and supplies used by contractors unless the property is to be incorporated into construction for the State of Maryland, its political sub-divisions or non-profit religious, charitable, educational, scientific or literary institutions. (Comptroller of the Treasury, Baltimore).

USE TAX: 3% on equipment, materials and supplies bought outside the State, to be used within the State. Receipts from rentals of tangible personal property, including equipment, subject to both sales and use tax. (Comptroller of the Treasury, Baltimore).

INCOME TAX: Corporations, 5% on net income allocable to State; formula gives equal weight to payrolls, property and sales, in state and elsewhere, less franchise tax credit to domestic corporations. (Section 292, Article 81, Maryland Code). Non-resident individuals, 5% on investment income over $500, 3% below; 3% on compensation earned in State, from any trade, business or profession carried on within the State.

General withholding law requires withholding of income tax by employers from both resident and non-resident employees, according to tables, and the filing of declarations of estimated tax on income not subject to withholding. Specials provisions for Va. and D. C. residents.

Withholding forms, tax forms and copy of the Law furnished upon request by the Comptroller of the Treasury, Income Tax Division, State Treasury Building, Annapolis, Maryland. Employers' withholding returns due April 30, July 31, October 31 and January 31. Declarations due April 15, June 15, September 15 and January 15. Annual return due April 15 or 15th day of fourth month following close of fiscal year. (Income Tax Division).

FUEL TAX: 6c per gallon on gasoline, but refund for non-highway use is obtainable if claim is filed within six months of date of purchase. There is a tax of 6c per gallon on diesel fuel used in motor vehicles operated on the highways, but such fuel consumed in off-the-highway use is tax-exempt. (Comptroller of the Treasury, Gasoline Tax Division, State Treasury Building, Annapolis, Maryland).

MASSACHUSETTS

PREQUALIFICATION *not* required in order to bid.

CONTRACTORS LICENSE *not* required.

PREQUALIFICATION

There is no State law requiring prequalification on any State work. Low bidders on State highway work are usually required to submit a financial statement, statement of experience, list of equipment and references, attesting to their ability to complete the required work within the stipulated time.

LICENSING

There is no State law requiring contractors on either public or private work to be licensed.

TAXES AND FEES

ADMISSION OF FOREIGN CORPORATIONS: Filing fee, $75. Foreign contracting corporations must be registered before doing business and must appoint the Commissioner as process attorney to represent them in State. (State Tax Commission).

ANNUAL CORPORATION EXCISE TAX: For foreign corporations, $5 per $1,000 on the corporate excess employed within Massachusetts, or $5 per $1,000 on the value of tangible property situated in Massachusetts not subject to local taxation, whichever is higher, plus 5½% of the net income allocable to Massachusetts.

There are three minimum measures:

(1) 1/20 of 1% of such proportion of the fair value of the capital stock of the corporation as the assets, both real and personal, employed within Massachusetts, bear to the total assets of the corporation, plus 3% of income allocable to Massachusetts.

(2) 1/20 of 1% of the gross receipts from business assignable to Massachusetts, plus 3% of income allocable to Massachusetts.

(3) $25, plus 3% of income allocable to Massachusetts.

Surtax is 3% with an additional 20% levy currently in effect. (Commissioner of Corporations and Taxation).

FUEL TAX: Gasoline, 5½c per gallon, with refund, upon proper application, for off-highway use. Diesel fuel, 5½c per gallon. (Commissioner of Corporations and Taxation).

MICHIGAN

PREQAULIFICATION *required* in order to bid on highway work.

LICENSE *required* in order to bid on residential building in counties of 300,000 or more or other counties that may so require.

PREQUALIFICATION

Under Act. No. 170, Public Acts of 1933, as amended, contractors proposing to bid on State Highway construction are require, at least 10 days prior to date set for opening of bids, to file with the State Highway Department an "Experience Questionnaire and Financial Statement," using standard forms provided by the Department.

Statements should be sworn to and, if possible, carry certification of a C.P.A. They should develop bidders' financial resouces, equipment, facilities, experience and qualifications to satisfactorily carry out work to be performed. Information will be kept confidential.

Bidders who qualify will be rated to show the maximum amount of work allowable at any one time. Following criteria will be used in determining the rating.

(a) Net liquid assets multiplied by 7½

(b) Approved letter of credit multiplied by 5

(c) Net equipment value multiplied by 4

Descriptions of all major rating factors and their method of application are set forth in Division I and II, Rules and Regulations of the booklet "Prequalification Regulations" published Feb., 1959, by State Highway Department. Joint bidding on a single contract may be engaged in by two or three contractors. See page 7 of booklet.

A low bidder on two or more projects totalling more than his financial rating will be awarded such project or projects as will be to best advantage of Highway Department.

Statements must be revised once a year or as often as Highway Commission requires.

LICENSING

Under Act 208, Public Acts of Michigan, 1953, as amended, general residential building contractors and maintenance and alteration contractors operating in counties with a population of 300,000 or more, or other counties that may elect to come within the provisions of this Act,

are required to obtain a license before bidding upon or contracting work of a residential or combination residential and commercial character. Application must be made in writing, on forms prescribed by the Commission. The applicant must undergo a written examination or submit satisfactory evidence of five years' experience in the type of work to be undertaken. The Act prescribes various regulations, exemptions, and penalties.

Fee for a residential builders' license is $25; for a residential maintenance and alteration contractor, $15.

No license will be issued to a foreign corporation until such corporation has been duly authorized to do business in Michigan by the Coroporation and Securities Commission. Foreign corporations must maintain place of business in Michigan.

Administration of the building division of the law is under the supervision of:

Michigan Corporation and Securities Commission.

Cadillac Square Building
Detroit 26, Michigan

Contractors desiring to do business in the residential building field in the counties affected should obtain the full text of the statute.

License is NOT required where contractor performs contracts for Federal Government, on Government land.

TAXES AND FEES

ADMISSION OF FOREIGN CORPORATIONS: Fee of 50c on each $1,000 of authorized capital stock employed or to be employed in Michigan. Minimum, $25. Also filing fee, $10. (Corporation and Securities Commission).

ANNUAL CORPORATION FRANCHISE TAX (Privilege Fee): 4 mills on each dollar of capital and surplus employed in State. Filing fee $2. Minimum Tax $10. (Corporation and Securities Commission).

BUSINESS ACTIVITIES TAX: 7¾ mills per dollar of "adjusted receipts" allocable to State based on formula utilizing property, payrolls and receipts in Michigan and everywhere. Adjusted receipts arrived at by deducting from gross receipts (1) practically all business costs—interest, rents, taxes, reasonable allowance for depreciation, cost of merchandise sold—*except* labor costs or (2) 50% of the gross receipts, plus $12,500. Businesses with less than $20,000 gross receipts exempt. Quarterly and annual returns required. (Department of Revenue, Specific Tax on Business Income Div.).

SALES AND USE TAX: 3% on sales in state of tangible personal property whether or not used in connection with fulfillment of construction contract, and on such property imported into state for use, storage or consumption. Contracts for State of Michigan and Federal Government *not* exempt. Contracts for Michigan political subdivisions and educational, religious, charitable institutions generally exempt. (Department of Revenue).

FUEL TAX: 6c per gallon on gasoline, with refund, upon proper application, for non-highway use. Diesel fuel, 6c per gallon, applied at time of sale if such fuel is placed in the fuel supply tanks of diesel-burning equipment, with no refund. (Motor Fuel Tax Division, Secretary of State).

MINNESOTA

PREQUALIFICATION *not* required.

CONTRACTOR'S LICENSE *not* required.

PREQUALIFICATION

No formal prequalification is required in order to submit bids on State highway work.

Each bidder shall furnish to the State upon request a statement showing the experience of the bidder and the amount of capital and equipment he has available for performance of the proposed work. (Department of Highways).

LICENSING

There is no law requiring contractors bidding on or executing work within the State to be licensed. Bids submitted by foreign or non-resident corporations will not be considered unless these corporations have furnished evidence to the State that they have met all legal requirements for transacting business in Minnesota.

TAXES AND FEES

ADMISSION OF FOREIGN CORPORATIONS: License fee of $100 and incidental fees of $16. (Secretary of State).

ANNUAL CORPORATION FRANCHISE FEE: Foreign corporation report filing fee, $10. (Commissioner of Taxation).

INCOME TAX: Corporations, 7½% on taxable net income derived from business in State, plus a "privilege and income" tax of 1.8% on all net taxable income, minimum tax, $10; nonresident individuals, graduated scale, 1½%-10½%, on net income from a trade or business carried on within the State, and on net income from personal services. A surtax equal to 10% of the normal tax is also imposed. (Commissioner of Taxation).

FUEL TAX: Gasoline, 5c per gallon, with refund, upon proper application, for non-highway use. Other fuels, when used on the highways, are taxed at 5c per gallon, as "Special Fuel." (Commissioner of Taxation).

MISSISSIPPI

PREQUALIFICATION *may be required* on all public work costing in excess of $25,000, and on highway work regardless of amount.

LICENSE *required* before bidding on public or private work costing over $10,000.

PREQUALIFICATION

Contractors may be required to prequalify before being awarded jobs involving public funds of more than $25,000.

Application for Certificate of Responsibility must be made to Board of Public Contractors, on forms provided for the purpose. Information furnished must include statement of applicant's experience, competency of personnel and adequacy of equipment. Board not authorized to inquire into financial status of applicant nor to fix maximum amounts of work permissible to have under contract at any one time.

If application approved Board will issue Certificate within 30 days after filing and after payment of $100 fee for "special privilege license." Certificates valid for one year after issue. Applications made and fees paid to

Board of Public Contractors,
105 Woolfolk State Office Building,
Jackson, Mississippi

Certificate *not required* where receiver of contract award holds certificate issued by similiar board of another State which recognizes certificates issued by Mississippi.

Bidders on jobs for Highway Department, regardless of amounts, must file financial statement and supply other information regarding themselves in advance of letting date. There is a fee of $5 for proposals, which is not refundable.

LICENSING

Before bidding on public or private work, on jobs costing in excess of $10,000, contractor must obtain a State-Wide Privilege License. Application must be on form prescribed by State Tax Commissioner. Cost of license $25. (State Tax Commissioner).

A further license is required for contractor before beginning work on jobs costing in excess of $10,000. This involves a fee of 50c for each $1,000 or fractional part thereof, of amount of contract. Where amount of contract cannot be determined in advance contractor shall obtain a license, costing $10, and, in addition, at end of each month shall obtain a supplementary license, paying therefor an amount based upon value of work performed during month, at rate of 50c per $1,000. (State Tax Commission).

License IS required where contractor bids on or performs contracts for Federal Government, on Government land.

TAXES AND FEES

ADMISSION OF FOREIGN CORPORATIONS: Fee, based on authorized capital stock, $20 for first $5,000, plus $2 for each additional $1,000 or fractional part thereof. Minimum, $20; maximum $500. Also, $5 fee for filing designation of Resident Agent. (Secretary of State).

ANNUAL CORPORATION FRANCHISE TAX: $2 per $1,000 of outstanding capital stock, surplus and undivided profits employed in the State, or on assessed value of real and tangible property within the State if this is greater. (State Tax Commission).

SALES TAX: 1/8 of 1% on construction materials which become a component or integral part of the building erected, constructed or repaired. 3% on equipment and supplies. 1½% of the gross amount on contracts in excess of $10,000. On contracts in excess of $10,000, contractors must post bond or pay tax in advance. (State Tax Commission).

USE TAX: 3% on equipment and supplies brought into State. Credit for sales tax paid another state against use tax due computed by applying rate of sales or use tax so paid to value of property at time of import into Mississippi. No credit on motor vehicles. (State Tax Commission).

INCOME TAX: Corporations and individuals, on graduated scale, beginning at 2% on first $5,000 and extending to 6% on taxable net income above $25,000 earned within the State. (State Tax Commission).

FUEL TAX: Gasoline, 7c per gallon, with refund of 6c per gallon obtainable, upon proper application, for non-highway use. Diesel fuel, 8c per gallon for highway use, ½c per gallon when used for other purposes. (Motor Vehicle Comptroller).

MISSOURI

PREQUALIFICATION *required* in order to bid on highway work.

CONTRACTOR'S LICENSE *not* required.

PREQUALIFICATION

Contractors desiring to bid on State Highway work must prequalify with the Missouri State Highway Department, Jefferson City, Mo. Bidders doing business under fictitious trade names are required to file with the Highway Commission a certified copy of registration of fictitious name, as issued by the Secretary of State. Foreign corporations which are successful bidders must furnish the Highway Department a certified copy of authority and license to do business in Missouri.

LICENSING

There is no State law requiring contractors on public or private work to be licensed.

TAXES AND FEES

ADMISSION OF FOREIGN CORPORATIONS: Fee, based on stated capital and surplus represented by property and business in State, is $50 on $30,000 or less and $5 on each $10,000 above $30,000. Minimum, $50 plus $10 license fee and $3 for two certificates, or a total of $63. (Secretary of State).

ANNUAL CORPORATION FRANCHISE TAX: 1/20 of 1% on amount of capital stock and surplus or total assets employed in State, whichever is greater. (Department of Revenue).

SALES AND USE TAX: 2% on retail sales of tangible personal property and on imports of same into state for use, consumption or storage, except where sales tax equal to or greater than Missouri tax has been paid. Sales to construction contractors for permanent installation in performance of contract *not* exempt. Sales to Federal government, State, political subdivisions and educational, religious or charitable organizations, exempt. (Department of Revenue).

INCOME TAX: Corporate, 2% on net income; individual, graduated rate, 1%-4%. (Department of Revenue).

FUEL TAX: 3c per gallon on gasoline and special fuels used as motor fuel on the highways. For special fuels, user must hold license, if he maintains own storage, and pay tax on such fuel used on highways. (Department of Revenue).

MONTANA

PREQUALIFICATION *required* in order to bid on highway work.

LICENSE *required* of contractors bidding on public work costing over $1,000, except where federal aid funds are involved.

PREQUALIFICATION

Contractors desiring to bid on State highway work must qualify, before bids are accepted, with the State Highway Commission, Helena, Mont. Prequalification statements required annually, by April 1st. New bidders wishing to qualify for specific letting must submit qualifications not later than 7 days prior to opening of bids.

LICENSING

Contractors bidding on public works in excess of $1,000 required to obtain license.

In the language of the revised statute:

"There shall be three classes of licenses issued

under the provisions of this act; and such classes of licenses are hereby designated as Classes A, B, and C. Any applicant for a license under the provisions hereof, shall specify in his application the class of license applied for.

"The holder of a Class A license shall be entitled to engage in the public contracting business within the State of Montana without any limitation as to the value of a single public contract project, subject however, to such prequalification requirements as may be imposed and at the time of making the application for such license the applicant shall pay to the registrar a fee in the sum of Two Hundred Dollars ($200).

"The holder of a Class B license shall be entitled to engage in the public contracting business within the State of Montana, but shall not be entitled to engage in the construction of any single public contract project of a value in excess of Fifty Thousand Dollars ($50,000); and shall pay unto the registrar as a license fee the sum of One Hundred Dollars ($100) for such Class B license at the time of making application therefor.

"The holder of a Class C license shall be entitled to engage in the public contracting business within the State of Montana, but shall not be entitled to engage in the construction of any single public contract project of a value in excess of Twenty-Five Thousand Dollars ($25,000); and shall pay unto the registrar as a license fee the sum of Ten Dollars ($10) at the time of making application therefor."

Licenses are issued by:
State Board of Equalization
Capitol Building
Helena, Montana

Since application must be held 10 days by Board before issuance of license, submission should be made sufficiently in advance to permit receipt of license before opening of bid on any project which contractor wishes to bid. Renewal, 50% of the original fee. Must be renewed before March 1st of each year.

License is NOT required where contractor bids on or performs contracts for Federal Government, on Government land or where Federal funds are involved from Bureau of Public Roads or Dept. of Agriculture.

TAXES AND FEES

ADMISSION OF FOREIGN CORPORATIONS: Fee, based on amount of issued capital stock employed in Montana, is on graduated scale, ranging from $1 per $1,000 on first $100,000 to $570 plus 20c per $1,000 on amounts over $1,000,000. Minimum qualification fee, including incidentals, $65. No par value stock of foreign corporations determined by market value of stock if readily ascertainable, otherwise assessment is made on book value of stock, which includes surplus. No par value stock of Montana corporations assigned value of $1 per share. There is an annual adjustment for any additional capital employed in business in the State. (Secretary of State).

ANNUAL CORPORATION FRANCHISE TAX (Corporation License Tax): 5% on net income from Montana sources. Minimum $10. (State Board of Equalization).

FUEL TAX: Gasoline of 46 test or over taxed at 6c per gallon. Tax for non-highway use refundable on application. Diesel fuel taxed at 9c per gallon. Exemption for non-highway use. (State Board of Equalization).

NEBRASKA

PREQUALIFICATION *required* in order to bid on State Highway work, except on maintenance or repair jobs costing below $2,500, and emergency work.

CONTRACTOR'S LICENSE *not* required on any work.

PREQUALIFICATION

On September 20, 1957, Department of Roads issued Rules and Regulations from which following is taken.

Contractors desiring to bid on State Highway work must prequalify with the Department of Roads. Excepted are maintenance and repair jobs costing less than $2,500, and emergency jobs.

Detailed statements, certified to by a C.P.A., must be submitted on forms provided for the purpose, at least seven days before date of contract letting on which it is desired to bid. Statements must provide information relating to finances, equipment, organization and experience, and must be sworn to. Statements will hold good for 15 months but Department may call for additional or new information at any time.

Department will rate applicants on basis of information supplied. Ratings will designate types and amounts of work for which applicants have been qualified to bid. Ratings will be of two types. A "maximum qualification" rating will first be issued to qualified applicants. After receiving same, applicant may request proposal forms for any specific letting, and state, on forms provided by Department, amounts and types of work then under contract, in Nebraska and elsewhere, and amounts thereof still uncompleted. Department may then grant a "current qualification" rating showing amount of work for which applicant may be qualified to bid at the particular letting. He may, however, be awarded contracts exceeding the "current" rating by 25%.

Proposal forms will be labelled with bidder's name and will not be transferable. They will not be issued after 5:00 P.M. of day preceding day of letting. Two or more qualified bidders may bid jointly.

LICENSING

There is no State law requiring contractors on either public or private work to be licensed.

TAXES AND FEES

ADMISSION OF FOREIGN CORPORATIONS: Fee of $50 for appointment of agent, certificate fee, $1. (Secretary of State).

ANNUAL CORPORATION FRANCHISE TAX (Annual Occupation Tax): Based on property and credits employed in State, fee is on graduated scale beginning at $5 on amounts up to $10,000 and increasing to $400 on $1,000,000. Maximum tax is $2,500 on property and credits exceeding $25,000,000. (Secretary of State).

FUEL TAX: Gasoline, 7c per gallon for all highway use, with 6c per gallon refundable for all non-highway use, except when used in a motor vehicle which is licensed or subject to licensing for highway use. Special fuels, including Diesel, taxable when used in motor vehicles. (Division of Motor Fuels).

NEVADA

PREQUALIFICATION *required* in order to bid on highway work.

LICENSE *required* of contractors in order to bid on both public and private work.

PREQUALIFICATION

Pursuant to statutes 1956-57 Chapter 370, Section 172, the State Highway Engineer shall, before furnishing any person proposing to bid on any duly advertised work with the plans and specifications for such work, require from such person a statement, under oath, in the form of answers to questions contained in a standard form of questionnaire and financial statement, which shall include a complete statement of the person's financial ability and experience in performing public work of a similar nature. Such statements shall be filed in ample time to permit the department to verify the information contained therein in advance of furnishing proposal form and plans and specifications to any such person proposing to bid on any such duly advertised public work, in accordance with the department's rules and regulations.

LICENSING

State of Nevada State Contractors' License Law, approved March 31, 1941 (as amended, 1959), requires all contractors, including subcontractors, doing business in the State to be licensed.

Application for license shall be on form prepared by State Contractor's Board, and shall be accompanied by fee prescribed by law. Form calls for exhaustive information as to applicant's past record, competence, character, finances (supported by CPA report) and other pertinent data. All bids and contracts must carry license number.

Licensing supervised by:
State Contractors Board
P.O. Box 405
Reno, Nevada

The board, in its discretion, is authorized to fix application and annual license fees to be paid by applicants and licensees under the terms of this act; provided, however, that the application fee shall not exceed $50 and the annual license fee shall not exceed $50.

A bond or deposit, of $1,000, is required from any applicant for a new license (as distinguished from renewal of an existing license). Renewal date, January 15th.

License is NOT required where contractor performs contracts for Federal Government, on Government land.

TAXES AND FEES

QUALIFICATION OF ALL CORPORATIONS: The fees for filing articles of incorporation or agreements of consolidation providing for shares shall be as provided below (Chap. 275, Mar. 22, 1951):

Amount represented by the total number of shares provided for in the articles of incorporation or the agreement of consolidation:

$25,000 or less	$ 25.
Over $25,000 and not over $75,000	$ 40.
Over $75,000 and not over $200,000	$ 75.
Over $200,000 and not over $500,000	$100.
Over $500,000 and not over $1,000,000	$150.
Over $1,000,000—	
(1) For the First $1,000,000	$150.
(2) For each additional $500,000 or fraction thereof	$ 75.

For the purposes of computing the filing fees according to the above schedule, the amount represented by the local number of shares provided for in the articles of incorporation or the agreement of consolidation shall be:

(1) The aggregate par value of the shares, if only shares with a par value are therein provided for, or

(2) The product of the number of shares multiplied by ten dollars ($10), if only shares without par value are therein provided for, or

(3) The aggregate par value of the shares with a par value plus the product of the number of shares without par value multiplied by ten dollars ($10), if shares with and without par value are therein provided for.

In addition to the above, there are various miscellaneous fees.

The foregoing fees apply equally to foreign and domestic corporations.

ANNUAL CORPORATION FRANCHISE TAX: There is no corporation franchise tax in Nevada. However, corporations, domestic or foreign, are required to file with the Secretary of State a list of officers, directors, and designation of resident agent, and his acceptance thereof, and pay a filing fee of $10. Failure to file such list before the first Monday of March, after the preceding July 1st, when lists are required to be filed, results in the revocation of the charters of domestic corporations, and the forfeiture of the right of foreign corporations to do business in the State. (Secretary of State).

Foreign corporation doing business in the State must publish, not later than March of each year, financial statement covering preceding year's business, in some Nevada newspaper.

SALES TAX: 2% on retail sales. (Nevada Tax Commission).

USE TAX: 2% on tangible personal property brought into state for use, consumption or storage. (Nevada Tax Commission).

FUEL TAX: Gasoline 4½c per gallon subject to refund for non-highway use. Also a county gas tax of 1½c per gallon, applicable in all counties. Diesel and miscellaneous fuels subject to a "use tax" of 6c per gallon, for highway use only. (Motor Vehicle Dept., Carrier Div.).

NEW HAMPSHIRE

PREQUALIFICATION *required* in order to bid on public work.

CONTRACTOR'S LICENSE *not* required.

PREQUALIFICATION

All contractors desiring to bid upon work under jurisdiction of Department of Public Works and Highways must prequalify by filing therewith, on forms provided for the purpose, information setting forth their qualifications. Such filing must be made once a year and, in any event, not less than eight days prior to opening of bids on any work on which they desire to bid. Department may require a prospective bidder to bring his statement up to date as of the last day of the month preceding that in which bids are to be opened.

Forms may be obtained from:
Department of Public Works and Highways
State House Annex
Concord, New Hampshire

LICENSING

There is no State law requiring contractors on either public or private work to be licensed.

TAXES AND FEES

ADMISSION OF FOREIGN CORPORATIONS: Registration fee of $50. Annual renewal (yearly maintenance) fee, due January 1st, $35. Fees for changing registration, changing name, withdrawal from state, $10 each. Foreign corporations must maintain registered agents in state authorized to act in name of corporation. (Secretary of State).

ANNUAL CORPORATION FRANCHISE TAX: Under House Bill 358, Chapter 171, approved June 1, 1955, every corporation, domestic and foreign, must submit, on or before April 1st of each year, a statement of condition, on blanks to be supplied by State, accompanied by a filing fee of $15. Failure to submit on time will involve an additional fee of $15 as well as necessity of making the return. Every domestic corporation, upon filing its annual return, shall pay a fee equal to one-fourth the amount paid upon filing its original record of organization, plus one-fourth of additional payments for increases in capital stock, if any. Maximum fee, $500. Minimum fee, $15. (Secretary of State).

FUEL TAX (ROAD TOLL): Gasoline, 7c per gallon, with refund for non-highway use available upon proper application. Users of diesel fuel required to hold User's License and pay tax of 7c per gallon for gallonage used in propulsion of motor vehicles on the highways. (Road Toll Administrator, Motor Vehicle Department).

NEW JERSEY

PREQUALIFICATION *required* of all bidders on State work, either new construction or repair, where same is advertised.

CONTRACTOR'S LICENSE *not* required on any work.

PREQUALIFICATION

Persons proposing to bid on any State work must prequalify.

Those proposing to bid on State Highway work must furnish a statement under oath in response to a Questionnaire to be submitted by the State Highway Commissioner. Such statement shall fully develop the Financial Ability, Adequacy of Plant and Equipment, Organization and Prior Experience, including complete record of work done in past two years and such other pertinent and material facts as may be desirable. New statements required every seven months, or oftener, if deemed necessary by Commissioner.

Applicant should indicate class or classes of work for which qualification is sought. Commissioner will classify approved applicants with respect to construction or repair work in grading, paving, bridge construction, general construction and miscellaneous work, involving test borings, landscaping, painting, demolition, utilities installation, plumbing, heating, ventilating and other activities. He will also stipulate dollar volume of work upon which applicants may bid, according to an alphabetical classification grading from A, for jobs between $50,000 and $75,000, to Y, jobs with a value of $6,000,000. For jobs over $6,000,000 special prequalification will be necessary at least 20 days prior to date set to receive bids. For jobs under $50,000 a special classification also will be accorded.

In addition to prequalification requirements above, prospective bidders on jobs up to and including $2,000,000 must submit new statements bearing date as of the end of the month preceding month during which bids will be accepted, except that bids submitted between the 1st and the 15th of any month may carry statements dated one month earlier than here stipulated.

Bidders on jobs with a value above $2,000,000 may resubmit statements previously filed if accompanied by affidavit that financial condition remains substantially unchanged.

BIDDING: Included in the bid envelope, in addition to the proposal being offered, should be (1) Revised Questionnaire, parts one and three, unless previously filed in accordance with the regulations (2) Certified check for 10% of the amount bid (minimum $500, maximum $20,000) (3) Proposal Bond in a sum not less than 50% of the bid (4) Non-collusion Affidavit, in duplicate (5) Appointment of local agent, by nonresident contractors.

LICENSING

There is no State law requiring contractors on either public or private work to be licensed.

TAXES AND FEES

ADMISSION OF FOREIGN CORPORATIONS: The laws of the State require the Secretary of State to exact from foreign corporations, for qualifying to transact business in New Jersey, the same fee that the foreign state exacts from a New Jersey corporation for qualifying in the foreign state. (Secretary of State).

ANNUAL CORPORATION FRANCHISE (CORPORATION BUSINESS) TAX: Based on net worth and net income allocable to State. Net worth allocable according to greater of (1) entire net worth multiplied by percentage of total assets allocable to New Jersey (2) entire net worth multiplied by the average of the three percentages of tangible assets, receipts and payrolls, respectively, allocable to New Jersey. Net income allocable by application of these three percentages.

Rates, based on net worth, from 2 mills per dollar on first $100,000,000 allocable to New Jersey to 2/10 mill per dollar on excess of $300,000,000. Minimum tax, the greater of 5/10 mill per dollar on net assets allocable to State of $100,000,000, and $25 for domestic corporations and $50 for foreign.

In lieu of above tax on net worth, corporations having total assets everywhere (less reasonable reserves for depreciation) of less than $150,000, may choose to pay tax according to scale, ranging, for domestic corporations, from $25 for less than $18,000 assets to $223 on assets of $146,000 to less than $150,000, and, for foreign corporations, from $50 on less than $34,000 assets to $223 on assets of $146,000 but less than $150,000.

Rate on that portion of tax based on net income, 1¾%. (Taxation Division, Corporation Tax Bureau, Department of Treasury).

FUEL TAX: 5c per gallon on all fuel, with refund, upon proper application, for most nonhighway uses. (See Sec. 54:-39-66, Revised Stat.) (Department of the Treasury, Division of Taxation, Motor Fuels Tax Bureau).

NEW MEXICO

PREQUALIFICATION *required* in order to bid on highway work.

LICENSE *required* in order to bid on both public and private work.

PREQUALIFICATION

By regulation of the State Highway Department, each bidder will be required to present satisfactory evidence that he has had sufficient experience and that he is fully prepared with the necessary capital, materials, machinery and skilled workmen to carry out the contract. Such evidence must be on forms which will be furnished upon request and must be on file in the office of the Chief Highway Engineer not later than five (5) days prior to the date of opening of bids. All bidders are required to properly and completely fill out the Contractor's Questionnaire found in the Proposal and to affix signatures as indicated.

LICENSING

License must be obtained before bids are made or work executed, public or private.

Fee is $40 for resident contractors doing an annual volume of $12,000 or more and for all non-resident contractors, regardless of volume. Licenses expire annually, on July 30th.

Applicant, in order to receive a license, must satisfy the Board that he is of good reputation and experienced and qualified to do the kind of work he proposes to do. He must have registered with the School Tax Div., Bureau of Revenue, and have received both a School Tax and Use Tax number. Familiarity with rules and regulations promulgated by the Board, concerning kind of work applicant proposes to contract, must also be demonstrated by taking and passing written examination.

The law provides that no license shall be issued to a foreign corporation which has not complied with the State laws requiring qualification to do business in the State, nor to any foreign corporation which has not maintained an office in the State for at least ninety days preceding application for license, and no license shall be issued to any individual, firm, partnership, or company, that has not been a resident of the State of New Mexico for ninety days, unless such person, firm, partnership or company has maintained an office in the State of New Mexico for at least ninety days preceding the filing of the application. No application will be accepted for consideration from any foreign corporation until after such corporation shall have registered with the New Mexico State Corporation Commission, Santa Fe, New Mexico, and obtained authority to do business in the State.

Contractors contemplating work in New Mexico should write to the Board for the official publication, "Provisions of Laws Relating to Contractors and the State Building Code."

Meetings of the Board are held quarterly.

Communications regarding licenses and application forms for same should be addressed to:

New Mexico Contractors' License Board
P. O. Box 1179
Santa Fe, New Mexico

License is not required where contractor bids on or performs contracts for Federal Government, on Government land.

TAXES AND FEES

ADMISSION OF FOREIGN CORPORATIONS: Filing fees, for certificate of authority, are on graduated scale, based on total capital stock; $250,000 or less, $25; over $250,000 but not over $5,000,000, 10c per $1,000, with maximum of $250; over $5,000,000 but not over $10,000,000, $500; over $10,000,000 but not over $20,000,000, $750; over $20,000,000 but not over $30,000,000, $1,000; over $30,000,000 but not over $75,000,000, $1,500; over $75,000,000 but not over $100,000,000, $2,000; over $100,000,000, $3,000. Corporations with no par value stock, $50.

At time of receiving certificate of authority foreign corporation shall pay an amount equal to annual minimum franchise tax. Such payment, however, will be credited against franchise tax due for that year. (State Corporation Commission).

ANNUAL CORPORATION FRANCHISE TAX: Based on issued capital stock employed in State, rate is $1 per $1,000 or fraction thereof, minimum $10. Filing annual report, rate 5c per $1,000 issued capital stock, or fraction thereof, represented by property and business in state; minimum $5. No par stock is based on book value as determined from certified balance sheet. (State Corporation Commission).

SALES TAX (EMERGENCY SCHOOL TAX): Contractors pay 2% on gross income from contracts, plus 1% School Tax. Materials becoming component part of structures not exempt.

Contracts for Federal government, State of New Mexico, its political subdivisions, and for non-profit organizations or societies, religious, hospital or fraternal, exempt. (Bureau of Revenue).

USE TAX: Purchases outside State taxed at same rate as domestic purchases, 2%. (Bureau of Revenue).

INCOME TAX: Corporations, 2% on taxable net income from business and property within State; individuals, 1% on first $10,000 net income, graduating upward thereafter. (Bureau of Revenue).

FUEL TAX: Gasoline, liquified petroleum gases and diesel fuel, 6c per gallon when used to propel motor vehicles on the highways. (Bureau of Revenue).

NEW YORK

PREQUALIFICATION *not* required.
CONTRACTOR'S LICENSE *not* required.

PREQUALIFICATION

There is no law or regulation of any department letting contracts for public construction that requires contractors to prequalify in advance of bidding. However, Chapter 480 of the Laws of 1947 provides that non-resident contractors, partnerships, having one or more partners who is a non-resident, or corporate contractors not organized under the laws of New York State must prove that all taxes due the State of New York have been paid before receiving payments due under a contract for highway construction. A certificate from the State Tax Commission to the effect that all such taxes have been paid constitutes proof of such fact.

LICENSING

There is no State law requiring contractors on either public or private work to be licensed.

TAXES AND FEES

ADMISSION OF FOREIGN CORPORATIONS: Filing fee for issuance of certificate of authority, $110. Application for same should give full names and addresses of corporation and principle officers, purpose of incorporation, state of incorporation, and statement that business to be done in New York is authorized by charter or articles of incorporation, also designation of Secretary of State as agent against whom action may be taken. (Division of Corporations, Secretary of State).

Fee of 1/8 of 1% on face value of issued stock employed in the State; 6c per issued no par value share employed in the State; taxes recomputed at same rates in case of any change of capital share structure or increase in the amount of capital stock employed; minimum tax, $10. (Department of Taxation & Finance).

ANNUAL CORPORATION FRANCHISE TAX: The greatest of (a) 5½% on all or an allocation of the entire net income or (b) 5½% on all or an allocation of a base which takes account of entire net income and compensation paid officers and holders of more than 5% of issued stock or (c) one mill for each dollar of its total or allocated portion of business and investment capital or (d) $25. Also ½ mill on each dollar of allocated subsidiary capital up to fifty million dollars; ¼ mill on the next fifty million dollars; and 1/8 mill on more than one hundred million dollars.

Entire net income is total net income, presumably as returned for Federal Income Tax purposes, with certain adjustments (such as the exclusion of all subsidiary income except recovery in respect of any war loss, 50% of nonsubsidiary dividends and without certain exclusions). Allocation of business income and capital on basis of real and tangible property, receipts and payroll. Allocation of investment income and capital and subsidiary capital on basis of allocation percentage of obligor or issuing corporation. (Department of Taxation and Finance).

INCOME TAX: Corporations, as above. Unincorporated businesses, 4%, on net income derived within the State. Exemptions, $5,000 per annum. Certain professions exempt. Deductions include salary allowance for proprietor or each partner not in excess of $5,000 for each active partner or an aggregate of 20% of net income, with remission of 15% of first $100 of tax and 10% of next $200, for 1959. Individuals, sliding scale from 2% or first $1,000 to 10% on amounts in excess of $15,000. Withholding of taxes and quarterly returns required from employers. Instruction pamphlet available from State Tax Commission. (Department of Taxation and Finance).

NEW YORK CITY TAXES: Various taxes on business enterprises, such as general business tax, retail sales tax, compensating use tax and occupancy tax on business premises are imposed in New York City. The contractor contemplating work within the city should familiarize himself with all taxes involved by obtaining full information from the City Treasurer.

OTHER LOCAL TAXATION: A number of other cities also, including Syracuse, Auburn and Poughkeepsie, and several counties, including Monroe and Erie, levy sales and use taxes about which contractors should seek information locally.

TRUCK MILEAGE TAX: Vehicular units, in excess of 18,000 pounds, maximum gross weight, taxed on a sliding scale. (Department of Taxation and Finance).

FUEL TAX: Gasoline, 6c per gallon, with refund, upon proper application, for non-highway use. Registered owners of vehicles using diesel fuel on the highways are required to register as distributors in accordance with section 282-a of the Tax Law. The tax is 9c per gallon, with refund available on fuel employed in non-highway use. (See Article 12A, Tax Law, as amended). (Department of Taxation and Finance).

NORTH CAROLINA

PREQUALIFICATION *required* in order to bid on all highway work, and on all other nonfederal jobs costing $20,000 or more, through operation of State licensing laws.

LICENSE *required* of contractors bidding on or undertaking projects over $20,000, both public and private.

PREQUALIFICATION

In order to bid on any highway project, federal or non-federal, contractor must prequalify with State Highway Commission. On all other types of work, except federal, costing $20,000 or more, contractor must prequalify, in effect, by obtaining a license.

LICENSING

Licensing is supervised by:
North Carolina Licensing Board for Contractors
607 First Citizens Bank Building
Raleigh, North Carolina

Contractors desiring to bid on or execute projects of $20,000 or more must make formal application and file with the Board at least thirty days prior to any regular or special meeting, a written application on such forms as may then be prescribed by the Board. Board meets in April and October, oftener when deemed necessary. The Board either issues or denies a license, basing its decision upon experience, organization, financial condition and performance record of the applicant, and successful completion of written examination.

Licenses are issued in three groups; jobs up to $75,000, fee, $40, renewal, $20; jobs from $75,000 to $300,000, fee, $60, renewal, $40; no limit on value of job, fee, $80, renewal, $60.

Licensing year begins June 1st. Fee may be prorated where bidder places first bid between Jan. 1st and May 31st. Licenses expire Dec. 31st, renewals required within one month.

License is NOT required where contractor performs contracts financed wholly or in part by the Federal Government. (Commissioner of Revenue).

Separate from the above, Section 122, Chapter 105-54, N. C. General Statutes, requires "Every person, firm or corporation who, for a fixed price, commission, fee, or wage, offers or bids to construct within the State of North Carolina any building, highway, street, sidewalk, bridge, culvert, sewer or water system, drainage or dredging system, electric or steam railway, reservoir or dam, hydraulic or power plant, transmission line, tower, dock, wharf, excavation, grading or other improvement or structure, or any part thereof, the cost of which exceeds the sum of ten thousand dollars ($10,000) shall apply for and obtain from the Commissioner of Revenue an annual Statewide license, and shall pay for such license a tax of one hundred dollars ($100) at the time of or prior to offering or submitting any bid on any of the above enumerated projects.

TAXES AND FEES

PROJECT TAX: Levied on total contract price of largest single job during year, scaled from $25 on jobs of $5,000, but not more than $10,000, to $625 on jobs of $1,000,000 or over. (Commissioner of Revenue).

ADMISSION OF FOREIGN CORPORATIONS: Based on total authorized capital stock, tax is 40c per $1,000. No par stock value at $100 per share. Minimum tax, $40, maximum, $500. Filing fee, $5. (Secretary of State).

ANNUAL CORPORATION FRANCHISE TAX: Franchise tax of $1.50 per $1,000 on the greatest of (1) outstanding capital stock and surplus allocable to State (2) investment in tangible personal property in State (3) assessed value for ad valorem tax purposes of property in State. Minimum, $10. (Commissioner of Revenue).

SALES TAX: 3% on retail sales, including building materials, equipment and supplies except where same become part of structures erected for state or political subdivisions or agencies, or for non-profit organizations of charitable, religious or educational nature. Forms, scaffolding, etc., not exempt. Tax on motor vehicles, as defined in statute, designated for use on highways 1%, maximum $80. (Commissioner of Revenue).

USE TAX: (Excise Tax) 3% on all purchases brought into state for use or storage. Tax on contractors' vehicles, machinery and equipment, brought into state for use, computed on basis of such proportion of original cost as duration of time of use in state bears to total useful life. Owner, or, if property is leased, lessee liable for tax. (Commissioner of Revenue).

INCOME TAX: (Corporations): 6% on net taxable income applicable to North Carolina. Net taxable income allocated to North Carolina by applying gross receipts ratio formula to total net income earned everywhere. No taxes based on or measured by income deductible in arriving at total net income.

(Individuals): 3% on first $2,000 taxable income, graduating to maximum, 7% over $10,000, with personal exemptions. (Commissioner of Revenue).

FUEL TAX: Gasoline, 7c per gallon, with refund, upon proper application, of 6c per gallon for non-highway use. Diesel fuel taxed only for use in diesel-driven motor vehicles. (Commissioner of Revenue).

FUEL USE TAX: Object, to require operators of heavy motor vehicles, having three or more axles, to purchase within state as much fuel as used in state. Such operators must register each vehicle keep records and make quarterly reports of gasoline purchased in state and used in state. Fuel bought outside North Carolina and used in North Carolina must bear full 7c North Carolina tax to extent that total use of fuel in state exceeds total fuel bought in state. If use within state is less than total fuel bought in state refund of tax on this difference is obtainable *provided* state where purchased has levied a tax on fuel bought in North Carolina but used within that other state. To obtain credit all purchases in state should be invoiced with name and address of both buyer and seller, amount of fuel bought, and date. Registration fee, $1.00 per vehicle. Reports and payments to Gasoline Tax Division, Department of Revenue, Raleigh, N. C.

NORTH DAKOTA

PREQUALIFICATION *required* in order to bid on highway work.

LICENSE *required* in order to bid on all public work costing $2,000 or more.

PREQUALIFICATION

North Dakota State Highway Department requires that "Names of bidders must be on the Commissioner's qualified list of Contractors, with a sufficient rating shown thereon to entitle them to bid upon the particular job in question before bid will be accepted. Each bidder shall furnish the Commissioner with satisfactory evidence of his competency to perform the work contemplated. At least ten (10) days prior to requesting his first proposal in each calendar year, to perform work for the Department, he shall submit a statement to the Commissioner showing his financial condition. This statement shall show the condition of his business as of December 31st of the preceding year. New statements shall be submitted annually thereafter for as long a period as the bidder continues to offer proposals for work advertised for letting by the Commissioner unless specifically requested oftener. With each financial statement filed with the Commissioner, the bidder shall also submit a questionnaire relating to his experience in performing construction work similar to that for which he is offering a proposal, and a list of machinery, plant and other equipment available for the proposed work. Experience questionnaire herein referred to shall be submitted on forms furnished by the Commissioner based on such statement, and the confirmation or verification of the fact set forth therein, together with such other material and pertinent data as the Commissioner may have to acquire relative to the competency of such a bidder. A rating will be assigned such bidder within six (6) days after receipt of the financial statement and questionnaire, which rating will set forth the maximum and type of work which will be awarded to him or which he will be permitted to have under contract and incomplete at any time."

LICENSING

North Dakota Public Contractors' Act, Chapter 43-07 North Dakota Revised Code of 1943, requires "public contractors" to be licensed.

A public contractor is defined as one who submits a proposal to or enters into a contract with the State or any political subdivision of the State, including all boards, commissions or departments thereof for the construction or reconstruction of public work, when the contract cost exceeds $2,000.

The provisions of the Act shall not apply to any road construction or road repair contract financed in whole, or in part, through Federal Aid furnished by the Bureau of Public Roads, except such Bureau approve the provisions of the Act.

There are four classes of licenses issued. License (A) costs $250, covers contracts of any amount. License (B) costs $150, covers contracts up to and including $125,000. License (C) costs $100, covers contracts up to and including $60,000. License (D) costs $15, covers contracts up to and including $15,000.

Applications must be on file with the Secretary of State ten days before license may be issued and license must be held by the contractor at least ten days prior to the date set for receiving bids.

License may be renewed from year to year at ten percent of original cost, if renewed between Jan. 1st and April 1st. (Secretary of State).

License is NOT required where contractor bids on or performs contracts for Federal Government, on Government land.

TAXES AND FEES

ADMISSION OF FOREIGN CORPORATIONS: Entrance fee of $93. (Secretary of State).

ANNUAL CORPORATION FRANCHISE TAX (Annual License Fee): Foreign corporations must file report each year for adjustment of license fee to equal that of domestic corporations of same character and capitalization. Fee for filing report is $10. (Secretary of State).

SALES TAX: 2% on retail sales. (State Tax Commissioner).

USE TAX: 2% on purchases brought into State. (State Tax Commissioner).

INCOME TAX: Foreign corporations, graduated scale beginning at 3% on first $3,000 of income earned in State; bonds must be furnished guaranteeing payment of amounts that may be due; individuals, graduated scale on taxable income earned in State, beginning at 1% on first $3,000. (State Tax Commissioner).

FUEL TAX: Gasoline and special fuels, 6c per gallon. Refund available, upon proper application, for some non-highway uses. Preparation of gravel and other road surfacing materials not construed as non-highway use. Special fuels defined as all combustible gases and liquids other than gasoline, as defined in Sec. 57-4101, 1943 Code Revised, suitable for generation of power for propulsion of motor vehicles. Heating fuel, delivered into tank connected with heating appliance, exempt. However on July 1, 1959 excise tax of 2% imposed on fuel used for heating purposes, not refundable. (State Auditor).

OHIO

PREQUALIFICATION *required* in order to bid on highway work.

CONTRACTOR'S LICENSE *not* required.

PREQUALIFICATION

Contractors desiring to bid on State highway work are required to file a financial statement and complete an equipment and experience questionnaire to be reviewed by the Credit Examiner.

For general work, statements must be submitted at least 10 days prior to date set for opening bids. For structures built to carry railroad traffic, whether permanent or temporary in nature, special prequalification necessary, with statements submitted at least 30 days before opening of bids.

Upon favorable action by the Examiner, a prequalification certificate indicating the amount and type of work the contractor is qualified to have under contract, is issued. Communications concerning prequalification should be addressed to the Credit Examiner, Department of Highways, Columbus, Ohio.

LICENSING

There is no State law requiring contractors on either public or private work to be licensed.

TAXES AND FEES

ADMISSION OF FOREIGN CORPORATIONS: Two types of licenses are granted foreign corporations—temporary and permanent. The temporary license carries a fee of $100 and is good for six months, with no surrender fee. However, only two will be granted in a period of three years. The permanent license carries an initial filing fee of $50, plus an annual fee on a graduated scale based on capital employed in Ohio; fees determined by prescribed formula, payable on an increase only over prior years. Rates from 10c per share on 1,000 shares or less to 1/4c per share on more than 500,000 shares. Minimum fee, $5, first year only. Fee for Surrender of License, $10. (Secretary of State).

ANNUAL CORPORATION FRANCHISE TAX: 3/10 of 1% on outstanding capital stock allocated to Ohio. Minimum $50. Annual report and payment due March 31st. No filing fee for same. (State Treasurer).

SALES TAX: 3% on retail sales of tangible personal property and on rentals. No exemptions for tangible personal property incorporated into structure, except where construction performed for state, political sub-divisions, hospitals, or religious, charitable or educational organizations. (Tax Commissioner).

USE TAX: 3% on purchases brought into State for "use" or "storage," as defined in Section 5741.01 of Act. (Tax Commissioner).

PERSONAL PROPERTY TAXES: All tangible property located and used in business in Ohio is subject to taxation at a rate which varies annually. All intangible property of residents and certain intangible property of non-residents is subject to taxation. (Tax Commissioner).

FUEL TAX: 7c per gallon. Refundable, upon proper application, for non-highway use except in certain classes of surface vessels. (Tax Commissioner).

TRUCK MILEAGE TAX: On all commercial vehicles and vehicular combinations, with three or more axles, graduated scale, from 1/2c to 2 1/2c per mile travelled within State and based on number of axles. Terms such as commercial car, commercial tandem and commercial tractor specifically described in law, which should be consulted. (Tax Commissioner).

LOCAL TAXES: A number of Ohio cities impose municipal income taxes about which contractors should seek information locally.

OKLAHOMA

PREQUALIFICATION *required* in order to bid on highway work.

CONTRACTOR'S LICENSE *not* required.

PREQUALIFICATION

At least 30 days prior to date upon which they expect to offer bids, contractors shall submit to Commission, under oath, on forms provided, complete information on themselves. This shall include: (1) complete financial statement, verified by C.P.A. (holding valid certificate as public accountant issued by State in which he is licensed), showing resources and liabilities, current and fixed, with all questions on this subject answered by figures or word "none;" if cash value of life insurance is included as asset, letters from insurance companies validating such claims shall also be presented: (2) Schedule of equipment; (3) Experience questionnairs; (4) Statement of past record; (5) Personnel of organization and; (6) At least six references as to character and quality of previous work done.

Commission will rate and notify applicants making such submissions, not later than 4 days prior to bid opening date. Applicants will not be rated to perform work at a value in excess of 10 times their net current assets. If additional information is asked it must be submitted at least 24 hours before time fixed for opening of bids.

Financial statements, experience records and equipment schedules shall be kept up to date by annual submissions made within 60 days of the close of each fiscal year.

Prior to engaging in any business for Department an out-of-state contractor must appoint and maintain an agent upon whom service or process may be had in any action to which said person or firm may become liable. Agent shall not be an official of Federal government or any political subdivision thereof or a bondsman, surety or material supplier. There shall be a filing with the Secretary of State of this person's name and place of residence, or business address in Oklahoma. A certified copy of this filing shall be furnished to the Prequalification Officer of the Commission.

There shall also be filed with the Highway Commission, prior to any work performance, by a foreign corporation, certified copies of the Articles of Incorporation or similar documents attesting to the nature of its formation.

Proposal blanks shall be obtained from the Prequalification Officer and from no other source.No proposal shall be issued to a contractor after 2:00 P.M. of the day preceding receiving of bids for any contract.

Should a contractor be low bidder on contracts totalling more than the amount for which he is qualified the Commission reserves the right to make awards on any part or all of such low bids as it appears to the best interest of the State to do. In event any award is not made to the low bidder it shall be made to the next lowest responsible bidder. The Prequalification Officer may, if it appears in the best interest of the State to do so, limit the amount of work for which any new contractor may qualify to 4 times the amount of the net current assets or working capital as shown on applicant's financial statement, until such time as he has satisfactorily completed at least one contract with the Commission.

The prospective bidder should familiarize himself with the Oklahoma State Highway Commission Standard Specifications, Edition 1937 and 1954 Revision thereof.

For copy write to State Highway Commission
Capitol Office Building
Oklahoma City 5, Oklahoma.

LICENSING

There is no State law requiring contractors on either public or private work to be licensed.

TAXES AND FEES

ADMISSION OF FOREIGN CORPORATIONS: $1 per $1,000 on capital employed in Oklahoma in excess of $10,000. Minimum, $18. (Secretary of State).

ANNUAL CORPORATION FRANCHISE TAX (License Tax): $1.25 on each $1,000 of actual capital employed in Oklahoma. Minimum, $10. Maximum, $20,000. (Oklahoma Tax Commission).

INCOME TAX: Corporations, 4% on taxable net income; individuals, 1% on first $1,500 to 6% on excess over $7,500. (State Tax Commission).

SALES TAX: 2% on retail sales. (Oklahoma Tax Commission).

USE TAX: 2% on purchases brought into State, including mail order purchases. (Oklahoma Tax Commission).

FUEL TAX: Gasoline, all sales and uses, 6½c per gallon; Special Fuels, highway use, 6½c per gallon. (Oklahoma Tax Commission).

OREGON

PREQUALIFICATION *required* in order to bid on public works projects in excess of $10,000.

CONTRACTOR'S LICENSE *not* required.

PREQUALIFICATION

Prospective bidders on all public works projects costing over $10,000 shall prequalify with awarding authority, not later than 10 days prior to date set for opening bids. Information required, on forms to be furnished, shall be sworn to and shall include financial statement, outline of experience, description of equipment available and other pertinent data that may be called for. Bidders once qualified before any public officer need not separately qualify for each contract later to be advertised, unless required to do so by said officer. Decisions as to prequalification submissions will be announced at least 8 days prior to time set for opening bids.

Prequalifying authorities include: State Highway Commission, Salem, Oregon; State Department of Finance and Administration, Salem, Oregon; State Board of Higher Education, Comptroller, University of Oregon, Eugene, Oregon; and Oregon State Board of Control, Salem, Oregon.

LICENSING

There is no State law requiring contractors, as such, to be licensed on either public or private work.

TAXES AND FEES

ADMISSION OF FOREIGN CORPORATIONS: Entrance fee of $50, in payment for Certificate of Authority to do business in State. Also prepayment of annual license fee. (Corporation Commission).

ANNUAL CORPORATION FRANCHISE TAX (Annual License Fee): Foreign corporations pay flat fee of $200 on anniversary date of Certificate of Authority. (Corporation Commissioner).

CORPORATION EXCISE TAX: Based on taxable net income from Oregon business. Rates, banks, 9%, public utilities, 6%, mercantile and manufacturing establishments, 6% with offset, not to exceed 33-1/3%, for non-depreciable personal property taxes paid by manufacturing corporations. Minimum, $10. (State Tax Commission).

INCOME TAX: Corporations, 6% on net income from Oregon business. Corporations pay *either* the Excise or Income Tax depending upon their methods of doing business. Non-resident individuals, graduated scale on income earned in State. (State Tax Commission).

FUEL TAX: All motor fuels ordinarily used by contractors, including diesel fuel, taxed at 6c per gallon, when used to propel vehicles on highways or public roads. Tax on gasoline collected at point of distribution or sale. Tax on gasoline used off highways refundable but, on contracts with State Highway Commission, any such refunds obtained must be repaid to Commission.

Tax on fuel, other than gasoline, collected by seller who places fuel in vehicles, or paid by user who obtains fuel not taxed. Refunds obtainable: (1) when used in another state if user pays to such state additional tax on same fuel or (2) when used on private road or (3) when used on any road other than state or county road pursuant to agreement with an agency or licensee of U.S., if agreement imposes upon user of the road the obligation either to construct the road at his own expense or to pay such agency or licensee a reasonable consideration for the use or right of way of such road. (Department of Motor Vehicles).

PENNSYLVANIA

PREQUALIFICATION *not* required in order to bid.

CONTRACTOR'S LICENSE *not* required.

PREQUALIFICATION

Contractors are not required to prequalify in order to bid, but low bidders on State Highway work must qualify as to finances and experience and file a plan questionnaire.

LICENSING

There is no State law requiring contractors on

either public or private work to be licensed. Fictitious business styles and foreign corporations must be registered. (Secretary of the Commonwealth.)

TAXES AND FEES

ADMISSION OF FOREIGN CORPORATIONS: (Excise) 1/3 of 1% on capital employed in commonwealth. (Department of Revenue).

ANNUAL EXCISE: 1/3 of 1% on any increase in capital employed in Pennsylvania. (Department of Revenue).

ANNUAL CORPORATION FRANCHISE TAX: 5 mills on value of capital stock allocated to Pennsylvania. (Department of Revenue).

SALES AND USE TAX: 4% on sales at retail of tangible personal property transferred for a consideration for consumption, use or storage, and on rentals.

Construction activities include furnishing of necessary materials and equipment and their installation or incorporation in constructing, reconstructing, improving or repairing a road, bridge, building or other structure. Contractor subject to tax on all tangible personal property used or consumed in such activities, whether or not performed for otherwise exempt entity such as Federal government or State of Pennsylvania. Contractor not permitted to collect tax from his customers on property used in performance of construction activities. On materials or equipment supplied by contractor but not so installed or incorporated into, or permanently affixed to structure, such as furniture, blinds, awnings, portable appliances, refrigerators, contractor functions as vendor, must register with Department as vendor, obtain license and remit taxes collected on such sales.

On tangible personal property brought into State if owner has paid a sales, use or occupation tax thereon, in another State, in an amount less than tax imposed in Pennsylvania, the difference between the two rates must be paid in Pennsylvania.

Persons covered as vendors of tangible personal property must have a Certificate of Registration and keep appropriate records. For further details consult local Sales Tax offices or State Bureau of Sales and Use Tax, 1846 Brookwood Street, Harrisburg. (Department of Revenue).

INCOME TAX: 6% on taxable net income allocated to Pennsylvania. (Department of Revenue).

LOCAL TAXES: One or more Pennsylvania cities, including Philadelphia, levy income or sales taxes about which contractors should obtain information locally.

FUEL TAX: 5c per gallon. Gasoline is taxable, regardless of purpose for which used. Diesel fuel is subject to the tax only when used on highways. (Department of Revenue).

PUERTO RICO

PREQUALIFICATION *required* in order to bid

on public work under jurisdiction Puerto Rico Industrial Development Co., and on public housing costing over $100,000 under jurisdiction Housing Authority Administration. Also *required* on all other public work prior to receiving contract awards.

CONTRACTOR'S LICENSE *not* required.

PREQUALIFICATION

Prospective bidders on "Pridco" work must file "statement of bidder's qualification" attesting to experience, equipment available, personnel and financial condition. Information bringing statement up to date must accompany each bid submitted.

Similar presentation required by Housing Authority Administration, prior to being awarded public housing contracts costing over $100,000, and by all other public agencies at time bids are submitted covering work under their jurisdiction. These include Department of Public Works, Aqueducts and Sewerage Authority and University of Puerto Rico.

LICENSING

There is no Puerto Rico law requiring that contractors be licensed.

TAXES AND FEES

ADMISSION OF FOREIGN CORPORATIONS: Filing must include certified copy of charter or certificate of incorporation, sworn statement of assets and liabilities and identification of authorized agent or agents who can be sued. Fee, $30. (Secretary of State).

ANNUAL REPORTS: Filing required, not later than April 15 of each year, of balance sheets showing existing financial condition and condition at close of preceding calendar or fiscal year, profit and loss statement for latest operating year, names and addresses and terms of office of all officers and directors. (Secretary of Treasury.) (Secretary of State).

INCOME TAX: Corporations, normal rate, 20% of net income derived from Puerto Rican sources, after certain credits; surtax rate, sliding scale, from 5% on net income above $25,000, but not above $50,000, with certain credits, to 20% on net income above $100,000, plus flat $7,500 levy. Also levy equal to 5% of combined normal and surtax. Individuals; U. S. Citizens, non-resident; normal rate, 7% on net income from Puerto Rican sources; surtax rate, sliding scale, from 5% on net income below $2,000 to 72% on net income above $200,000, after personal exemptions and credits.

Construction firms *may* qualify for accelerated depreciation allowances on depreciable property at rates up to 100% in a single year.

U.S. mainland residents, located in Puerto Rico

for periods of one year or longer *may*, under certain circumstances, qualify as Puerto Rican residents and become exempt from U.S. income taxes on income derived from Puerto Rican sources during such periods. (Department of Treasury).

EXCISE TAXES: Imposed on selected products imported from U.S. including trucks, above 3/4-ton, 10% on appraised value, truck-tractors, 17%, passenger cars, 20% on first $2,000 appraised value, 60% to $2,500, 80% above $2,500. In appraising, consideration given to original cost, depreciation, cost of transportation to Puerto Rico, other factors. (Department of Treasury).

FUEL TAX: Gasoline, 8c a gallon; gas oil or diesel oil, 4c a gallon. No refund for non-highway use. (Department of Treasury).

RHODE ISLAND

PREQUALIFICATION *not* required in order to bid.

CONTRACTOR'S LICENSE *not* required.

PREQUALIFICATION

A registered list of bidders, for highway and bridge work, is maintained and when there is work to be bid, contractors on the registered list are notified. The job is also listed in local newspapers. To be placed on the list, it is necessary only to make request to the State Department of Public Works Division of Roads and Bridges, Providence, R. I. Out-of-State contractors are requested to file authority for officers authorized to sign or, if a partnership, to give names and addresses of partners. A contractor is not required to file statement of finances and performance history until and unless he is a low bidder.

LICENSING

There is no State law requiring contractors on either public or private work to be licensed.

TAXES AND FEES

ADMISSION OF FOREIGN CORPORATIONS: Entrance fee of $25 (payable to General Treasurer), plus filing fee of $5. (Secretary of State).

ANNUAL CORPORATION TAX (Business Corporation Tax): All corporations must pay a tax at rate of 5½% of net income allocable to State. Thereafter State calculates tax, based on corporate excess, at 40c per $100, and if greater than amount paid, corporation is billed for the difference. (Department of Administration, Division of Taxation).

SALES AND USE TAX: 3% on sales at retail of tangible personal property. Use tax, 3% on

tangible personal property brought into state by a resident for use, consumption or storage. Tangible personal property purchased by a non-resident and brought into State for his own use, exempt.

Regulation C, September 17, 1951, specifies that, with exceptions noted, contractors are liable on all purchases of equipment, materials and supplies and on all rentals of equipment; where contractor contracts to furnish materials at agreed price and to render services in connection therewith he functions as a retailer, must have permit to make sales at retail and must collect tax from person to whom he sells.

Sales to state, any political subdivision thereof, non-profit hospitals and educational institutions, and to institutions operated exclusively for religious or charitable purposes, exempt.

On federal contracts for construction, improvement or repair, which are written on a lump sum, guaranteed or upset price basis, contractor is liable for tax on all purchases. On federal contracts written on cost-plus or fixed fee basis, where contractor is authorized to act as agent for government, and government agrees to make payment *direct* to suppliers for materials, equipment or supplies purchased, contractor is exempt from sales tax. On contracts such as foregoing where government agrees simply to reimburse contractor for cost of purchases made, contractor is liable for sales tax. (Sales and Use Tax Section, Division of Taxation, Department of Administration).

FUEL TAX: All motor fuels ordinarily used by contractors, including diesel fuel, used either on or off the highways, subject to tax of 6c per gallon. (Department of Administration, Division of Taxation).

SOUTH CAROLINA

PREQUALIFICATION *required* in order to bid on highway work.

LICENSES of two types *required* of general contractors in order to bid on jobs of $20,000 or more.

If contractor has prequalified for highway work license will automatically issue, after 7 days following date of application.

PREQUALIFICATION

Under "Rules and Regulations for Prequalification," established in accordance with Section 33-223 of the 1952 Code entitled "State Highway Commission to Fix Eligibility of Bidders on State Highway Work," contractors desiring to bid on State highway work are required to submit a financial statement, an experience questionnaire and an equipment questionnaire to the South Carolina State Highway Commission, Columbia, South Carolina. The required forms may be obtained from the Commission.

LICENSING

Two types of licenses required for general contractors desiring to bid on jobs costing $20,000 or over. Both licenses under jurisdiction, Licensing Board for Contractors, 1628 Laurel St., Columbia, S. C.

General Contractor's license obtainable by applying to Board, on form provided, at least 30 days in advance of regularly scheduled meeting and, upon being notified, submitting to examination. Approved applicants will appear on roster maintained by Board. Fee for this license $60, paid at time of application. Expiration date, December 31. Renewal required during month of January, same fee.

This license will be issued without examination, seven days after application, to those presenting bidder's or contractor's certificate issued by Highway Department. Same fees.

Bidder's license, issuable without examination to those holding General Contractor's license upon payment of fee of $100. This license also expires in December and is renewable either the following January or prior to submitting the next bid. Renewal fee, $100.

All bids by licensed contractor must show license numbers, for both types of licenses held.

License is NOT required where contractor bids on or performs contracts for Federal Government, on Government land. Projects involving public or private funds augmented by Federal Aid are subject to license requirements.

TAXES AND FEES

ADMISSION OF FOREIGN CORPORATIONS: Entrance fee of $50. Annual statement required. Filing fee, $10. (Secretary of State).

ANNUAL CORPORATION FRANCHISE TAX (Corporation License Fee).

Under Franchise Tax Law, effective January 1, 1955, tax of one mill on a proportion of total capital stock and paid in surplus is levied on both Foreign and Domestic corporations. Proportion will be based on application of statutory formulae. Minimum tax, $10. (State Tax Commissioner).

INCOME TAX: Corporations, 5% on taxable net income earned in State.

Indivduals, graduated scale on taxable income from sources within State. (State Tax Commission).

SALES AND USE TAX: 3% on retail sales within State whether for use, storage or consumption, and on retail purchases outside State when brought into State for use irrespective of whether or not sales tax has been paid on same in State where purchased. Contractors' tangible personal property, such as contruction equipment, vehicles, tools, and machinery, where bought outside State and used before being brought into State, taxed on basis of life expectancy in relation to time used in

State. Where imported new, and used for the first time in State, taxed on full purchase price.

Tax applies to gross proceeds or receipts from rentals.

Where there is transfer of title, under lease or rental agreement, tax applies to gross proceeds derived. Sales of machines and parts therefor, used in mining, quarrying and manufacturing exempt. Exemptions do not include such items as screens and bins used in rock quarry. (State Tax Commission, Sales and Use Tax Division).

FUEL TAX: Gasoline, 7c per gallon for highway or non-highway use. Diesel fuel subject to 7c per gallon tax only when used to propel motor vehicles on the highways. (State Tax Commission).

SOUTH DAKOTA

PREQUALIFICATION *required* **in order to bid on highway work.**

CONTRACTOR'S LICENSE *not* **required.**

PREQUALIFICATION

All persons proposing to bid on State highway work must furnish a statement under oath on a form prescribed and provided by the Highway Commission. Such statement shall fully develop the financial ability, adequacy of equipment, organization, prior experience and other pertinent and material facts as may be desirable. Such information furnished shall be used to determine a contractor's classification and rating in accordance with the regulations (of the Highway Commission).

In case the same shall be demanded by the Commission the financial statement shall be supported by a certificate as to the correctness thereof by a certified public accountant.

Statements upon which a qualification is desired shall be on file with the Commission in sufficient time for the Committee to properly analyze and consider the same, and to obtan confirmation or verification of the Statements set forth therein.

Ratings are given for one or more of the following classifications:

Grading
Concrete Paving
Bituminous Paving—Plant Mix Work
Bituminous Paving—Road Mix Work
Gravel and Crushed Rock Servicing Base Course
Bridges and Grade Separations
Guard Rail
Roadside Improvement

A rating on "maximum capacity," or largest amount that contractor may have under contract at any one time is also asigned.

Also assigned, for any particular letting, will be a "current rating," bearing upon contractor's eligibility to be considered for that job.

Necessary forms for making application together with a copy of "Regulations Governing the Classification and Rating of Prospective Bidders" can be obtained from:

South Dakota State Highway Commission
Pierre, South Dakota.

LICENSING

There is no State law requiring contractors on either public or private work to be licensed.

TAXES AND FEES

ADMISSION OF FOREIGN CORPORATIONS: $1 per $1,000 of issued capital stock allocable to State, plus incidental fees. Minimum fee, $25. In addition there is a 50c minimum fee for filing and recording appointment of Secretary of State as resident agent. Filed as separate instrument. (Secretary of State).

SALES TAX: 2% on retail sales. (Director of Taxation).

USE TAX: 2% on purchases brought into State. (Director of Taxation).

FUEL TAX: Gasoline and liquified petroleum, 6c per gallon; diesel fuel, 7c per gallon.

Refund of tax on fuel used in stationary power plants may be obtained by holders of proper permits. No refund of gasoline tax for non-highway use in connection with construction or maintenance work paid for with public funds. (Director, Motor Fuel Tax Division).

TENNESSEE

PREQUALIFICATION *required* in order to bid on highway work.

LICENSE *required* of contractors undertaking work in excess of $10,000, public or private.

PREQUALIFICATION

The Department of Highways and Public Works requires prequalification by contractors bidding on work under the supervision of that Department, regardless of the size of the project and irrespective of whether or not the contractor has been licensed by the State Board for Licensing General Contractors.

Prequalification questionnaires prepared by a Certified or Licensed Public Accountant, must be filed once a year and must reach office of State Construction Engineer at least five days prior to date of contract letting in which contractor is interested. Forms may be obtained from:

Department of Highways and Public Works
804 Cotton States Building
Nashville, Tennessee.

LICENSING

Chapter 135, Public Acts of 1945, as amended March 13, 1947, requires general contractors to be licensed before undertaking any project costing more than $10,000.

Any person, firm or corporation desiring to be licensed under this Act must make application on the prescribed form thirty (30) days prior to any regular or special meeting of the Board for Licensing General Contractors. Regular meetings of the Board are held in January, April, July and October of each year.

Written application accompanied by a remittance of $25 must be filed with the Board on such forms as the Board may prescribed. If the application is satisfactory to the Board, the applicant shall be entitled to an examination to determine his qualifications. If the results of the examination are satisfactory to the Board a certificate will be issued authorizing the applicant to operate as a general contractor in the State of Tennessee. Anyone failing to pass such examination may be re-examined at a regular meeting of the Board without additional fee.

Licenses expire on the last day of December following issuance or renewal. Renewal may be effected by payment of $15 to the Secretary of the Board.

Issuance of a certificate of license by the Board is evidence that the person, firm or corporation named therein is entitled to all the rights and privileges of a licensed general contractor, while the license remains unrevoked, provided that the contractor has paid his privilege tax (outlined below) and that the license is recorded in the office of the County Clerk in each County in which the contractor engages in business. Fee for recording of the certificate by the County Clerk of any county is $1.

Licensing is under the supervision of:
State Board of Licensing General Contractors
101 Cotton States Building
Nashville 3, Tennessee.

TAXES AND FEES

ADMISSION OF FOREIGN CORPORATIONS: Entrance fee of $300. (Secretary of State).

CORPORATION FRANCHISE TAX: 15c per $100 on the greater of the following three bases:

(1) Capital stock, surplus and undivided profits allocated to business within the State.

(2) Assessed value of real and tangible personal property, owned or used in the State.

(3) Book value of real and tangible personal property, owned or used in the State. (Commissioner of Revenue).

ALSO: Corporation Filing Fee—Paid annually. Option of: (1) ½ of 1% on gross amount of receipts from Tennessee business or (2) Graduated rate on outstanding capital stock. Minimum $5; maximum, $150. (Commissioner of Revenue).

ALSO: Corporation Excise Tax—3.75% on taxable net earnings from Tennessee business. Specified accounting basis for contractors. (Commissioner of Revenue).

CONTRACTORS' PRIVILEGE TAX: Any person engaged in the business of constructing railroads and public roads in the State as prime contractor is subject to a privilege tax of $250.

Sub-contractors engaged in railroad or highway work must pay an annual privilege tax of $150.

Contractors engaged in construction other than railroads or highways, pay an annual privilege tax, imposed by the State, ranging from a minimum of $10 on contracts with a value of less than $10,000 to a maximum of $200 on contracts over $500,000.

Unless he prefers to pay applicable tax on full amount of contract at beginning of contract a contractor or sub-contractor engaged in business in the State must pay to the County Court Clerk in the county in which application is made the minimum tax of $10 and enter into bond satisfactory to the County Court Clerk covering maximum that could be due both the county and the State for one year's privilege tax. Such contractor will be liable, at the end of a 12-month period, for the additional fee based upon the aggregate contracts taken, without deduction for any part of the work subcontracted.

In addition to the State privilege tax, counties and cities have the authority to assess like taxes. (Commissioner of Revenue).

SALES AND USE TAX: 3% on materials and equipment purchased in Tennessee; on rentals of tangible personal property within State; and on purchase price of or rental paid for tangible personal property, materials or supplies imported for use within State, minus, however, any Sales Tax paid to another State on such property, provided such other State gives credit for tax paid to Tennessee. Otherwise full Tennessee levy applies. Applicable on both private and public work including work done for Federal Government and including any materials, supplies or equipment furnished in connection therewith; however, property which becomes part of an electric generating or transmission system belonging to Federal, State or local government or a political subdivision thereof exempt. Machinery for a new or expanded industry may be subject to a 1% instead of 3% tax. (Commissioner of Revenue).

FUEL TAX: 7c per gallon on all motor fuels, including diesel, ordinarily used by contractors, when placed in fuel tank of motor vehicle licensed to operate on public highways. (Gasoline Tax Division, Commissioner of Revenue).

TEXAS

PREQUALIFICATION *required* in order to bid on highway work.

CONTRACTOR'S LICENSE *not* required.

PREQUALIFICATION

Contractors desiring to bid on State highway work must submit to the Texas Highway Department, Austin 14, Texas, on forms provided for the purpose, a confidential questionnaire containing financial, equipment and experience data. The financial statement must afford complete information on contractors' financial resources, including a balance sheet, and must be certified to by an independent C.P.A., registered and in good standing in any State, an independent non-certified public accountant authorized to practice before U.S. Treasury Department, or a public accountant registered under the Texas Public Accountancy Act.

LICENSING

There is no State law requiring contractors on either public or private work to be licensed.

TAXES AND FEES

ADMISSION OF FOREIGN CORPORATIONS: Filing of new charters or permit to do business in Texas, fee, $50. (Secretary of State).

ANNUAL CORPORATION FRANCHISE TAX: Based on that proportion of the stated capital, surplus and undivided profits, and long-term notes, bonds and debentures payable, which the gross receipts from business done in Texas bear to the total gross receipts from the entire business, or if higher, on the assessed value of corporation's property in State. Rate of tax is $2.25 for each $1,000 or fractional part thereof; minimum, $25.

For period Sept. 1, 1959-April 30, 1960, temporary additional tax levied, equal to one-third of tax already computed. For years beginning May 1, 1960 and May 1, 1961, temporary additional tax will be 22.22% of regular computed tax. (Comptroller of Public Accounts).

FUEL TAX: Motor fuel (gasoline), 5c per gallon. Motor fuel used off highway subject to tax refund, except on work paid for out of such tax. Special fuels (liquefied gas, butane propane), 5c per gallon. Distillate fuel (diesel, kerosene), 6½c per gallon. Tax refunds paid to non-bonded users and dealers(retailers) on fuel used or sold for use off highway.

Bonded users and retail dealers, whose purchases are predominantly for use off highway, purchase tax-free and pay tax to State only on fuel used or sold for use on highway. (Comptroller of Public Accounts).

UTAH

PREQUALIFICATION *required* to bid on projects under the jurisdiction of State Road Commission, and on State Building Board projects in excess of $15,000.

LICENSE *required* in order to bid on both public and private work except that bidder on a *single undertaking,* involving less than $1,000, exempt.

PREQUALIFICATION

Under regulations of State Road Commission contractors desiring to bid on State projects must submit a Confidential Financial Statement, Equipment and Experience Questionnaire, on forms provided by the Department of Contractors, 307 State Capitol, Salt Lake City, Utah. State Building Board requirements similar, with respect to projects in excess of $15,000.

Financial statements must be of a date not more than nine months prior to date of application for prequalification. Financial statements must be audited by independent C.P.A., registered and in good standing in any State, or independent public accountant, licensed and in good standing in Utah, *If* applicant desires qualification in excess of $100,000. Accountant's audit must not contain nullifying qualifications.

New statements required annually at least 30 days prior to expiration of current qualification.

Prior to being prequalified each applicant must have been licensed in compliance with Utah laws, and, if an out-of-State corporation, must have certificate from Secretary of State showing that corporation is duly qualified to transact business in Utah.

LICENSING

All contractors and sub-contractors on jobs costing $1,000 or more, with exceptions noted below, must be licensed by Department of Contractors. Applications for license must be on forms provided by Department and afford pertinent information regarding applicant's experience, financial responsibility and general competence.

Licenses to be issued will be of three types: (1) General Engineering (2) General Building (3) Specialty. Applicants should state type of license or licenses desired. A qualified applicant may be licensed in more than one classification.

Exceptions include such work as irrigation or drainage ditches, agricultural construction, quarrying, well drilling, metal and coal mining, construction for public utilities subject to regulation by Public Utilities Commission, and any *single undertaking* involving less than $1,000.

License is required on jobs for Federal Government on government land.

Licenses expire December 31 and applications for renewal must be made on or before that date. Fee, $50, renewal, $15. (Department of Contractors).

TAXES AND FEES

ADMISSION OF FOREIGN CORPORATIONS: Based on proportion of authorized capital stock to be represented by property and business in the State, fee is 25c per $1,000 plus $5 for Certificate of Compliance and $3 for filing acceptance of provisions of constitution and appointment of resident agent. (Secretary of State).

ANNUAL CORPORATION FRANCHISE TAX (Income Tax): Rate, 4% on income attributable to Utah sources, or 1/20 of 1% on fair average value of tangible property within the State, whichever is greater. Minimum, $10. (State Tax Commission).

SALES TAX: 2% on retail sales, including installation of tangible personal property and repairs to or renovation thereof. (State Tax Commission).

USE TAX: 2% on purchases brought into State. (State Tax Commission).

MOTOR FUEL TAX: 6c per gallon tax on all motor fuels with refund for *agricultural* nonhighway use, except that tax on special fuel is applicable only when used for the propulsion of motor vehicles. (State Tax Commission).

VERMONT

PREQUALIFICATION *required* in order to bid on highway and bridge work estimated at $25,000 or over. Statement also required of contractors desiring to bid on smaller jobs under jurisdiction of Highway Dept., at time first bid of year is submitted.

CONTRACTOR'S LICENSE *not* required.

PREQUALIFICATION

Contractors proposing to bid on State Highway projects must furnish a statement under oath on forms prescribed and provided by the Department of Highways at least once each year.

Prequalification year ends April 1.

Proposals for contract work of $25,000 or over will not be issued to any contractor prior to establishing of prequalification status.

Request for prequalification, including properly executed statement, must be submitted on or before the eighth day before the opening of bids in order to receive proposals for a particular bid opening.

Contractors desiring to bid on contract work involving less than $25,000 are required to submit with their first bid of the year a properly executed financial statement and questionnaire, using forms prescribed and provided by the Department of Highways.

LICENSING

There is no State law requiring contractors on either public or private work to be licensed.

TAXES AND FEES

ADMISSION OF FOREIGN CORPORA-

TIONS: A foreign corporation is required by State law to obtain a certificate of authority from the Secretary of State. Certified copy of charter or articles of incorporation must accompany application for certificate. Fee is $25. Annual renewal or certificate of extension of authority must be obtained each year at a cost of $10. All certificates of authority obtained from the Secretary of State, who is ex-officio Commissioner of Foreign Corporations, are effective until the first day of the following April. (Secretary of State.) In order to do business subject to regulation by the Vermont Public Service Commission a foreign corporation must have certificate of authority issued by that Commission. (Public Service Commission).

ANNUAL CORPORATION FRANCHISE TAX: 5% on taxable net income or earnings allocable to Vermont. (Commissioner of Taxes).

REGISTRATION OF BUSINESS TITLES: Partnerships and individuals doing business under names other than their own must register with the Commissioner of Taxes in accordance with Sections 1621 and 1623 of the Vermont Statutes. Registration fee is $5. (Commissioner of Taxes).

INCOME TAX: Individuals, graduated scale based on taxable business income earned in the State. Ranges from 2% on less than $1,000 net income to $220 plus 7½% on more than $5,000. Employers are required to withhold taxes on the wages of all persons, resident and nonresident, receiving compensation for work and services performed in Vermont. (Commissioner of Taxes).

FUEL TAX: 6½c per gallon on all motor fuels except diesel fuel, with no refund for non-highway use. On motor trucks or motor vehicles using diesel fuel, the registration is 1¾ times that of gasoline-powered motor vehicles. (Commissioner of Motor Vehicles).

VIRGINIA

PREQUALIFICATION *required* with State Highway Commission in order to bid on State Highway work.

REGISTRATION and possession of registration certificate, issued by State Registration Board for Contractors, required to bid on or undertake all other work, costing $20,000 or more, whether private or public, and including highway work under jurisdiction of counties or local authorities.

PREQUALIFICATION

Contractors desiring to bid on State highway work in Virginia must prequalify. Application is by standard form, "Contractors' Financial and Experience Statements," affidavits No's 1, 2 and 3, supplied by Commonwealth of Virginia, Department of Highways, Richmond, Virginia.

LICENSING

General contractors and sub-contractors bidding on or undertaking projects of $20,000 or more within the State, except those under the supervision of the State Highway Commission, must be licensed and registered.

Title 54, Chapter 7, Virginia Code, 1950, as amended, 1958, provides substantially as follows:

Any person, firm, association or corporation desiring to be registered as a general contractor or a subcontractor in this State shall make and file with the State Registration Board for Contractors 30 days prior to any regular or special meeting thereof a written application on such form as may then be prescribed for examination by the Board, which application shall be accompanied by thirty ($30) dollars.

Regular meetings are held in January, April, July and October and special meetings as determined by Board.

The Board may require the applicant to furnish evidence of his ability, character and financial responsibility, and if said application is satisfactory to the Board, then the applicant shall be entitled to an examination to determine his qualifications. If the result of the examination of any applicant shall be satisfactory to the Board, then the Board shall issue to the applicant a certificate to engage as a general contractor or a subcontractor in this State as provided in the certificate in any one or more of four classifications of work: (1) building, (2) highway, (3) public utilities, (4) specialty; and it shall be the responsibility of the Board, or the members of said Board, to ascertain from reliable sources whether or not the past performance record of the applicant is good, and whether or not he has the reputation of paying his labor and material bills, as well as carrying out other contracts that he may have entered into.

Before issuing certificate Board shall ascertain whether applicant has complied with laws respecting foreign corporations and/or any other laws of State affecting contractors or subcontractors as defined by statute. Any applicant failing to pass such examination may be re-examined at any regular meeting of the Board without additional fee. Certificate of registration shall expire on the last day of December following its issuance or renewal and shall become invalid on that date unless renewed, subject to the approval of the Board. Renewal may be effected anytime during the month of January by the payment of a fee of fifteen ($15) dollars to the secretary-treasurer of the Board.

Registration supervised by:
State Registration Board for Contractors
504 Lyric Building, 9th and Broad Streets
Richmond 19, Virginia.

BIDDING: Revised bidding procedure announced September 20, 1955, by State Department of Highways, permits bidder, if he so chooses, to withdraw proposals submitted but not yet opened if it apears that he is successful on one or more projects already opened and read. Bidders will furnish lists of proposals which they may wish to withdraw. Such projects will then be numbered on slips of paper, placed in container, withdrawn one at a time and number posted. Acceptable proposals on each project will then be read and, before other proposals are opened, opportunity afforded apparent low bidder to withdraw any or all other unopened proposals he has submitted, should he so desire.

Opening and reading of proposals which have not been listed with requests for possible withdrawals will follow opening and reading of all proposals where such requests have been submitted.

TAXES AND FEES

ADMISSION OF FOREIGN CORPORATIONS: Based on authorized capital stock, fee is 60c per $1,000 for capitalization up to $1,000,000. Over $1,000,000 fees range from $1,000 to $5,000. No par value shares taken at par of $100. Minimum, $30. Foreign corporations must maintain registered local agents who may be sued. (State Corporation Commission).

ANNUAL CORPORATION REGISTRATION FEE: Graduated scale based on authorized capital stock. Minimum, $5. Maximum, $25. (State Corporation Commission).

CONTRACTORS' LICENSE TAX (Privilege Tax): The Tax Code of Virginia, Section 58-297 to 58-303.1, imposes a license tax, measured by the gross amount of all orders or contracts accepted during the preceding year. The tax is ascertained in the following manner:

$1,000 or under	no license required
$1,000 to $5,000	$ 5
$5,000 to $10,000	$ 10
$10,000 to $20,000	$ 15
$20,000 to $50,000	$ 20
$50,000 to $100,000	$ 50
$100,000 to $150,000	$100
$150,000 to $300,000	$150
$300,000 or above	$250

Licenses expire December 31. They may be procured from the Commissioner of Revenue for the city or county in which the contractor has his office, or in the city or county in which he conducts his business. (Department of Taxation).

Contractors' registration and license are NOT required where contractor performs contracts for Federal Government, on Govenment land over which Virginia has ceded exclusive jurisdiction to Federal Government.

INCOME TAX: Corporations, 5% on taxable net income allocable to State; resident and nonresident individuals on sliding scale, beginning at 2% on first $3,000 of taxable net income and extending to 5% on income in excess of $5,000 allocable to State. (Corporations: Department of Taxation. Individuals: Local Commissioner of Revenue).

FUEL TAX: 6c per gallon on all fuels ordinarily used by contractors including diesel fuel. Tax refundable for non-highway use. (Commissioner of Motor Vehicles).

SALES TAX PROPOSAL of 3% under consideration by 1960 State legislature.

WASHINGTON

PREQUALIFICATION *required* in order to bid on highway work.

CONTRACTOR'S LICENSE *not* required.

PREQUALIFICATION

Chapter 53, Session Laws, 1937, provides substantially as follows, in connection with bid proposals for construction or improvement of Primary State Highways. A call for bid proposals, which has been published by the Director of Highways, shall be made upon contract proposal form supplied by the Department of Highways, and in no other manner. The Director of Highways shall, before furnishing any person, firm or corporation desiring to bid upon any work for which a call or bid proposal has been published, with a contract proposal form, require from such person, firm, or corporation answers to questions contained in a standard form of Questionnaire and Financial Statement, including a complete statement of the financial ability and experience of such person, firm, or corporation in performing State highway, road or other public work. Such questionnaire shall be sworn to before a notary public or other person authorized to take acknowledgment of deeds. Whenever the Director of Highways is not satisfied with the sufficiency of the answers contained in such Questionnaire and Financial Statement he may refuse to furnish such person, firm or corporation with a contract proposal form and any bid proposal of such person, firm or corporation must be disregarded. The Questionnaire and Financial Statement will be used to establish the classified rating of the applicant for an annual fiscal period, or remaining portion thereof, of the fiscal year following date of financial statement.

New filings required annually within 90 days of expiration of current qualification.

Applications for prequalification are filed with the Director of Highways of the State of Washington, Olympia, Washington.

LICENSING

There is no State law requiring contractors on either public or private work to be licensed.

TAXES AND FEES

ADMISSION OF FOREIGN CORPORATIONS: Fee is on graduated scale, based on capital stock represented by property and capital employed in State.

First $50,000 stock, $50, plus $1 per $1,000 additional stock, up to and including $1,000,000. First $1,000,000 stock, $1,000, plus 40 cents per $1,000 additional stock, up to and including $4,000,000. First $4,000,000 stock, $2,200, plus 20 cents per $1,000 additional stock. Maximum, $5,000. Same rate for domestic corporations. (Secretary of State).

ANNUAL CORPORATION FRANCHISE TAX: (License Fee): Graduated scale, based on capital stock represented by property and business in State. Same brackets as admission fee, but fee starts at $30 for first $50,000 stock, with 50 cents for each additional $1,000 stock, up to and including $1,000,000, and highest bracket starting at $1,105 for first $4,000,000 stock, plus 10 cents each additional $1,000 stock; maximum fee, $2,500. (Secretary of State).

BUSINESS PRIVILEGE TAX: 44/100% on building contractors' gross receipts, including receipts from construction or improvement of public roads. (State Tax Commission).

SALES TAX: 4% on retail sales which include full contract price of construction contracts, tangible property installed and labor and services performed. Razing and moving of buildings and moving of earth are classified as construction. (State Tax Commission).

FUEL TAX: 6½c per gallon on motor fuels ordinarily used by contractors, including diesel fuel. Refund allowed on gasoline used in any equipment not licensed to operate on the highways. (Department of Licenses).

WEST VIRGINIA

PREQUALIFICATION *required* in order to bid on highway work.

CONTRACTOR'S LICENSE *not* required.

PREQUALIFICATION

Standard Specifications for Roads issued by State Road Commission, 1952, pursuant to Acts of the Legislature, 1937, Revised July 1, 1941, provide that contractors desiring to bid on State highway work must prequalify.

To obtain a Certificate of Qualification, the contractor must file, under Oath, a "Contractor's Prequalification Statement," containing Financial Statements, and Experience Records with detailed information as to available financial resources, equipment, property and other assets, together with an account of past experience, a record of work accomplished, personnel of organization and other facts called for in the Prequalification Statement, and with such other information the Commissioner may desire for consideration in issuing a Certificate.

Financial Statement must be filed in January of each year.

Application will be accepted by the Commissioner until 15 calendar days prior to date set for opening of bids. Form may be obtained from:

The State Road Commissioner of West Virginia
Charleston, West Virginia.

For the issuance of a Certificate, net current assets shall be not less than 20% of the amount the contractor may be permitted to have under contract, for a base rating. Premium rating up to 8 times net current assets allowed on experience in State. Sub-contractors must qualify on the same basis as general contractors.

Based upon an analysis of the Financial Statement and consideration of the Experience Questionnaire, contractor will be rated and Certificate issued for the following classifications:

General Construction
Portland Cement Concrete Paving
Bituminous Paving
Gravel, Crushed Rock and Knapped Stone Construction
Grading
Structures
Guard Rail, Sodding, Tiling, and Culverts
Hauling and Placing
Building, Housemoving, Reconstruction

A contractor may not be awarded contracts to a value in excess of his rating. In cases where a contractor is found to be low bidder on projects whose value exceeds his rating he may be allowed to withdraw those bids which account for the excess.

Two or more persons, partnerships or corporations may combine for a joint bid by obtaining joint qualification under the same regulations governing individual concerns.

Before being awarded any contract an out-of-State corporation must have obtained from the Secretary of State authority to do business in West Virginia.

LICENSING

There is no State law requiring contractors on either public or private work to be licensed.

TAXES AND FEES

ADMISSION OF FOREIGN CORPORATIONS: Filing fee of $10, plus $5 for certificate for Workman's Compensation Commission, plus pro rata share of annual license tax, which totals $250, and of State Auditor's fee of $10. (Secretary of State).

ANNUAL CORPORATION FRANCHISE TAX: (LICENSE TAX): For domestic corporations, tax is on graduated scale and ranges from $20 on authorized capital stock of $5,000 or less to $2,500 on capital stock of more than $15,000,000.

For foreign corporations, tax is based on the same percentage of total issued capital stock as is represented by the percentage of corporate property owned in West Virginia. Tax ranges from

$250 minimum to $4,375 maximum. Also, there is a statutory attorney fee of $10 for foreign corporations, or domestic corporations having their main place of business or chief works outside the State. (State Auditor).

SALES TAX: Consumers' sales tax of 1c on purchases from 6c to 50c; 2c on purchases from 51c to $1; with 1c on each additional 50c or part thereof. Contractors are exempt from this tax on purchases of materials and supplies to be used in the original construction of buildings or substantial repairs to same, if they pay the Business and Occupation Privilege Tax. (State Tax Commissioner).

BUSINESS AND OCCUPATION PRIVILEGE TAX (Gross Sales Tax): Contractor's tax of 2.6% on gross amount of all West Virginia contracts. (State Tax Commissioner).

FUEL TAX: Gasoline, 7c per gallon with refund available for certain non-highway uses. Diesel fuel, 7c per gallon. (State Tax Commissioner).

WISCONSIN

PREQUALIFICATION *may be required* on highway and other public work.

CONTRACTOR'S LICENSE *not* required.

PREQUALIFICATION

Prequalification optional with awarding authority. Where required prospective bidders must submit sworn, confidential statement containing information on finances, equipment and experience, not less than five days prior to time set for opening bids. Authority will rate applicants on the basis of information supplied. Once rated, applicants need not separately qualify on each contract let unless so required by awarding authority.

On State highway work, applications to State Highway Commission, State Office Building, Madison, Wisconsin.

On other public work, to State Chief Engineer.

LICENSING

There is no State law requiring general contractors to be licensed.

TAXES AND FEES

ADMISSION OF FOREIGN CORPORATIONS: Entrance fee of $25, plus $1 for each $1,000 of capital in excess of $25,000 represented in the State. (Secretary of State).

ANNUAL CORPORATION REPORT: Filing fees, Jan. 1-March 31, $5; April 1-May 31, $30. All reports due by June 1. However, grace period of 30 days is allowed, with notices sent in early June advising of impending revocation. Those

responding within 30 days of date of notice will be granted renewal of certificate of authority, at cost of $55. Failure to file report within the 30 days will result in revocation of certificate.

Also required payment of $1 per $1,000 on capital in excess of the amount upon which fee has theretofore been paid. (Secretary of State).

INCOME TAX: Corporations, graduated scale beginning at 2% on the first $1,000 and increasing to 7% on excess above $6,000, on net income earned within the State.

Individuals, graduated scale, from $10, plus 1¼% on net income above $1,000, to $622.50, plus 8½% on net income over $14,000.

All non-residents, corporations and individuals engaged in construction contracting on a single job of $50,000 or two or more jobs totalling $50,000 or more must file surety bonds of 1% to guarantee payment of income taxes. (Department of Taxation).

FUEL TAX: 6c per gallon on all motor fuels ordinarily used by contractors, including gasoline and special fuels (diesel fuel and liquefied petroleum gas). Refund, on proper application, for non-highway use of gasoline including use in mobile equipment and machinery. Special fuels used in mobile machinery and equipment not taxable. (Department of Taxation).

WYOMING

PREQUALIFICATION required in order to bid on highway work.

CONTRACTOR'S LICENSE *not* required.

PREQUALIFICATION

Prospective bidders on highway work must prequalify with Wyoming Highway Department, Cheyenne, Wyoming, using form provided, which is designed to set forth applicant's financial status, adequacy of plant and equipment and past experience.

Applications required sufficiently far in advance of contract letting date as to allow Department time to act thereon at least four days before said date. Bids of contractors not prequalified at least one day in advance will not be considered.

Department will rate all approved applicants as to character of work for which qualified and amount allowed to have under contract at any one time. Decision will be taken within ten days of receipt of application.

LICENSING

There is no State law requiring contractors on either public or private work to be licensed.

TAXES AND FEES

ADMISSION OF FOREIGN CORPORATIONS: Entrance fee of $10, plus $1 per $1,000 of capital, property and assets allocable to State, plus $2.50 for designation of agent who may be sued and $2.50 for filing Acceptance of Constitution. (Secretary of State).

ANNUAL CORPORATION FRANCHISE TAX: Corporate capital, property and assets located in and employed in Wyoming are taxed on graduated scale, beginning at $5 for $50,000 or less and extending to $50 for each $1,000,000. Annual report, duly notarized, and payment of tax due July 1. (Secretary of State).

SALES TAX: 2% on retail sales. (State Board of Equalization).

USE TAX: 2% on purchases brought into State. (State Board of Equalization).

FUEL TAX: 5c per gallon on gasoline used on or off the highways. Diesel fuel taxed at 7c for highway use; exempt from tax for off-the-highway use. Special fuels, including propane and butane, 5c a gallon for highway use. (Department of Revenue).

Appendix **B**

Standard federal forms

In this appendix are reproduced Standard Forms 19, 19A, 20, 21, 22, 24, 25, 25A, 27, 27A, 27B, 28, used for federal construction work.

Standard Forms 23, and 23A appear in Chap. 9.

For a discussion of the application of these forms, please see sections 9.3 and 9.4 of this book.

STANDARD FORM 19
JANUARY 1959 EDITION
GENERAL SERVICES ADMINISTRATION
FED. PROC. REG. (41 CFR) 1–16.401
19–104

INVITATION, BID, AND AWARD
(Construction, Alteration or Repair)

REFERENCE

(Include in correspondence)

INVITATION FOR BIDS

DATE ISSUED:

ISSUING OFFICE

BID RECEIVING OFFICE

INFORMATION REGARDING BIDDING MATERIAL MAY BE OBTAINED FROM THE ISSUING OFFICE

SEALED BIDS in covering work described in specifications, schedules, drawings and conditions entitled and dated as follows:

will be received at the Bid Receiving Office until ____________ *(Time)* ____________ *(Date)*
and then publicly opened.
Sealed envelopes containing bids shall be addressed to the Bid Receiving Office and shall be marked to show: Bidder's Name and Address; Reference ; Time and Date of Opening;

BID *(This Section to be completed by Bidder)*

→ DATE BID SUBMITTED:

The undersigned agrees, if this bid is accepted within calendar days after date of opening, to complete all work specified in strict accordance with the above-identified documents and the General Provisions on the reverse hereof, within calendar days after receipt of notice to proceed, for the following amount ____________

including all applicable Federal, State, and local taxes.

BIDDER REPRESENTS: *(Check appropriate boxes)*
1. That he ☐ is, ☐ is not, a small business concern. For this purpose, a small business concern is one that (a) is independently owned and operated, (b) is not dominant in its field of operation, and (c) with affiliates, had average annual receipts for the preceding three years of $5,000,000.00 or less. *(See Code of Federal Regulations, Title 13, Part 121, as amended, for additional information.)*
2. *That he operates as an* ☐ *individual,* ☐ *partnership,* ☐ *corporation, incorporated in the State of* ____________.

NAME AND ADDRESS OF BIDDER *(Street, city, zone, and State. Type or print.)*

SIGNATURE OF PERSON AUTHORIZED TO SIGN THIS BID
→

TYPE OR PRINT SIGNER'S NAME AND TITLE

AWARD *(This Section for Government only)*

DATE OF AWARD:

The above Bid is accepted in the amount of $
☐ You are directed to proceed with the work upon receipt of this Award.
☐ Notice to proceed will be issued upon receipt of acceptable payment and performance bonds.

THE UNITED STATES OF AMERICA

BY ____________
(Contracting Officer)

____
(Title)

Standard form 19 (face)

GENERAL PROVISIONS

1. CHANGES AND CHANGED CONDITIONS.—
(a) The Contracting Officer may, in writing, order changes in the drawings and specifications within the general scope of the contract. *(b)* The Contractor shall, before proceeding further, notify the Contracting Officer in writing of subsurface or latent conditions differing materially from those indicated in this contract or unknown unusual physical conditions at the site.
(c) If changes under *(a)* or conditions under *(b)* increase or decrease the cost of, or time required for, performing the work, upon assertion of a claim by the Contractor before final settlement of the contract, a written equitable adjustment shall be made; except that no adjustment under *(b)* shall be made unless the notice required therein was given or unless the Contracting Officer determines the facts justify its waiver. If the adjustment cannot be agreed upon, the dispute shall be decided pursuant to Clause 3.

2. TERMINATION FOR DEFAULT—DAMAGES FOR DELAY—TIME EXTENSIONS.—
(a) If the Contractor does not prosecute the work so as to insure completion, or fails to complete it, within the time specified, the Government may, by written notice to the Contractor, terminate his right to proceed. Thereafter, the Government may have the work completed and the Contractor shall be liable for any resulting excess cost to the Government. If the Government does not terminate the Contractor's right to proceed, he shall continue the work and shall be liable to the Government for any actual damages occasioned by such delay unless liquidated damages are stipulated.
(b) The Contractor's right to proceed shall not be terminated nor the Contractor charged with actual or liquidated damages under *(a)* above because of any delays in completion of the work due to causes other than normal weather, beyond his control and without his fault or negligence, including but not restricted to, acts of God, acts of the public enemy, acts of the Government (in either its sovereign or contractual capacity), acts of another contractor in the performance of a contract with the Government, fires, floods, epidemics, quarantine restrictions, strikes, freight embargoes, and unusually severe weather, or delays of subcontractors or suppliers due to causes beyond their control and without their fault or negligence: *Provided*, That the Contractor shall within 10 days from the beginning of any such delay, unless the Contracting Officer shall grant a further period of time prior to the date of final settlement of the contract, notify the Contracting Officer in writing of the causes of delay. The Contracting Officer shall ascertain the facts and the extent of the delay and extend the time for completing the work when in his judgment the findings of fact justify such an extension, and his findings of fact shall be final and conclusive on the parties hereto, subject only to appeal as provided in Clause 3 hereof.

3. DISPUTES.—
Any dispute concerning a question of fact arising under this contract, not disposed of by agreement, shall be decided by the Contracting Officer, who shall reduce his decision to writing and furnish a signed copy to the Contractor. Such decision shall be final and conclusive unless, within 30 days from the date of receipt thereof, the Contractor mails or otherwise furnishes to the Contracting Officer a written appeal, addressed to the head of the Federal agency. The Contractor shall be afforded an opportunity to be heard and to offer evidence. The decision of the head of the Federal agency or his authorized representative, shall be final and conclusive unless fraudulent, or capricious, or arbitrary, or so grossly erroneous as necessarily to imply bad faith, or not supported by substantial evidence. Pending final decision of a dispute hereunder, the Contractor shall proceed diligently with the performance of the contract and in accordance with the Contracting Officer's decision.

4. RESPONSIBILITY OF CONTRACTOR.—
At his own expense the Contractor shall: *(a)* obtain all required licenses and permits; *(b)* provide competent superintendence; *(c)* take precautions necessary to protect persons or property against injury or damage and be responsible for any such injury or damage that occurs as a result of his fault or negligence; *(d)* perform the work without unnecessarily interfering with other contractors' work or Government activities; *(e)* be responsible for all damage to work performed and materials delivered (including Government-furnished items), until completion and final acceptance.

5. MATERIALS AND WORKMANSHIP.—
The work shall be under the general direction and subject to the inspection of the Contracting Officer or his duly authorized representative who may require the Contractor to correct defective workmanship and materials without cost to the Government.

6. PAYMENTS.—
Progress payments equal to 90 percent of the value of work performed may be made monthly on estimates approved by the Contracting Officer. Upon payment therefor, title to the property shall vest in the Government. The Contractor will notify the Government when all work is complete. Final payment will be made after final acceptance.

7. OFFICIALS NOT TO BENEFIT.—
No Member of or Delegate to Congress, or Resident Commissioner, shall be admitted to any share or part of this contract, or to any benefit that may arise therefrom; but this provision shall not be construed to extend to this contract if made with a corporation for its general benefit.

8. BUY AMERICAN ACT.—
The Contractor, subcontractors, material men, and suppliers must comply with the Buy American Act of March 3, 1933 (41 U.S.C. 10a–10d) and Executive Order 10582 of December 17, 1954 (19 Fed. Reg. 8723). (In substance the above require use of domestic materials except as otherwise authorized by the Act and Executive Order.)

9. NONDISCRIMINATION IN EMPLOYMENT.—
(a) In connection with the performance of work under this contract, the Contractor agrees not to discriminate against any employee or applicant for employment because of race, religion, color, or national origin. The aforesaid provision shall include, but not be limited to, the following: Employment, upgrading, demotion, or transfer; recruitment or recruitment advertising; layoff or termination; rates of pay or other forms of compensation; and selection for training, including apprenticeship. The Contractor agrees to post hereafter in conspicuous places, available for employees and applicants for employment, notices to be provided by the Contracting Officer setting forth the provisions of the nondiscrimination clause.
(b) The Contractor further agrees to insert the foregoing provision in all subcontracts hereunder, except subcontracts for standard commercial supplies or raw materials.

10. LABOR STANDARDS PROVISIONS APPLICABLE TO CONTRACTS NOT IN EXCESS OF $2,000.
(a) Eight-Hour Laws—Overtime Compensation.—The Eight-Hour Laws (40 U.S.C. 321–326) are applicable to this contract. (In substance they provide that laborers and mechanics employed by the Contractor or his subcontractors shall be paid not less than time and a half for work in excess of 8 hours a day. Violations are punishable as prescribed in 40 U.S.C. 322–324.)
(b) Nonrebate of Wages.—The Regulations issued by the Secretary of Labor (29 CFR, Part 3) pursuant to the Anti-Kickback Act, as amended (40 U.S.C. 276 (c), 18 U.S.C. 874), are applicable to this contract. (In substance they provide that no deductions may be made from wages except those required by law or permitted by the Regulations, that contractors and subcontractors shall preserve for 3 years after completion of the work payrolls which contain for each employee, his name, address, correct classification, rate of pay, daily and weekly number of hours worked, deductions made, and actual wages paid, and shall submit weekly a statement of compliance, the form of which is stated in the Regulations.)

(c) Subcontractors—Termination.—The Contractor agrees to insert paragraphs *(a)* and *(b)* immediately above in all subcontracts. The term "Contractor" as used in such paragraphs in any subcontract shall be deemed to refer to the subcontractor. Breach of the requirements of this Clause 10 may be grounds for the termination of this contract.

11. CONVICT LABOR.—
In connection with the performance of work under this contract, the Contractor agrees not to employ any person undergoing sentence of imprisonment at hard labor.

January 1959. Reverse of Standard Form 19

U.S. GOVERNMENT PRINTING OFFICE : 1959 O—518309

Standard form 19 (reverse)

STANDARD FORM 19–A
JANUARY 1959 EDITION
GENERAL SERVICES ADMINISTRATION
FED. PROC. REG. (41 CFR) 1–16.401
19–202

LABOR STANDARDS PROVISIONS

APPLICABLE TO CONTRACTS IN EXCESS OF $2,000

1. DAVIS-BACON ACT (40 U.S.C. 276a–a(7))

(a) All mechanics and laborers employed or working directly upon the site of the work will be paid unconditionally and not less often than once a week, and without subsequent deduction or rebate on any account (except such payroll deductions as are permitted by the Copeland Act (Anti-Kickback) Regulations (29 CFR, Part 3)) the full amounts due at time of payment, computed at wage rates not less than those contained in the wage determination decision of the Secretary of Labor which is attached hereto and made a part hereof, regardless of any contractual relationship which may be alleged to exist between the Contractor or subcontractor and such laborers and mechanics; and a copy of the wage determination decision shall be kept posted by the Contractor at the site of the work in a prominent place where it can be easily seen by the workers.

(b) In the event it is found by the Contracting Officer that any laborer or mechanic employed by the Contractor or any subcontractor directly on the site of the work covered by this contract has been or is being paid at a rate of wages less than the rate of wages required by paragraph (a) of this clause, the Contracting Officer may (1) by written notice to the Government Prime Contractor terminate his right to proceed with the work, or such part of the work as to which there has been a failure to pay said required wages, and (2) prosecute the work to completion by contract or otherwise, whereupon such Contractor and his sureties shall be liable to the Government for any excess costs occasioned the Government thereby.

(c) Paragraphs (a) and (b) of this clause shall apply to this contract to the extent that it is (1) a prime-contract with the Government subject to the Davis-Bacon Act or (2) a subcontract under such prime contract.

2. EIGHT-HOUR LAWS–OVERTIME COMPENSATION

No laborer or mechanic doing any part of the work contemplated by this contract, in the employ of the Contractor or any subcontractor contracting for any part of said work contemplated, shall be required or permitted to work more than eight hours in any one calendar day upon such work, except upon the condition that compensation is paid to such laborer or mechanic in accordance with the provisions of this clause. The wages of every laborer and mechanic employed by the Contractor or any subcontractor engaged in the performance of this contract shall be computed on a basic day rate of eight hours per day and work in excess of eight hours per day is permitted only upon the condition that every such laborer and mechanic shall be compensated for all hours worked in excess of eight hours per day at not less than one and one-half times the basic rate of pay. For each violation of the requirements of this clause a penalty of five dollars shall be imposed for each laborer or mechanic for every calendar day in which such employee is required or permitted to labor more than eight hours upon said work without receiving compensation computed in accordance with this clause, and all penalties thus imposed shall be withheld for the use and benefit of the Government: *Provided,* That this stipulation shall be subject in all respects to the exceptions and provisions of the Eight-Hour Laws as set forth in 40 U.S.C. 321, 324, 325, 325a, and 326, which relate to hours of labor and compensation for overtime.

3. APPRENTICES

Apprentices will be permitted to work only under a bona fide apprenticeship program registered with a State Apprenticeship Council which is recognized by the Federal Committee on Apprenticeship, U.S. Department of Labor; or if no such recognized Council exists in a State, under a program registered with the Bureau of Apprenticeship, U.S. Department of Labor.

4. PAYROLL RECORDS AND PAYROLLS

(a) Payroll records will be maintained during the course of the work and preserved for a period of three years thereafter for all laborers and mechanics working at the site of the work. Such records will contain the name and address of each such employee, his correct classification, rate of pay, daily and weekly number of hours worked, deductions made and actual wages paid. The Contractor will make his employment records available for inspection by authorized representatives of the Contracting Officer and the U.S. Department of Labor, and will permit such representatives to interview employees during working hours on the job.

(b) A certified copy of all payrolls will be submitted weekly to the Contracting Officer. The Government Prime Contractor will be responsible for the submission of certified copies of the payrolls of all subcontractors. The certification will affirm that the payrolls are correct and complete, that the wage rates contained therein are not less than the applicable rates contained in the wage determination decision of the Secretary of Labor attached to this contract, and that the classifications set forth for each laborer or mechanic conform with the work he performed.

5. COPELAND (ANTI-KICKBACK ACT)–NONREBATE OF WAGES

The regulations of the Secretary of Labor applicable to Contractors and subcontractors (29 CFR, Part 3), made pursuant to the Copeland Act, as amended (40 U.S.C. 276c) and to aid in the enforcement of the Anti-Kickback Act (18 U.S.C. 874) are made a part of this contract by reference. The Contractor will comply with these regulations and any amendments or modifications thereof and the Government Prime Contractor will be responsible for the submission of statements required of subcontractors thereunder. The foregoing shall apply except as the Secretary of Labor may specifically provide for reasonable limitations, variations, tolerances, and exemptions.

6. WITHHOLDING OF FUNDS TO ASSURE WAGE PAYMENT

There may be withheld from the Contractor so much of the accrued payments or advances as may be considered necessary to pay laborers and mechanics employed by the Contractor or any subcontractor the full amount of wages required by this contract. In the event of failure to pay any laborer or mechanic all or part of the wages required by this contract, the Contracting Officer may take such action as may be necessary to cause the suspension, until such violations have ceased, of any further payment, advance, or guarantee of funds to or for the Government Prime Contractor.

7. SUBCONTRACTS–TERMINATION

The Contractor agrees to insert Clauses 1 through 7 hereof in all subcontracts and further agrees that a breach of any of the requirements of these clauses may be grounds for termination of this contract. The term "Contractor" as used in such clauses in any subcontract shall be deemed to refer to the subcontractor except in the phrase "Government Prime Contractor."

* U.S. GOVERNMENT PRINTING OFFICE: 1959 OF—512724

Standard form 19A

STANDARD FORM 20
REVISED MARCH 1953
GENERAL SERVICES ADMINISTRATION
GENERAL REGULATION NO. 13

REFERENCE

INVITATION FOR BIDS
(CONSTRUCTION CONTRACT)

DATE

NAME AND LOCATION OF PROJECT

DEPARTMENT OR AGENCY

BY (*Issuing office*)

Sealed bids in for furnishing all labor, equipment, and materials and performing all work for the project described herein will be received until

in

and then publicly opened.

Information regarding bidding material, bid guarantee, and bonds

Description of work

Information regarding liquidated damages (if any), payments, etc., is attached or made a part of the specifications. Bids shall be submitted on the forms furnished or copies thereof.

U. S. GOVERNMENT PRINTING OFFICE 16—58475-2

Standard form 20

STANDARD FORM 21
JANUARY 1959 EDITION
GENERAL SERVICES ADMINISTRATION
FED. PROC. REG. (41 CFR) 1–16.401
21–104

BID FORM
(CONSTRUCTION CONTRACT)

REFERENCE

Read the Instructions to Bidders (Standard Form 22)
This form to be submitted in

DATE OF INVITATION

NAME AND LOCATION OF PROJECT

(Date)

TO:

In compliance with your invitation for bids of the above date, the undersigned hereby proposes to furnish all labor, equipment, and materials and perform all work for

at

in strict accordance with the specifications, schedules, drawings, and conditions for the consideration of the following amount(s)

and agrees that, upon written acceptance of this bid, mailed, or otherwise furnished, within calendar days (calendar days unless a shorter period be inserted by the bidder) after the date of opening of bids, he will within calendar days (unless a longer period is allowed) after receipt of the prescribed forms, execute Standard Form 23, Construction Contract, and give performance bond and payment bond on Government standard forms, if these forms are required, with good and sufficient surety or sureties.

(Continue on other side)

Standard form 21 (face)

The undersigned agrees that if awarded the contract, he will commence the work after the date of receipt of notice to proceed, and that he will complete the work within calendar days after the date of receipt of notice to proceed.

The undersigned acknowledges receipt of the following addenda to the drawings and/or specifications (*Give number and date of each*):

The undersigned represents (*Check appropriate boxes*):

(1) That he ☐ is, ☐ is not, a small business concern. For this purpose, a small business concern is one that (a) is independently owned and operated, (b) is not dominant in its field of operation, and (c) with affiliates, had average annual receipts for the preceding three years of $5,000,000.00 or less.. (See Code of Federal Regulations, Title 13, Part 121, as amended, for additional information.)

(2) (a) That he ☐ has, ☐ has not, employed or retained any company or person (*other than a full-time bona fide employee working solely for the bidder*) to solicit or secure this contract; and

(b) That he ☐ has, ☐ has not, paid or agreed to pay any company or person (*other than a full-time bona fide employee working solely for the bidder*) any fee, commission, percentage or brokerage fee contingent upon or resulting from the award of this contract, and agrees to furnish information relating thereto as requested by the Contracting Officer. (*For interpretation of this representation, including the term "bona fide employee," see Code of Federal Regulations, Title 44, Chapter I, Part 150.*)

(3) That he operates as an ☐ individual, ☐ partnership, ☐ corporation, incorporated in State of __________

Enclosed is bid guarantee, consisting of

in the amount of

NAME OF FIRM OR INDIVIDUAL (*Type or print*)	FULL NAME OF ALL PARTNERS (*Type or print*)
BUSINESS ADDRESS (*Type or print*)	
BY (*Signature in ink. Type or print name under signature*)	
TITLE (*Type or print*)	

DIRECTIONS FOR SUBMITTING BIDS

Envelopes containing bids, guarantee, etc., must be sealed, marked, and addressed as follows:

CAUTION: ***Do not include in the envelope any bids for other work.***
Bids should not be qualified by exceptions to the bidding conditions.

21–104

* U.S. GOVERNMENT PRINTING OFFICE : 1959 O—518313

Standard form 21 (reverse)

STANDARD FORM 22
MARCH 1953 EDITION
GENERAL SERVICES ADMINISTRATION
FED. PROC. REG. (41 CFR) 1-16.401
22-102

INSTRUCTIONS TO BIDDERS

(CONSTRUCTION CONTRACTS)

(These instructions are not to be incorporated in the contract)

1. *Explanation to Bidders.* Any explanation desired by bidders regarding the meaning or interpretation of the drawings and specifications must be requested in writing and with sufficient time allowed for a reply to reach them before the submission of their bids. Oral explanations or instructions given before the award of the contract will not be binding. Any interpretation made will be in the form of an addendum to the specifications or drawings and will be furnished to all bidders and its receipt by the bidder shall be acknowledged.

2. *Conditions at Site of Work.* Bidders should visit the site to ascertain pertinent local conditions readily determined by inspection and inquiry, such as the location, accessibility and general character of the site, labor conditions, the character and extent of existing work within or adjacent thereto, and any other work being performed thereon.

3. *Bidder's Qualifications.* Before a bid is considered for award, the bidder may be requested by the Government to submit a statement of facts in detail as to his previous experience in performing similar or comparable work, and of his business and technical organization and financial resources and plant available and to be used in performing the contemplated work.

4. *Bid Guaranty.* Where security is required, failure to submit the same with the bid may be cause for rejection. The bidder, at his option, may furnish a bid bond, postal money order, certified check, or cashier's check, or may deposit, in accordance with Treasury Department regulations, bonds or notes of the United States (at par value) as security in the amount required: *Provided,* That where the total amount of the bid is $2,000 or less, the contracting agency may declare a bid bond unacceptable by so stating in the specifications or Invitation for Bids.

In case security is in the form of postal money order, certified check, cashier's check, or bonds or notes of the United States, the Government may make such disposition of the same as will accomplish the purpose for which submitted. Checks may be held uncollected at the bidder's risk. Checks, or the amounts thereof, and bonds or notes of the United States deposited by unsuccessful bidders will be returned as soon as practicable after the opening.

5. *Preparation of Bids.* (*a*) Bids shall be submitted on the forms furnished, or copies thereof, and must be manually signed. If erasures or other changes appear on the forms, each such erasure or change must be initialed by the person signing the bid.

(*b*) The form of bid will provide for quotation of a price, or prices, for one or more items which may be lump sum bids, alternate prices, scheduled items resulting in a bid on a unit of construction or a combination thereof, etc. Where required on the bid form, bidders must quote on all items and *they are warned* that failure to do so may disqualify the bid. When quotations on all items are not required, bidders should insert the words "no bid" in the space provided for any item on which no quotation is made.

(*c*) Alternative bids will not be considered unless called for.

(*d*) Unless specifically called for, telegraphic bids will not be considered. Modification by telegraph of bids already submitted will be considered if received prior to the time fixed in the Invitation for Bids. Telegraphic modifications shall not reveal the amount of the original or revised bid.

6. *Submission of Bids.* Bids must be submitted as directed on the bid form.

7. *Receipt and Opening of Bids.* (*a*) Bids will be submitted prior to the time fixed in the Invitation for Bids. Bids received after the time so fixed are late bids; and the exact date and hour of mailing such bids, as shown by the cancellation stamp or by the stamp of an approved metering device will be recorded. Such late bids will be considered, *Provided,* They are received before the award has been made, *And provided further,* The failure to arrive on time was due solely to a delay in the mails for which the bidder was not responsible; otherwise late bids will not be con-

Standard form 22 (face)

sidered but will be held unopened until the time of award and then returned to the bidder, unless other disposition is requested or agreed to by the bidder.

(*b*) Subject to the provisions of paragraph 5(*d*) of these instructions, bids or bid modifications which were deposited for transmission by telegraph in time for receipt, by normal transmission procedure, prior to the time fixed in the Invitation for Bids and subsequently delayed by the telegraph company through no fault or neglect on the part of the bidder, will be considered if received prior to the award of the contract. The burden of proof of such abnormal delay will be upon the bidder and the decision as to whether or not the delay was so caused will rest with the officer awarding the contract.

(*c*) No responsibility will attach to any officer for the premature opening of, or the failure to open, a bid not properly addressed and identified.

8. *Withdrawals of Bids.* Bids may be withdrawn on written or telegraphic request received from bidders prior to the time fixed for opening. Negligence on the part of the bidder in preparing the bid confers no right for the withdrawal of the bid after it has been opened.

9. *Bidders Present.* At the time fixed for the opening of bids, their contents will be made public for the information of bidders and others properly interested, who may be present either in person or by representative.

10. *Bidders Interested in More than One Bid.* If more than one bid be offered by any one party, by or in the name of his or their clerk, partner, or other person, all such bids will be rejected. A party who has quoted prices to a bidder is not thereby disqualified from quoting prices to other bidders or from submitting a bid directly for the work.

11. *Award of Contract.* (*a*) The contract will be awarded as soon as practicable to the lowest responsible bidder, price and other factors considered, provided his bid is reasonable and it is to the interest of the Government to accept it.

(*b*) The Government reserves the right to waive any informality in bids received when such waiver is in the interest of the Government. In case of error in the extension of prices, the unit price will govern.

(*c*) The Government further reserves the right to accept or reject any or all items of any bid, unless the bidder qualifies such bid by specific limitation; also to make an award to the bidder whose aggregate bid on any combination of bid items is low.

12. *Rejection of Bids.* The Government reserves the right to reject any and all bids when such rejection is in the interest of the Government; to reject the bid of a bidder who has previously failed to perform properly or complete on time contracts of a similar nature; and to reject the bid of a bidder who is not, in the opinion of the Contracting Officer, in a position to perform the contract.

13. *Contract and Bonds.* The bidder to whom award is made shall, within the time established in the bid and when required, enter into a written contract with the Government and furnish performance and payment bonds on Government Standard Forms. The bonds shall be in the amounts indicated in the specifications or the Invitation for Bids.

U.S. GOVERNMENT PRINTING OFFICE: 1960—O-517509

Standard form 22 (reverse)

STANDARD FORM 24
NOVEMBER 1950 EDITION
GENERAL SERVICES ADMINISTRATION
FED. PROC. REG. (41 CFR) 1-16.801
24-102

BID BOND

(See Instructions on Reverse)

DATE BOND EXECUTED

PRINCIPAL

SURETY

PENAL SUM OF BOND *(express in words and figures)*

DATE OF BID

KNOW ALL MEN BY THESE PRESENTS, That we, the PRINCIPAL and SURETY above named, are held and firmly bound unto the United States of America, hereinafter called the Government, in the penal sum of the amount stated above, for the payment of which sum well and truly to be made, we bind ourselves, our heirs, executors, administrators, and successors, jointly and severally, firmly by these presents.

THE CONDITION OF THIS OBLIGATION IS SUCH, that whereas the principal has submitted the accompanying bid, dated as shown above, for

NOW THEREFORE, if the principal shall not withdraw said bid within the period specified therein after the opening of the same, or, if no period be specified, within sixty (60) days after said opening, and shall within the period specified therefor, or, if no period be specified, within ten (10) days after the prescribed forms are presented to him for signature, execute such further contractual documents, if any, as may be required by the terms of the bid as accepted, and give bonds with good and sufficient surety or sureties, as may be required, for the faithful performance and proper fulfillment of the resulting contract, and for the protection of all persons supplying labor and material in the prosecution of the work provided for in such contract, or in the event of the withdrawal of said bid within the period specified, or the failure to enter into such contract and give such bonds within the time specified, if the principal shall pay the Government the difference between the amount specified in said bid and the amount for which the Government may procure the required work, supplies, and services, if the latter amount be in excess of the former, then the above obligation shall be void and of no effect, otherwise to remain in full force and virtue.

IN WITNESS WHEREOF, the above-bounden parties have executed this instrument under their several seals on the date indicated above, the name and corporate seal of each corporate party being hereto affixed and these presents duly signed by its undersigned representative, pursuant to authority of its governing body.

In Presence of:

	WITNESS		INDIVIDUAL PRINCIPAL	
1.	....	as to	....	[SEAL]
2.	....	as to	....	[SEAL]
3.	....	as to	....	[SEAL]
4.	....	as to	....	[SEAL]

	WITNESS		INDIVIDUAL SURETY	
1.	....	as to	....	[SEAL]
2.	....	as to	....	[SEAL]

Attest	CORPORATE PRINCIPAL	
	BUSINESS ADDRESS	
	BY	AFFIX CORPORATE SEAL
	TITLE	

Attest:	CORPORATE SURETY	
	BUSINESS ADDRESS	
	BY	AFFIX CORPORATE SEAL
	TITLE	

STANDARD FORM 24
NOVEMBER 1950 EDITION

Standard form 24 (face)

The rate of premium on this bond is .... per thousand.

Total amount of premium charged, $..............................

(The above must be filled in by corporate surety)

CERTIFICATE AS TO CORPORATE PRINCIPAL

I, ..., certify that I am the .... secretary of the corporation named as principal in the within bond; that ...., who signed the said bond on behalf of the principal, was then .... of said corporation; that I know his signature, and his signature thereto is genuine; and that said bond was duly signed, sealed, and attested for and in behalf of said corporation by authority of its governing body.

.... [CORPORATE SEAL]

INSTRUCTIONS

1. This form shall be used for construction work or the furnishing of supplies or services whenever a bid bond is required. There shall be no deviation from this form except as authorized by the General Services Administration.

2. The surety on the bond may be any corporation authorized by the Secretary of the Treasury to act as surety, or two responsible individual sureties. Where individual sureties are used, this bond must be accompanied by a completed Affidavit of Individual Surety for each individual surety (Standard Form 28).

3. The name, including full Christian name, and business or residence address of each individual party to the bond shall be inserted in the space provided therefor, and each such party shall sign the bond with his usual signature on the line opposite the scroll seal, and if signed in Maine or New Hampshire, an adhesive seal shall be affixed opposite the signature.

4. If the principals are partners, their individual names shall appear in the space provided therefor, with the recital that they are partners composing a firm, naming it, and all the members of the firm shall execute the bond as individuals.

5. If the principal or surety is a corporation, the name of the State in which incorporated shall be inserted in the space provided therefor, and said instrument shall be executed and attested under the corporate seal as indicated in the form. If the corporation has no corporate seal the fact shall be stated, in which case a scroll or adhesive seal shall appear following the corporate name.

6. The official character and authority of the person or persons executing the bond for the principal, if a corporation, shall be certified by the secretary or assistant secretary, according to the form herein provided. In lieu of such certificate there may be attached to the bond copies of so much of the records of the corporation as will show the official character and authority of the officer signing, duly certified by the secretary or assistant secretary, under the corporate seal, to be true copies.

7. The date of this bond must not be prior to the date of the instrument in connection with which it is given.

U.S. GOVERNMENT PRINTING OFFICE : 1959—O-529754

Standard form 24 (reverse)

Standard Form 25
November 1950 Edition
General Services Administration
Fed. Proc. Reg. (41 CFR) 1-16.801
25-102

PERFORMANCE BOND

(See Instructions on Reverse)

DATE BOND EXECUTED

PRINCIPAL

SURETY

PENAL SUM OF BOND *(express in words and figures)*	CONTRACT NO.	DATE OF CONTRACT

KNOW ALL MEN BY THESE PRESENTS, That we, the PRINCIPAL and SURETY above named, are held and firmly bound unto the United States of America, hereinafter called the Government, in the penal sum of the amount stated above, for the payment of which sum well and truly to be made, we bind ourselves, our heirs, executors, administrators, and successors, jointly and severally, firmly by these presents.

THE CONDITION OF THIS OBLIGATION IS SUCH, that whereas the principal entered into a certain contract with the Government, numbered and dated as shown above and hereto attached;

NOW THEREFORE, if the principal shall well and truly perform and fulfill all the undertakings, covenants, terms, conditions, and agreements of said contract during the original term of said contract and any extensions thereof that may be granted by the Government, with or without notice to the surety, and during the life of any guaranty required under the contract, and shall also well and truly perform and fulfill all the undertakings, covenants, terms, conditions, and agreements of any and all duly authorized modifications of said contract that may hereafter be made, notice of which modifications to the surety being hereby waived, then, this obligation to be void; otherwise to remain in full force and virtue.

IN WITNESS WHEREOF, the above-bounden parties have executed this instrument under their several seals on the date indicated above, the name and corporate seal of each corporate party being hereto affixed and these presents duly signed by its undersigned representative, pursuant to authority of its governing body.

In Presence of:

WITNESS		INDIVIDUAL PRINCIPAL	
1.	as to	....	[SEAL]
2.	as to	....	[SEAL]
3.	as to	....	[SEAL]
4.	as to	....	[SEAL]

WITNESS		INDIVIDUAL SURETY	
1.	as to	....	[SEAL]
2.	as to	....	[SEAL]

Attest:	CORPORATE PRINCIPAL	
	BUSINESS ADDRESS	
	BY	AFFIX CORPORATE SEAL
	TITLE	

Attest:	CORPORATE SURETY	
	BUSINESS ADDRESS	
	BY	AFFIX CORPORATE SEAL
	TITLE	

STANDARD FORM 25
NOVEMBER 1950 EDITION

Standard form 25 (face)

The rate of premium on this bond is .... per thousand.

Total amount of premium charged, $......................................

(The above must be filled in by corporate surety)

CERTIFICATE AS TO CORPORATE PRINCIPAL

I, ..., certify that I am the .... secretary of the corporation named as principal in the within bond; that ...., who signed the said bond on behalf of the principal, was then .... of said corporation; that I know his signature, and his signature thereto is genuine; and that said bond was duly signed, sealed, and attested for and in behalf of said corporation by authority of its governing body.

.... [CORPORATE SEAL]

INSTRUCTIONS

1. This form shall be used for construction work or the furnishing of supplies or services, whenever a performance bond is required. There shall be no deviation from this form except as authorized by the General Services Administration.

2. The surety on the bond may be any corporation authorized by the Secretary of the Treasury to act as surety, or two responsible individual sureties. Where individual sureties are used, this bond must be accompanied by a completed Affidavit of Individual Surety for each individual surety (Standard Form 28).

3. The name, including full Christian name, and business or residence address of each individual party to the bond shall be inserted in the space provided therefor, and each such party shall sign the bond with his usual signature on the line opposite the scroll seal, and if signed in Maine or New Hampshire, an adhesive seal shall be affixed opposite the signature.

4. If the principals are partners, their individual names shall appear in the space provided therefor, with the recital that they are partners composing a firm, naming it, and all the members of the firm shall execute the bond as individuals.

5. If the principal or surety is a corporation, the name of the State in which incorporated shall be inserted in the space provided therefor, and said instrument shall be executed and attested under the corporate seal as indicated in the form. If the corporation has no corporate seal the fact shall be stated, in which case a scroll or adhesive seal shall appear following the corporate name.

6. The official character and authority of the person or persons executing the bond for the principal, if a corporation, shall be certified by the secretary or assistant secretary, according to the form herein provided. In lieu of such certificate there may be attached to the bond copies of so much of the records of the corporation as will show the official character and authority of the officer signing, duly certified by the secretary or assistant secretary, under the corporate seal, to be true copies.

7. The date of this bond must not be prior to the date of the instrument in connection with which it is given.

☆ U.S. GOVERNMENT PRINTING OFFICE: 1959 O—512918

Standard form 25 (reverse)

STANDARD FORM 25A
NOVEMBER 1950 EDITION
GENERAL SERVICES ADMINISTRATION
FED. PROC. REG. (41 CFR) 1-16.801
25-202

PAYMENT BOND
(See Instructions on Reverse)

DATE BOND EXECUTED

PRINCIPAL

SURETY

PENAL SUM OF BOND *(express in words and figures)*	CONTRACT NO.	DATE OF CONTRACT

KNOW ALL MEN BY THESE PRESENTS, That we, the PRINCIPAL and SURETY above named, are held and firmly bound unto the United States of America, hereinafter called the Government, in the penal sum of the amount stated above, for the payment of which sum well and truly to be made, we bind ourselves, our heirs, executors, administrators, and successors, jointly and severally, firmly by these presents.

THE CONDITION OF THIS OBLIGATION IS SUCH, that whereas the principal entered into a certain contract with the Government, numbered and dated as shown above and hereto attached;

NOW THEREFORE, if the principal shall promptly make payment to all persons supplying labor and material in the prosecution of the work provided for in said contract, and any and all duly authorized modifications of said contract that may hereafter be made, notice of which modifications to the surety being hereby waived, then this obligation to be void; otherwise to remain in full force and virtue.

IN WITNESS WHEREOF, the above-bounden parties have executed this instrument under their several seals on the date indicated above, the name and corporate seal of each corporate party being hereto affixed and these presents duly signed by its undersigned representative, pursuant to authority of its governing body.

In Presence of:

WITNESS		INDIVIDUAL PRINCIPAL	
1.	as to	....	[SEAL]
2.	as to	....	[SEAL]
3.	as to	....	[SEAL]
4.	as to	....	[SEAL]

WITNESS		INDIVIDUAL SURETY	
1.	as to	....	[SEAL]
2.	as to	....	[SEAL]

Attest:	CORPORATE PRINCIPAL	
	BUSINESS ADDRESS	
	BY	AFFIX CORPORATE SEAL
	TITLE	

Attest:	CORPORATE SURETY	
	BUSINESS ADDRESS	
	BY	AFFIX CORPORATE SEAL
	TITLE	

STANDARD FORM 25A
NOVEMBER 1950 EDITION

16—49803-2

Standard form 25A (face)

The rate of premium on this bond is .... per thousand.

Total amount of premium charged, $........................

(The above must be filled in by corporate surety)

CERTIFICATE AS TO CORPORATE PRINCIPAL

I, ..., certify that I am the .... secretary of the corporation named as principal in the within bond; that ...., who signed the said bond on behalf of the principal, was then .... of said corporation; that I know his signature, and his signature thereto is genuine; and that said bond was duly signed, sealed, and attested for and in behalf of said corporation by authority of its governing body.

.... [CORPORATE SEAL]

INSTRUCTIONS

1. This form, for the protection of persons supplying labor and material, shall be used whenever a payment bond is required under the act of August 24, 1935, 49 Stat. 793, as amended (40 U. S. C. 270a–270e). It may also be used in any other case in which a payment bond is to be required. There shall be no deviation from this form except as authorized by the General Services Administration.

2. The surety on the bond may be any corporation authorized by the Secretary of the Treasury to act as surety, or two responsible individual sureties. Where individual sureties are used, this bond must be accompanied by a completed Affidavit of Individual Surety for each individual surety (Standard Form 28).

3. The name, including full Christian name, and business or residence address of each individual party to the bond shall be inserted in the space provided therefor, and each such party shall sign the bond with his usual signature on the line opposite the scroll seal, and if signed in Maine or New Hampshire, an adhesive seal shall be affixed opposite the signature.

4. If the principals are partners, their individual names shall appear in the space provided therefor, with the recital that they are partners composing a firm, naming it, and all the members of the firm shall execute the bond as individuals.

5. If the principal or surety is a corporation, the name of the State in which incorporated shall be inserted in the space provided therefor, and said instrument shall be executed and attested under the corporate seal as indicated in the form. If the corporation has no corporate seal the fact shall be stated, in which case a scroll or adhesive seal shall appear following the corporate name.

6. The official character and authority of the person or persons executing the bond for the principal, if a corporation, shall be certified by the secretary or assistant secretary, according to the form herein provided. In lieu of such certificate there may be attached to the bond copies of so much of the records of the corporation as will show the official character and authority of the officer signing, duly certified by the secretary or assistant secretary, under the corporate seal, to be true copies.

7. The date of this bond must not be prior to the date of the instrument in connection with which it is given.

U.S. GOVERNMENT PRINTING OFFICE: 1959—O-513801

Standard form 25A (reverse)

STANDARD FORM 27
NOVEMBER 1950
PRESCRIBED BY GENERAL SERVICES ADMINISTRATION
GENERAL REGULATION NO. 5

PERFORMANCE BOND
CORPORATE CO-SURETY FORM
(See Instructions on Last Page)

DATE BOND EXECUTED

PRINCIPAL

PENAL SUM OF BOND (*express in words and figures*)	CONTRACT NO.	DATE OF CONTRACT

KNOW ALL MEN BY THESE PRESENTS, That we, the PRINCIPAL above named, and the Corporations hereinafter designated as SURETY A to SURETY , inclusive, as SURETIES, are held and firmly bound unto the United States of America, hereinafter called the Government, in the penal sum of the amount stated above, for the payment of which sum well and truly to be made, we bind ourselves, our heirs, executors, administrators, and successors, jointly and severally, firmly by these presents: *Provided,* That we the sureties bind ourselves in such sum "jointly and severally" as well as "severally" only for the purpose of allowing a joint action or actions against any or all of us, and for all other purposes each surety binds itself, jointly and severally with the principal, for the payment of such sum only as is set forth opposite its name in the following schedule:

SURETY	NAME AND STATE OF INCORPORATION	LIMIT OF LIABILITY
A		
B		
C		
D		
E		
F		
G		
H		
I		
J		
K		

STANDARD FORM 27
NOVEMBER 1950 EDITION

16—47533-2

Standard form 27

THE CONDITION OF THIS OBLIGATION IS SUCH, that whereas the principal entered into a certain contract with the Government, numbered and dated as shown above and hereto attached;

NOW THEREFORE, if the principal shall promptly make payment to all persons supplying labor and material in the prosecution of the work provided for in said contract, and any and all duly authorized modifications of said contract that may hereafter be made, notice of which modifications to the sureties being hereby waived, then this obligation to be void; otherwise to remain in full force and virtue.

IN WITNESS WHEREOF, the above-bounden parties have executed this instrument under their several seals on the date indicated above, the name and corporate seal of each corporate party being hereto affixed and these presents duly signed by its undersigned representative, pursuant to authority of its governing body.

In Presence of:

WITNESS — INDIVIDUAL PRINCIPAL

1. .... as to .... [SEAL]

2. .... as to .... [SEAL]

3. .... as to .... [SEAL]

4. .... as to .... [SEAL]

Attest:	CORPORATE PRINCIPAL BUSINESS ADDRESS BY TITLE	AFFIX CORPORATE SEAL
Attest:	CORPORATE SURETY TITLE	AFFIX CORPORATE SEAL
Attest:	CORPORATE SURETY BY TITLE	AFFIX CORPORATE SEAL
Attest:	CORPORATE SURETY BY TITLE	AFFIX CORPORATE SEAL

2 16—63284-1

Standard form 27 (cont.)

Attest:	CORPORATE SURETY BY TITLE	AFFIX CORPORATE SEAL
Attest:	CORPORATE SURETY BY TITLE	AFFIX CORPORATE SEAL
Attest:	CORPORATE SURETY BY TITLE	AFFIX CORPORATE SEAL
Attest:	CORPORATE SURETY BY TITLE	AFFIX CORPORATE SEAL
Attest:	CORPORATE SURETY BY TITLE	AFFIX CORPORATE SEAL
Attest:	CORPORATE SURETY BY TITLE	AFFIX CORPORATE SEAL
Attest:	CORPORATE SURETY BY TITLE	AFFIX CORPORATE SEAL
Attest:	CORPORATE SURETY BY TITLE	AFFIX CORPORATE SEAL

3 16—47882-3

Standard form 27 (cont.)

The rate of premium on this bond is .... per thousand.

Total amount of premium charged, $...

(The above must be filled in by corporate surety)

CERTIFICATE AS TO CORPORATE PRINCIPAL

I, ..., certify that I am the .... secretary

of the corporation named as principal in the within bond; that ....,

who signed the said bond on behalf of the principal, was then .... of said corporation; that I know his signature, and his signature thereto is genuine; and that said bond was duly signed, sealed, and attested for and in behalf of said corporation by authority of its governing body.

.... [CORPORATE SEAL]

INSTRUCTIONS

1. This form shall be used for construction work or the furnishing of supplies or services whenever a performance bond is required and two or more corporate sureties authorized by the Secretary of the Treasury to act as surety, each limiting its liability to a portion only of the total penalty, are to execute the bond. There shall be no deviation from this form except as authorized by the General Services Administration.

2. The letter, such as B, F, or K, applicable to the last surety listed in the schedule, must be inserted in the space indicated in the body of the bond. In cases involving only two sureties the word "to" following "Surety A" should be changed to "and" by interlineation, and the word "inclusive" should be stricken. When more than 11 sureties are to execute the bond, attach Standard Form 27B, Continuation Sheet, and identify sureties beginning with "L" on the first sheet.

3. The name, including full Christian name, and residence of each individual party to the bond shall be inserted in the space provided therefor, and each such party shall sign the bond with his usual signature on the line opposite the scroll seal, and if signed in Maine or New Hampshire, an adhesive seal shall be affixed opposite the signature.

4. If the principals are partners, their individual names shall appear in the space provided therefor, with the recital that they are partners composing a firm, naming it, and all the members of the firm shall execute the bond as individuals.

5. The name of the State in which each surety is incorporated and the State of incorporation of the principal, if a corporation, shall be inserted in the appropriate place in the bond, and said instrument shall be executed and attested under the corporate seal as indicated in the form. If the corporation has no corporate seal the fact shall be stated, in which case a scroll or adhesive seal shall appear following the corporate name.

6. The official character and authority of the person or persons executing the bond for the principal, if a corporation, shall be certified by the secretary or assistant secretary, according to the form set forth above. In lieu of such certificate there may be attached to the bond copies of so much of the records of the corporation as will show the official character and authority of the officer signing, duly certified by the secretary or assistant secretary, under the corporate seal, to be true copies.

7. The date of this bond must not be prior to the date of the instrument in connection with which it is given.

☆ U. S. GOVERNMENT PRINTING OFFICE : 1952—O-211298 16—47532-2

4

Standard form 27 (cont.)

STANDARD FORM 27A
NOVEMBER 1950
PRESCRIBED BY GENERAL
SERVICES ADMINISTRATION
GENERAL REGULATION NO. 5

PAYMENT BOND

CORPORATE CO-SURETY FORM

(See Instructions on Last Page)

DATE BOND EXECUTED

PRINCIPAL

PENAL SUM OF BOND (*express in words and figures*)	CONTRACT NO.	DATE OF CONTRACT

KNOW ALL MEN BY THESE PRESENTS, That we, the PRINCIPAL above named, and the Corporations hereinafter designated as SURETY A to SURETY , inclusive, as SURETIES, are held and firmly bound unto the United States of America, hereinafter called the Government, in the penal sum of the amount stated above, for the payment of which sum well and truly to be made, we bind ourselves, our heirs, executors, administrators, and successors, jointly and severally, firmly by these presents: *Provided*, That we the sureties bind ourselves in such sum "jointly and severally" as well as "severally" only for the purpose of allowing a joint action or actions against any or all of us, and for all other purposes each surety binds itself, jointly and severally with the principal, for the payment of such sum only as is set forth opposite its name in the following schedule:

SURETY	NAME AND STATE OF INCORPORATION	LIMIT OF LIABILITY
A		
B		
C		
D		
E		
F		
G		
H		
I		
J		
K		

STANDARD FORM 27A
NOVEMBER 1950 EDITION

16—63284-1

Standard form 27A

THE CONDITION OF THIS OBLIGATION IS SUCH, that whereas the principal entered into a certain contract with the Government, numbered and dated as shown above and hereto attached;

NOW THEREFORE, if the principal shall well and truly perform and fulfill all the undertakings, covenants, terms, conditions, and agreements of said contract during the original term of said contract and any extensions thereof that may be granted by the Government, with or without notice to the sureties, and during the life of any guaranty required under the contract, and shall also well and truly perform and fulfill all the undertakings, covenants, terms, conditions, and agreements of any and all duly authorized modifications of said contract, that may hereafter be made, notice of which modifications to the sureties being hereby waived, then, this obligation to be void; otherwise to remain in full force and virtue.

IN WITNESS WHEREOF, the above-bounden parties have executed this instrument under their several seals on the date indicated above, the name and corporate seal of each corporate party being hereto affixed and these presents duly signed by its undersigned representative, pursuant to authority of its governing body.

In Presence of:

WITNESS		INDIVIDUAL PRINCIPAL	
1.	as to	....	[SEAL]
2.	as to	....	[SEAL]
3.	as to	....	[SEAL]
4.	as to	....	[SEAL]

Attest:	CORPORATE PRINCIPAL BUSINESS ADDRESS BY TITLE	AFFIX CORPORATE SEAL
Attest:	CORPORATE SURETY TITLE	AFFIX CORPORATE SEAL
Attest:	CORPORATE SURETY BY TITLE	AFFIX CORPORATE SEAL
Attest:	CORPORATE SURETY BY TITLE	AFFIX CORPORATE SEAL

2 16—47208-2

Standard form 27A (cont.)

Attest:	CORPORATE SURETY	
	BY	AFFIX CORPORATE SEAL
	TITLE	

Attest:	CORPORATE SURETY	
	BY	AFFIX CORPORATE SEAL
	TITLE	

Attest:	CORPORATE SURETY	
	BY	AFFIX CORPORATE SEAL
	TITLE	

Attest:	CORPORATE SURETY	
	BY	AFFIX CORPORATE SEAL
	TITLE	

Attest:	CORPORATE SURETY	
	BY	AFFIX CORPORATE SEAL
	TITLE	

Attest:	CORPORATE SURETY	
	BY	AFFIX CORPORATE SEAL
	TITLE	

Attest:	CORPORATE SURETY	
	BY	AFFIX CORPORATE SEAL
	TITLE	

Attest:	CORPORATE SURETY	
	BY	AFFIX CORPORATE SEAL
	TITLE	

3 16—63284-1

Standard form 27A (cont.)

The rate of premium on this bond is .... per thousand.

Total amount of premium charged, $............................

(The above must be filled in by corporate surety)

CERTIFICATE AS TO CORPORATE PRINCIPAL

I, ..., certify that I am the .... secretary of the corporation named as principal in the within bond; that ...., who signed the said bond on behalf of the principal, was then .... of said corporation; that I know his signature, and his signature thereto is genuine; and that said bond was duly signed, sealed, and attested for and in behalf of said corporation by authority of its governing body.

.... [CORPORATE SEAL]

INSTRUCTIONS

1. This form, for the protection of persons supplying labor and material for construction work, shall be used whenever a payment bond under the act of August 24, 1935, 49 Stat.793, as amended (40 U. S. C. 270a–270e), is required and two or more corporate sureties authorized by the Secretary of the Treasury to act as surety, each limiting its liability to a portion only of the total penalty, are to execute the bond. It may also be used in any other case in which a payment bond is to be required. There shall be no deviation from this form except as authorized by the General Services Administration.

2. The letter, such as B, F, or K, applicable to the last surety listed in the schedule, must be inserted in the space indicated in the body of the bond. In cases involving only two sureties the word "to" following "Surety A" should be changed to "and" by interlineation, and the word "inclusive" should be stricken. When more than 11 sureties are to execute the bond, attach Standard Form 27B, Continuation Sheet, and identify sureties beginning with "L" on the first sheet.

3. The name, including full Christian name, and residence of each individual party to the bond shall be inserted in the space provided therefor, and each such party shall sign the bond with his usual signature on the line opposite the scroll seal, and if signed in Maine or New Hampshire, an adhesive seal shall be affixed opposite the signature.

4. If the principals are partners, their individual names shall appear in the space provided therefor, with the recital that they are partners composing a firm, naming it, and all the members of the firm shall execute the bond as individuals.

5. The name of the State in which each surety is incorporated and the State of incorporation of the principal, if a corporation, shall be inserted in the appropriate place in the bond, and said instrument shall be executed and attested under the corporate seal as indicated in the form. If the corporation has no corporate seal the fact shall be stated, in which case a scroll or adhesive seal shall appear following the corporate name.

6. The official character and authority of the person or persons executing the bond for the principal, if a corporation, shall be certified by the secretary or assistant secretary, according to the form set forth above. In lieu of such certificate there may be attached to the bond copies of so much of the records of the corporation as will show the official character and authority of the officer signing, duly certified by the secretary or assistant secretary, under the corporate seal, to be true copies.

7. The date of this bond must not be prior to the date of the instrument in connection with which it is given.

U. S. GOVERNMENT PRINTING OFFICE 16—63284-1

Standard form 27A (cont.)

STANDARD FORM 27B
NOVEMBER 1950
PRESCRIBED BY GENERAL
SERVICES ADMINISTRATION
GENERAL REGULATION NO. 5

CONTINUATION SHEET
CORPORATE CO-SURETY BOND

.... Bond of ....
("Performance"—"Payment") (Name of Principal must be filled in)

.... under Contract No. ...., dated ....
(Must be filled in)

CONTINUATION OF PAGE 1

SURETY	NAME AND STATE OF INCORPORATION	LIMIT OF LIABILITY

CONTINUATION OF PAGE 3

Attest:	CORPORATE SURETY BY TITLE	AFFIX CORPORATE SEAL
Attest:	CORPORATE SURETY BY TITLE	AFFIX CORPORATE SEAL

STANDARD FORM 27B
NOVEMBER 1950 EDITION

16—63316-1

Standard form 27B (face)

Attest:	CORPORATE SURETY	
	BY	AFFIX CORPORATE SEAL
	TITLE	

Attest:	CORPORATE SURETY	
	BY	AFFIX CORPORATE SEAL
	TITLE	

Attest:	CORPORATE SURETY	
	BY	AFFIX CORPORATE SEAL
	TITLE	

Attest:	CORPORATE SURETY	
	BY	AFFIX CORPORATE SEAL
	TITLE	

Attest:	CORPORATE SURETY	
	BY	AFFIX CORPORATE SEAL
	TITLE	

Attest:	CORPORATE SURETY	
	BY	AFFIX CORPORATE SEAL
	TITLE	

Attest:	CORPORATE SURETY	
	BY	AFFIX CORPORATE SEAL
	TITLE	

Attest:	CORPORATE SURETY	
	BY	AFFIX CORPORATE SEAL
	TITLE	

16—63316-1 U. S. GOVERNMENT PRINTING OFFICE

Standard form 27B (reverse)

STANDARD FORM 28
NOVEMBER 1950 EDITION
GENERAL SERVICES ADMINISTRATION
FED. PROC. REG. (41 CFR) 1-16.801
28-102

AFFIDAVIT OF INDIVIDUAL SURETY

(See Instructions on Reverse)

STATE OF

COUNTY OF } SS:

I, the person whose signature appears below as surety, being duly sworn, depose and say that I am one of the sureties to the attached bond; that I am a citizen of the United States, and of full age and legally competent; that I am not a partner in the business of the principal on the bond or bonds on which I appear or may appear as surety; that the information herein below furnished is true and correct. This affidavit is made to induce the United States of America to accept me as surety on the attached bond.

MY NAME *(first, middle, last)*

MY ADDRESS *(street and number, city and State)*

TYPE AND DURATION OF MY OCCUPATION

NAME OF MY EMPLOYER

MY BUSINESS ADDRESS *(street and number, city and State)*

AMOUNT I AM WORTH IN REAL ESTATE AND PERSONAL PROPERTY OVER AND ABOVE (1) ALL MY DEBTS AND LIABILITIES OWING AND INCURRED, (2) ANY PROPERTY EXEMPT FROM EXECUTION, (3) ANY PECUNIARY INTEREST I HAVE IN THE BUSINESS OF THE PRINCIPAL ON SAID BOND, AND (4) ANY INTEREST I HAVE IN ANY SO-CALLED COMMUNITY PROPERTY

LOCATION AND DESCRIPTION OF REAL ESTATE OF WHICH I AM SOLE OWNER IN FEE SIMPLE *(not exempt from seizure and sale under any homestead law, community or marriage law, or upon attachment, execution, or judicial process)*

FAIR VALUE OF SUCH REAL ESTATE

ASSESSED VALUE OF SUCH REAL ESTATE FOR TAXATION PURPOSES

ALL MORTGAGES OR OTHER ENCUMBRANCES AGAINST ABOVE REAL ESTATE, THERE BEING NO OTHERS *(if none, so state)*

MY LIABILITIES OWING AND INCURRED DO NOT EXCEED THE AMOUNT OF

AMOUNT I AM WORTH, OVER AND ABOVE JUST DEBTS AND LIABILITIES, IN PERSONAL PROPERTY SUBJECT TO EXECUTION AND SALE, THIS AMOUNT BEING ADDITIONAL TO THE REAL ESTATE ABOVE DESCRIBED

THE ABOVE PERSONAL PROPERTY CONSISTS OF THE FOLLOWING

ALL OTHER BONDS ON WHICH I AM SURETY *(state character and amount of each bond; if none, so state)*

MY SIGNATURE AS SURETY

Subscribed and sworn to before me this date at ....

OFFICIAL SEAL

.... *(Signature)* .... *(Title of official administering oath)* .... *(Date)*

STANDARD FORM 28
NOVEMBER 1950 EDITION

Standard form 28 (face)

CERTIFICATE OF SUFFICIENCY

I HEREBY CERTIFY, That the surety named herein is personally known to me; that, in my judgment, said surety is responsible, and qualified to act as such; and that, to the best of my knowledge and belief, the facts stated by said surety in the foregoing affidavit are true.

NAME (*typewritten*)	SIGNATURE

OFFICIAL TITLE

ADDRESS

INSTRUCTIONS

1. This form shall be used whenever sureties on bonds to be executed in connection with Government contracts are individual sureties. There shall be no deviation from this form except as authorized by the General Services Administration.

2. A firm, as such, will not be accepted as a surety, nor a partner for copartners or for a firm of which he is a member. Stockholders of a corporate principal may be accepted as sureties provided their qualifications as such are independent of their stock holdings therein. Sureties, if individuals, shall be citizens of the United States, except that sureties on bonds executed in any foreign country, the Canal Zone, Puerto Rico, Hawaii, Alaska, Guam, or any possession of the United States, for the performance of contracts entered into in these places, need not be citizens of the United States, but if not citizens of the United States shall be domiciled in the place where the contract is to be performed.

3. The individual surety shall justify, under oath, in a sum not less than the penalty of the bond, according to the form appearing on the face hereof, before a United States commissioner, a clerk of a United States court, a notary public, or some other officer having authority to administer oaths generally. If the officer has an official seal, it shall be affixed, otherwise the proper certificate as to his official character shall be furnished. Where citizenship is not required, as provided in paragraph 2 of these instructions, the affidavit may be amended accordingly.

4. The certificate of sufficiency shall be signed by an officer of a bank or trust company, a judge or clerk of a court of record, a United States district attorney or commissioner, a postmaster, a collector or deputy collector of internal revenue, or any other officer of the United States acceptable to the department or establishment concerned. Further certificates as to the financial qualification of the sureties may be required from time to time. Such certificates must be based on the personal investigation of the certifying officers at the time of the making thereof, and not upon prior certificates.

U.S. GOVERNMENT PRINTING OFFICE : 1959 O—527523

Standard form 28 (reverse)

Abbreviations

Ala.	Alabama Supreme Court Reports
ALR.	American Law Reports
ALR. 2d	American Law Reports, Second Series
Am. Jur.	American Jurisprudence
App. Cas. D.C.	Appeal Cases, District of Columbia
App. Div.	New York Supreme Court, Appellate Division
App. Div. 2d	New York Supreme Court, Appellate Division, Second Series
Ariz.	Arizona Reports
Ark.	Arkansas Reports
At.	Atlantic Reporter, National Reporter System
At. 2d	Atlantic Reporter, Second Series
B. and Ad.	Barnewall & Adolphus (England)
Barb.	Barbour, Supreme Court Reports, New York
C.B. N.S.	Common Bench Reports, New Series (Scott)
C.J.	Corpus Juris
C.J.S.	Corpus Juris Secundum
Cal.	California Reports
Cal. 2d	California Reports, Second Series
Cal. App.	California Appeals Reports
Colo.	Colorado Reports
Conn.	Connecticut Reports
Ct. Cl. N.Y.	Court of Claims Reports, New York
Ct. Cl. U.S.	United States Court of Claims Reports
Cush.	Cushing's Massachusetts Reports
F.	Federal Reporter

F. 2d	Federal Reporter, Second Series
F. Supp.	Federal Supplement
Fla.	Florida Reports
Ga.	Georgia Reports
Gray	Gray's Massachusetts Reports
Heisk	Heiskell (Tennessee)
How. Pr.	Howard's Practice Reports, New York
Hun	Hun's New York Supreme Court Reports, also Appellate Division, Supreme Court, New York
Idaho	Idaho Reports
Ill.	Illinois Reports
Ill. App.	Illinois Appellate Court Reports
Ind.	Indiana Reports
Ind. App.	Indiana Appellate Court Reports
Iowa	Iowa State Reports
Kan.	Kansas Reports
Ky.	Kentucky Reports
L. ed.	Lawyers Edition, Supreme Court Reports, United States
L.R. App. Cas.	English Law Reports, Appeals Cases, House of Lords
La.	Louisiana Reports
Lans.	Lansing's Chancery Decisions, New York
Mass.	Massachusetts Reports
Md.	Maryland Reports
Me.	Maine Reports
Minn.	Minnesota Reports
Misc.	Miscellaneous Reports, New York
Misc. 2d	Miscellaneous Reports, New York, Second Series
Miss.	Mississippi Reports
M. & W.	Messon & Welsby, English Exchequer Reports
Mich.	Michigan Reports
Mo.	Missouri Reports
Mo. App.	Missouri Appeal Reports
Mont.	Montana Reports
N.C.	North Carolina Reports
N.E.	Northeastern Reporter, National Reporter System
N.E. 2d	Northeastern Reporter, Second Series
N.H.	New Hampshire Reports
N.J.	New Jersey Law Reports
N.J. Eq.	New Jersey Equity Reports
N.J. Super.	New Jersey Superior Courts
N.M.	New Mexico Reports
N.W.	Northwestern Reporter, National Reporter System

N.W. 2d	Northwestern Reporter, Second Series
N.Y.	New York Court of Appeals Reports
N.Y. 2d	New York Court of Appeals Reports, Second Series
N.Y.S.	New York Supplement Reports, National Reporter System
N.Y.S. 2d	New York Supplement Reports, National Reporter System, Second Series
N.Y.S.R.	New York State Reporter
Neb.	Nebraska Reports
N.D.	North Dakota Reports
Ohio C.C.	Ohio Circuit Court Reports
Ohio St.	Ohio State Reports
Okla.	Oklahoma Reports
Oregon	Oregon Reports
P.	Pacific Reporter, National Reporter System
P. 2d	Pacific Reporter, Second Series
Pa.	Pennsylvania State Reports
Pa. Sup. Ct.	Pennsylvania Superior Court Reports
Pet.	Peters, United States Supreme Court
Q.B.	Queen's Bench
R.I.	Rhode Island Reports
S.C.	South Carolina Reports
S.E.	Southeastern Reporter, National Reporter System
S.E. 2d	Southeastern Reporter, Second Series
So.	Southern Reporter, National Reporter System
So. 2d	Southern Reporter, Second Series
S.W.	Southwestern Reporter, National Reporter System
S.W. 2d	Southwestern Reporter, Second Series
So. Dak.	South Dakota Reports
Tenn.	Tennessee Reports
Tex.	Texas Supreme Court Reports
U.S.	United States Supreme Court Reports
Ut.	Utah Reports
Ut. 2d	Utah Reports, Second Series
Va.	Virginia Reports
Vt.	Vermont Reports
Wall.	Wallace, U.S. Supreme Court Reports
Wash.	Washington State Reports
Wend.	Wendell Reports, New York
West Va.	West Virginia Reports
Wis.	Wisconsin Reports
Wyoming	Wyoming Reports

Table of Cases

Abells v. *City of Syracuse,* 7 App. Div. 501 127
Adams v. *Brenan,* 52 N.E. 314 17
Aetna L. Ins. Co. v. *Middleport,* 124 U.S. 534 199
Albert v. *City of Salem,* 164 P. 567 25
Albert & Harrison v. *United States,* 68 F. Supp. 732 246
Matter of Albro Contracting Corp. v. *Dept. of Public Works, etc.,* 13 Misc. 2d 845 84
Allanion v. *Albany,* 43 Barb. 33 238
Allen v. *Labsap,* 188 Mo. 692 15
Alpina ex rel. Beaudrie v. *Murray Co.,* 123 N.W. 1128 205
Altoona Electric Co. v. *Kittaning Street Rwy.,* 126 F. 559 227
Alvarado v. *Davis, et al.,* 6 P. 2d 121 93
Ambaum v. *State,* 141 P. 314 131
American Bridge Co. v. *State of New York,* 245 App. Div. 535 231
American Pipe Co. v. *Westchester County,* 292 F. 941 134
American Store Equipt. & Constr. Corpn. v. *Jack Dempsey's Punch Bowl, Inc.,* 283 N.Y. 601 93
American Surety Co. v. *Billingham Natl. Bank,* 254 F. 54 134
American Surety Co. of New York v. *Lawrenceville Cement Co. et al.,* 110 F. 717 205
Amoskeag Manuf. Co. v. *United States,* 17 Wall. 592 227
Ampt ex rel. Cincinnati v. *Cincinnati,* 54 N.E. 1097 57
Matter of Anderson, 109 N.Y. 554 55

Anderson v. *Fuller,* 41 So. 684 3
Anderson v. *State of New York,* 103 Misc. 388 131
Andrews v. *Ada County,* 63 P. 592 83
Andrews v. *City of Detroit,* 206 N.W. 514 61
Antenoff v. *Basso,* 78 N.Y. 2d 604 186
Arensmeyer et al. v. *Wray et al.,* 118 Misc. 619 87
Armaniaco v. *Borough of Cresskill,* 163 At. 2d 379 54
Armour & Co. v. *Rinaker,* 202 F. 901 71
Armstrong v. *Ashley,* 204 U.S. 272 246
Armstrong v. *State of New York,* 111 Misc. 297 156
W. H. Armstrong & Co., v. *United States,* 98 Ct. Cl. U.S. 519 270
Arnold v. *United States,* 280 F. 338 203
The Arundel Corporation v. *United States,* 96 Ct. Cl. U.S. 77 269
275
The Arundel Corporation v. *United States,* 103 Ct. Cl. U.S. 688 275
Atlanta v. *Stein,* 36 S.E. 932 17
Atlantic Bitulithic Co. v. *Town of Edgewood,* 137 S.E. 223 145
Atlanta Construction Co. v. *State of New York,* 103 Misc. 233 137
Attorney-General ex rel. Cook et al. v. *City of Detroit,* 26 Mich. 263 57
59, 83
Atwood v. *Burpee,* 58 At. 237 164

Bacon v. *Kennedy,* 22 N.W. 824 25
Baird v. *Mayor,* 96 N.Y. 567 15
Matter of Baitinger Elec. Co. v. *Fones,* 170 Misc. 599 29
Baker Co. v. *State of New York,* 267 App. Div. 712 234
Baltimore v. *Clark,* 128 Md. 291 126
Baltimore v. *DeLuca-Davis Const. Co.,* 124 At. 2d 557 65
Baltimore v. *Flack,* 64 At. 702 57
Baltis v. *Westchester,* 121 N.E. 2d 495 177
Bangor v. *Ridley,* 104 At. 230 9
A. W. Banko, Inc. v. *State of New York,* 186 Misc. 491 152
Barash v. *Board of Education,* 226 App. Div. 249, affirmed 255 N.Y. 587 119
M. Barash v. *State of New York,* 2 Misc. 2d 680 152
Barber Asphalt Paving Co. v. *Garr,* 73 S.W. 1106 57
Barbier v. *Connolly,* 113 U.S. 27 177
Barker Lumber Co. v. *Marathon Paper Mills Co.,* 146 Wis. 12 206
Barkley v. *Oregon City,* 33 P. 978 189
W. E. Barling v. *United States,* 126 Ct. Cl. U.S. 34 283
Barlow v. *Jones,* 87 At. 649 73

Barnard Bakeshops v. *Dirig,* 173 Misc. 862 154
Barrett Mfg. Co. v. *New Orleans,* 63 So. 505 191
Bates v. *Rogers,* 274 F. 659 119
Bates v. *Santa Barbara County,* 27 P. 438 191
Bator v. *Ford Motor Co.,* 269 Mich. 648 166
Baumann v. *West Allis,* 204 N.W. 907 203
Beals v. *Fidelity & Deposit Co.,* 76 App. Div. 526 207
Bechthold et al. v. *City of Wauwatosa et al.,* 280 N.W. 320 23
Becker v. *City of New York,* 176 N.Y. 441 79
Bedford-Carthage Stone Co. et al. v. *Ramey et al.,* 34 S.W. 2d 387 227
Bell v. *New York,* 11 N.E. 495 191
Belmar Contr. Co. v. *State of New York,* 233 N.Y. 189 138
Bennecke v. *Insurance Co.,* 105 U.S. 355 135
Bentley v. *State,* 41 N.W. 388 116, 117, 122
Bigham v. *Wabash-Pittsburgh Term. Ry.,* 72 At. 318 187
Bintz v. *City of Hornel,* 268 App. Div. 742, affirmed 295 N.Y. 628 152
Birmingham v. *Thompson,* 200 F. 2d 505 251
Bjerkeseth v. *Lysnes,* 22 P. 2d 660 131
Blackmer v. *United States,* 284 U.S. 421 291
Blair v. *United States,* 147 F. 2d 840 135
Blair v. *United States et al.,* 66 F. Supp. 405 121
Blanchard v. *Burns,* 162 S.W. 63 197
Blanton et al., v. *Town of Wallins,* 291 S.W. 372 30
Bluffton v. *Miller,* 70 N.E. 989 83
Board of Education v. *Elliott,* 276 Ky. 790 93
Board of Regents v. *Cole,* 209 Ky. 761 73
Board of School Commissioners v. *Bender,* 72 N.E. 154, 36 Ind. App. 164 70, 73
Boden v. *Maher,* 105 Wis. 539 227
Boettler v. *Tendick,* 11 S.W. 497 160
Boren v. *Darke County Comrs.,* 21 Ohio St. 311 83
Born v. *Schrenkeisen,* 110 N.Y. 55 75
Borough Const. Co. v. *City of New York,* 200 N.Y. 149 132
Boston v. *McGovern,* 292 F. 705 134
Bowen v. *Kimball,* 203 Mass. 364 185
Boyden v. *Hill,* 188 Mass. 477 37
Braaten v. *Olsen,* 148 N.W. 829 5
Bradbury v. *Nagelhus,* 319 P. 2d 503 61
Brady v. *New York,* 20 N.Y. 312 8, 80
Brainard v. *New York Central R.R. Co.,* 242 N.Y. 125 151
Brandese v. *City of Schenectady,* 194 Misc. 150, affirmed 273 App. Div. 831 and 297 N.Y. 965 35

Brawley v. *United States,* 96 U.S. 168 77
Breath v. *City of Galveston,* 49 S.W. 575 24
Brennan Const. Co. v. *State of New York,* 117 Misc. 816 230
Bridge Co. v. *Hamilton,* 110 U.S. 108 117
Bright v. *Boyd,* 1 Story 478 246
Brown v. *Houston,* 48 S.W. 760 32
Browning v. *Bergen County,* 76 At. 1054 83
R.O. Brumagin & Co., v. *City of Bloomington,* 234 Ill. 114 73
Bucholz v. *Green Bros. Co.,* 172 N.E .101 239
Burke v. *Dunbar,* 128 Mass. 499 117
Burke v. *Mayor,* 7 App. Div. 128 160
Burland Printing Co., Inc. v. *LaGuardia,* 9 N.Y.S. 2d 616 19
Burr v. *Gardelle,* 200 P. 493 216
Burt v. *Los Angeles Olive Growers' Assn.,* 175 Cal. 668 73
Byrne v. *Bellingham Consol. School Dist. #301,* 108 P. 2d 791 232

Caldwell v. *Donaghey,* 156 S.W. 839 142
California Highway Comm. v. *Riley,* 218 P. 579 130, 131
Camdenton, etc., ex rel. W. H. Powell Lumber Co. v. *New York Casualty Co.,* 104 S.W. 2d 319 195
Campbell v. *City of New York,* 244 N.Y. 317 85
Campbell Bldg. Co. v. *State Road Commission,* 70 P. 2d 857 129
Canal Bank v. *Hudson,* 111 U.S. 66 246
Cardell v. *City of Perry,* 207 N.W. 775 187
Philip Carey Co. v. *Maryland Casualty Co.,* 206 N.W. 808 195
Carr v. *Fenstermacher,* 119 Neb. 172 26
Carr v. *State,* 26 N.E. 778 138
Carr, Auditor, etc., v. *State,* 127 Ind. 204 230
Carroll v. *O'Connor,* 35 N.E. 1006 114
Carson Opera House Assn. v. *Miller,* 16 Nev. 327 217
In Matter of Carter, 21 App. Div. 118 36
Carter v. *City and County of Denver,* 160 P. 2d 991 146
Carthage Tissue Paper Mills v. *Village of Carthage,* 200 N.Y. 1 151
Case v. *Fowler,* 65 Ind. 29 22
Cauldwell-Wingate Co. v. *State of New York,* 276 N.Y. 365 112, 235 236,
Cavanaugh v. *Globe Indemnity Co.,* 44 P. 2d 216 193
Central Trust Co. v. *Richmond, N.I. & B.R. Co.,* 54 F. 723 204
Chalender v. *United States,* 127 Ct. Cl. U.S. 557 282
Champlain Constr. Co. v. *O'Brien,* 177 F. 271 227
Chapman v. *State,* 104 Cal. 690 and 38 p. 547 140, 230

Charleston v. *Southeastern Constr. Co.,* 64 S.E. 2d 676 179
Chouteau v. *United States,* 95 U.S. 61 274
Christie v. *United States,* 237 U.S. 234 109, 110, 112, 114, 117, 120, 122
Cincinnati Quarries Co. v. *Hess et al.,* 162 N.E. 686 206
City of Baltimore v. *Clack,* 64 At. 702 60
City of Baton Rouge, La. v. *Robinson,* 127 F. 2d 693 121
City of Chicago v. *Duffy,* 218 Ill. 242 and 75 N.E. 912 78, 241
City of Earlington v. *Powell,* 10 S.W. 2d 1060 188
City of Hartford v. *King,* 249 S.W. 2d 13 48
City of Hattiesburg v. *Cobb Bros. Const. Co.,* 184 So. 630 66
City of Houston v. *L. J. Fuller, Inc.,* 311 S.W. 2d 284 116
City of Indianapolis v. *Indianapolis Water Co.,* 113 N.E. 369 230
City of Knoxville v. *Melvin E. Burgess,* 175 S.W. 2d 548 195
City of New York v. *Doud Lumber Co.,* 140 App. Div. 359 70, 74
City of New York v. *Seely-Taylor Co.,* 140 App. Div. 98, affirmed 208 N.Y. 548 35
City of Newport News v. *Potter,* 122 F. 321 24
City of Salisbury v. *Lynch-McDonald Const. Co.,* 261 S.W. 356 126
Clark v. *Clark,* 164 Minn. 201 166
Clark v. *Pope,* 70 Ill. 128 117
Clark County Const. Co. v. *State Highway Comm.,* 58 S.W. 2d 388 130, 131
Clayton v. *Taylor,* 49 Mo. App. 117 81
Clement v. *Cash,* 21 N.Y. 253 34
Cleveland Fire Alarm Teleg. Co. v. *Board of Fire Comrs.,* 55 Barb. 288 81
Cohen v. *City of New York et al.,* 205 Misc. 105 47
Cole v. *Arizona Edison Co., Inc.,* 53 Ariz. 141 166
Colella v. *County of Allegheny,* 137 At. 2d 265 68
Coley v. *Cohen,* 289 N.Y. 365 166
Collender v. *Dinsmore,* 55 N.Y. 200 157
Columbus Bldg. & Constr. Trades, etc., Council v. *Moyer, etc.,* 126 N.E. 2d 429 12
Comey v. *U.S. Surety Co.,* 217 N.Y. 268 226
Commercial Casualty Ins. Co. v. *Durham County,* 128 S.E. 469 201
Commonwealth v. *Mitchell,* 82 Pa. 343 81
Comstock v. *Droney Lumber Co.,* 71 S.E. 255 141
Comstock v. *Eagle Grove City,* 111 N.W. 51 22
Conduit & Foundation Corpn. v. *Atlantic City,* 2 N.J. Super, 433 73
Connolly v. *State,* 120 Misc. 854 229
Connolly v. *Sullivan,* 53 N.E. 143 239
Constable v. *National Steamship Co.,* 154 U.S. 73 164

Continental Illinois National Bank and Trust Co., v. *United States,* 101 F. Supp. 755 266, 268
Contra Costa Constr. Co. v. *Daly City,* 192 P. 178 131
Cook v. *Dean,* 11 App. Div. 123 163
Re. Cook County, 177 N.W. 103 5
Cooke et al. v. *United States,* 91 U.S. 389 243
Coonan v. *City of Cape Girardeau,* 129 S.W. 745 36
Cornell v. *Standard Oil Co.,* 91 App. Div. 345 227
Coryell v. *Dubois Borough,* 75 At. 25 121
Costa v. *Callanan,* 15 Misc. 2d 198 167
Matter of V.J. Costanzi, Inc. v. *Brandt, Albany County, N.Y.,* Special Term, Nov. 8, 1940. 27
Coster v. *Mayor, etc.,* 43 N.Y. 399 139
Coward v. *Bayonne,* 51 At. 490 32
Crane Const. Co. v. *Commonwealth,* 195 N.E. 110 131
Crawfordville v. *Braden,* 28 N.E. 849 140
Creedon v. *Automatic Voting Machine Corp.,* 243 App. Div. 339 166
Crook Co. v. *United States,* 270 U.S. 4 229, 274, 277, 278, 282
Crouse v. *Stanley,* 154 S.E. 40 216

Daddario v. *Wilford,* 296 Mass. 92 37
Daegling v. *Gilmore,* 49 Ill. 248 117
Daneis v. *M. DeMatteo Const. Co.,* 102 F. Supp. 874 202
Danolds v. *State,* 89 N.Y. 36 145
Davenport v. *Buffington,* 97 F. 234 139
Davidson County v. *Harmon,* 292 S.W. 2d 777 179
Davin v. *City of Syracuse,* 69 Misc. 285, affirmed 145 App. Div. 904 34
Davis v. *Patrick,* 122 U.S. 138 164
Davison-Nicholson Co. v. *Pound,* 94 S.E. 500 32
Day v. *United States,* 245 U.S. 159 117, 122
Matter of Delaware Co. Elec. Corp. v. *City of New York,* 257 App. Div., 526, affirmed 304 N.Y. 196 152, 152
Denton v. *Carey-Reed Co., et al.,* 183 S.W. 262 30
DeRiso Bros. v. *State,* 161 Misc. 934 239
Dermott v. *Jones,* 2 Wall. 1 117, 122
Detroit v. *Hosmer,* 79 Mich. 384 83
Detroit v. *Osborne,* 135 U.S. 492 134
Dever v. *Corwall,* 86 N.W. 227 25
Devin v. *Belt,* 17 At. 375 60
Matter of Dictaphone Corp. v. *O'Leary et al.,* 287 N.Y. 491 30
Dillingham v. *Mayor, etc. of City of Spartanburg,* 56 S.E. 381 8, 13

District of Columbia v. *Camden Iron Works,* 15 App. Cas. D.C. 198 — 227
Dixey v. *Atl. City & D. River Quarry & Constr. Co.,* 58 At. 370 — 57
Dodd v. *Churton,* 12 Q.B. 562 — 227
Dolman v. *Board of Comrs. etc.,* 226 P. 240 — 126
Douglass & Varnum v. *Village of Morrisville,* 95 At. 810 — 129
Dovel v. *Village of Lynbrook,* 213 App. Div. 571 — 9
Dow v. *United States,* 157 F. 2d 707 — 93
Drainage Dist. No. 1 v. *Rude,* 21 F. 2d 257 — 125, 149
Draper v. *Miller,* 140 P. 890 — 93
Dunavan v. *Caldwell & Northern R.R.,* 122 N.C. 999 — 227
Dunbar & Sullivan Dredging Co. v. *State of New York,* 259 App. Div. 440, affirmed 291 N.Y. 652 — 118
Dunn v. *Steubing,* 120 N.Y. 232 — 135
E. I. Dupont deNemours & Co., v. *Glenwood Springs,* 19 F. 2d 225 — 196

E. and F. Const. Co. v. *Town of Stamford,* 158 At. 551 — 109
Re Eager, 46 N.Y. 100 — 83
East River Gas Light Co. v. *Donnelly,* 93 N.Y. 55 — 87
Edwards v. *Hartshorn,* 82 P. 520 — 160
Edwards v. *Thoman,* 187 Mich. 361 and 153 N.W. 806 — 164
Eighth Ave. Coach Corp. v. *City of New York,* 286 N.Y. 84 — 149
Electric Reduction Co. v. *Colonial Steel Co.,* 120 At. 116 — 154
Ellis v. *Grand Rapids,* 82 N.W. 244 — 146
Elmer v. *U.S. F. & G. Co.,* 174 F. Supp. 437, affirmed 275 F. 2d 89 — 212
Emmerson v. *Hutchinson,* 63 Ill. App. 203 — 111
Empire State Surety Co. v. *Des Moines,* 132 N.W. 837 — 205
Empson Packing Co. v. *Clawson,* 95 P. 346 — 160
Emslie v. *Livingston,* 51 App. Div. 628 — 129
Endres Plumbing Corp. v. *State of New York,* 198 Misc. 546 — 232
English v. *Shelby,* 172 S.W. 817 — 152
Equitable Surety Co. v. *McMillan,* 234 U.S. 448 — 215
Erickson v. *Edmond School Dist.,* 125 P. 2d 275 — 235
Ertle v. *Leary,* 46 P. 1 — 83
Erving v. *Mayor etc., of New York,* 131 N.Y. 133 — 29, 87
Evans v. *Western Brass Mfg. Co.,* 24 S.W. 175 — 153

Faber v. *City of New York,* 222 N.Y. 255 — 112, 119, 151
Fairchild Engine & Airplane Corp. v. *Cox,* 50 N.Y.S. 2d 643 — 156

Faist v. *Mayor, etc., of City of Hoboken,* 60 At. 1120 46
Federal Crop Ins. Corp. v. *Merrill,* 332 U.S. 380 245
Feigel Construction Corp. v. *The City of Evansville, et al.* 150 N.E. 2d 263 26
Ference v. *State of New York,* 251 App. Div. 13 37
Ferris v. *Snively,* 172 Wash. 167 93
Fidelity & Deposit Co. v. *Auburn,* 272 P. 34 202
Fidelity & Deposit Co. v. *Hegewald,* 139 S.W. 975 206
Fidelity & Deposit Co. of Maryland v. *Jones,* 256 Ky. 181 228
Field v. *United States,* 16 Ct. Cl. U.S. 434 77
Filbert v. *City of Philadelphia,* 181 Pa. 530, 37 At. 545 117, 122
Findley v. *City of Pittsburgh,* 82 Pa. 351 81
Finlayson v. *Peterson,* 67 N.W. 953 25
First M.E. Church v. *Isenberg,* 246 Pa. 221 and 92 At. 141 164
First National Bank v. *City Trust, S.D. and Surety Co.* 114 F. 529 201
Fish v. *Hubbard,* 21 Wend. 651 154
Flint v. *Chicago Bonding & S. Co.,* 168 N.W. 528 217
Florence Oil & Refining Co. v. *Reeves,* 56 P. 674 238
Floyd County & Owego Bridge Co., 137 S.W. 237 8
Flynn v. *Smith,* 111 App. Div. 873 75
Flynn Const. Co. et al. v. *Leininger,* 257 P. 374 143
Foeller v. *Heintz,* 137 Wis. 169 185
Fogarty v. *Davis,* 264 S.W. 879 196
Foley Bros. Inc. v. *Filardo,* 336 U.S. 281 290
Fones Hardware Co. v. *Erb,* 54 Ark. 645 3, 59, 83
Fore v. *Feimster,* 88 S.E. 977 197
Fort Worth Independent School Dist. v. *Aetna Casualty & Surety Co.,* 48 F. 2d 1 201
Fosmire v. *National Surety Co.,* 229 N.Y. 44 194
Foster v. *Gaston,* 23 N.E. 1092 217
Benjamin Foster Co. v. *Com.,* 61 N.E. 147 113
Foundation Co. v. *State of New York,* 233 N.Y. 177 112
Fox Film Corp. v. *Springer,* 273 N.Y. 434 155
Frankfurt-Barnett Co. v. *William Prym Co.,* 237 F. 21 135
Franklin v. *Horton,* 116 At. 176 5
Frazier-David Construction Co. v. *United States,* 97 Ct. Cl. U.S. 1 266
Friederick v. *County of Redwood,* 190 N.W. 801 118, 123
Friedman v. *Hampden County,* 90 N.E. 851 191
Frye v. *State of New York,* 192 Misc. 264 154
Fuccy v. *Coal & Coke Rwy. Co.,* 83 S.W. 301 125
George A. Fuller Co. v. *Elderkin,* 154 At. 548 60
George A. Fuller v. *United States,* 108 Ct. Cl. U.S. 70 281, 283

Gadsen v. *Brown,* Speers, Equity 38, 41 199
Galbreath v. *Newton,* 45 Mo. App. 312 7
Ganley Bros. v. *Butler Bros. Bldg. Co.,* 212 N.W. 602 120
Garofano Constr. Co. v. *State of New York,* 206 Misc. 760, affirmed 4 N.Y. 2d 748 118
Gearty v. *City of New York,* 171 N.Y. 61 134
Gemsco, Inc. v. *United States,* 115 Ct. Cl. U.S. 209 77
General Bonding & Gas. Co. v. *Dallas,* 175 S.W. 1098 191
Geremia v. *Boyarsky,* 107 Conn. 387 73
Giant Powder Co. v. *Oregon Pac. R.R. Co.,* 42 F. 470 207
Gibbons v. *Bente,* 53 N.W. 756 141
Gibson County v. *Mothwell Iron & Steel Co.,* 24 N.E. 115 127
Gillet v. *Bank of America,* 160 N.Y. 549 151
Gilmore v. *Utica,* 29 N.E. 841 57
Gisel v. *City of Buffalo,* 48 Hun 615 111
Glass v. *Wiesner,* 238 P. 2d 712 123
Goddard v. *Lowell,* 61 N.E. 53 17
Goss v. *Lannin,* 152 N.W. 43 61
Grace Co. v. *Chesapeake Co.,* 281 F. 904 119
Grand Investment Co. v. *United States,* 102 Ct. Cl. U.S. 40 266, 267
Gravenhorst v. *Zimmerman,* 236 N.Y. 22 155, 156
Graveson v. *Tobey,* 75 Ill. 540 227
Gray v. *State,* 72 Ind. 567 139
Great Lakes Const. Co. v. *Republic Creosoting Co.,* 139 F. 2d 456 231
Great Lakes Const. Co. v. *United States,* 95 Ct. Cl. U.S. 479 259
Green v. *Okanogan County,* 111 P. 226 6
Green County v. *Monroe,* 87 N.W. 2d 827 179
Greene v. *City of New York et al.,* 283 App. Div. 485 76
Griffiths v. *United States,* 74 Ct. Cl. U.S. 245 269
Grogan et al. v. *City of San Francisco,* 18 Cal. 590 139, 230
Grymes v. *Sanders,* 93 U.S. 62 70
Guerini Stone Co. v. *P. J. Carlin Const. Co.,* 248 U.S. 334 231
Gustor v. *Clark,* 157 N.W. 49 186
Guttenberg v. *Vassel,* 65 At. 994 216

Hagerman v. *Norton,* 105 F. 996 158
Hagerman v. *United States,* 180 F. Supp. 181 279
Hall v. *Wisconsin,* 103 U.S. 5 139
Hall & Olswang v. *Aetna Casualty & Surety Co.,* 296 P. 162 202
Handy v. *Bliss,* 204 Mass. 513 185
Hannan v. *Bd. of Education,* 25 Okla. 372 3

Hardison v. *Yeaman*, 91 S.W. 1111 197
Harlem Gaslight Co. v. *Mayor of New York*, 33 N.Y. 309 3
Harnbach v. *Ward*, 125 P. 140 196
Harnett Co. Inc. v. *Thruway Authority*, 3 Misc. 2d 257 150
Harper, Inc., v. *City of Newburgh*, 159 App. Div. 695 34, 35, 69
Harris v. *City of Philadelphia*, 149 At. 722 91
Harris v. *Taylor*, 129 S.W. 995 217
Hart v. *Thompson*, 10 App. Div. 183 154
Hartford Mill Co. v. *Hartford Tobacco Warehouse Co.*, 121 S.W. 477 186
Hartigan v. *Casualty Co. of America*, 227 N.Y. 175 150
Hartman v. *Greenhow*, 102 U.S. 672 139, 230
Hartsville Oil Mill v. *United States*, 271 U.S. 43 247
Hathaway v. *Sabin*, 63 Vt. 527 36
Thomas Haverty Co. v. *Jones*, 197 P. 105 187
Hawkins v. *United States*, 96 U.S. 607 150
Hayden v. *Cook*, 62 N.W. 165 217
Hays v. *Port of Seattle et al.*, 251 U.S. 233 144
Hazard v. *Board of Education*, 75 At. 237 193
Hazelhurst Oil Mill Co. v. *United States*, 70 Ct. Cl. U.S. 335 247
Heaver v. *Lanahan*, 22 At. 263 141
Heilbrun v. *Cuthbert*, 23 S.E. 206 140
Heim v. *McCall*, 239 U.S. 175 16
Heisel v. *Volkmann*, 55 App. Div. 607 154
Henderson Bridge Co. v. *McGrath*, 134 U.S. 260 125
Hendrick v. *Lindsey*, 93 U.S. 143 164
Hendry v. *City of Salem*, 129 P. 531 189
Heninger v. *City of Akron*, 112 N.E. 77 92
Hennessy v. *Preston*, 106 N.E. 570 182
Henningsen v. *U.S. F. & G. Co.*, 143 F. 810, affirmed 208 U.S. 404 200
Hercules Powder Co. v. *Knoxville, L. & J. R. Co.*, 113 Tenn. 382 206
Herman Constr. Co. v. *Lyon*, 211 S.W. 68 32
Hetherington-Berner Co. v. *Spokane*, 135 P. 484 233
Hickey v. *Sutton*, 191 Wis. 313 93
Hillside Twp. v. *Sternin*, 136 At. 2d 265 48
Hinckley v. *Pittsburgh Bessemer Steel Co.*, 121 U.S. 264 141
Hines v. *City of Bellefontaine*, 57 N.E. 2d 164 44
Hobbs v. *Brick Co.*, 157 Mass. 109 36
Hodgeman v. *City of San Diego*, 128 P. 2d 412 81
Hoffman v. *McMullen*, 83 F. 372 62
Holcomb v. *American Surety Co.*, 42 S.W. 2d 765 203
Holden v. *Alton*, 53 N.E. 556 17

Hollerbach v. *United States,* 233 U.S. 165 — 109, 110, 118, 120, 122, 138, 148, 243, 245
Holme v. *Guppey,* 3 M. & W. 387 — 227
Holmes v. *Council Detroit,* 120 Mich. 226 — 15, 59
Home Bank v. *Drumgoole,* 109 N.Y. 63 — 227
Horgan v. *The Mayor,* 160 N.Y. 516 — 234
Horgan & Slattery v. *New York,* 114 App. Div. 555 — 5
Hornung et al. v. *Town of West New York et al.,* 81 At. 1116 — 48
Hoskins v. *Powder L. and I. Co.,* 176 P. 124 — 126
Hostetter v. *Park,* 137 U.S. 30 — 153
Houghe v. *Woodruff,* 19 F. 136 — 158
Howell v. *Dimock,* 15 N.Y. 102 — 157
Hunt v. *Treka Terrazzo & Mosaic Co.,* 11 P. 2d 521 — 131
Hunter v. *Whitaker,* 230 S.W. 1096 — 4
Huntington v. *Force,* 53 N.E. 443 — 126
Hyatt v. *Williams,* 84 P. 41 — 140
Hyde v. *Tanner,* 1 Barb. 75 — 75
Hydraulic Press Brick Co. v. *School Dist.,* 79 Mo. App. 655 — 197

Interstate Power Co. v. *Incorp. Town of McGregor et al.,* 296 N.W. 770 — 42
Island Nav. Co. v. *American Surety Co. of New York,* 227 S.W. 809 — 216

Jackson v. *State,* 210 App. Div. 115, affirmed 241 N.Y. 563 — 120, 150
Jacob & Youngs, Inc., v. *Kent,* 129 N.E. 889 — 183
Jahn Contracting Co. v. *Seattle,* 170 P. 549 — 114
Janus v. *Dravo Contracting Co.,* 302 U.S. 134 — 249
Jefferson v. *Asch,* 53 Minn. 446 — 166
Jefferson County v. *City of Birmingham,* 55 So. 2d 196 — 178
Jenkins v. *City of Bowling Green,* 251 Ky. 119 — 23, 26
Jenkins v. *C. & O. Ry. Co.,* 57 S.E. 48 — 162
Jenny v. *City of Des Moines,* 103 Iowa 347 — 26
Jewett Pub. Co. v. *Butler,* 159 Mass. 517 — 36
Johnson v. *Maryland,* 254 U.S. 51 — 251
Johnson County Savings Bank et al., v. *City of Creston et al.,* 231 N.W. 705 — 8
Johns-Manville v. *Lander County,* 240 P. 925 — 196
Jones v. *Salm,* 123 P. 109 — 24

Matter of Kaelber v. *Sahm et al.,* 305 N.Y. 858 82
Kansas City Bridge Co. v. *State,* 250 N.W. 344 121, 127, 131
Kansas Turnpike Authority v. *Abramson,* 272 F. 2d 711 123
A. Kaplen & Son, Ltd., v. *Housing Authority,* 126 At. 2d 13 237
Kasbo Const. Co. v. *Minto School Dist.,* 184 N.W. 1029 185
Kawananakoa v. *Polyblank,* 205 U.S. 349 248
Kelly v. *United States,* 31 Ct. Cl. U.S. 361 235
Kemp v. *United States,* 38 F. Supp. 568 71
Kemper v. *City of Los Angeles,* 235 P. 2d 7 71
Kenny v. *Monahan,* 169 N.Y. 591 227
Peter Kiewit Sons' Co. v. *United States,* 151 E. Supp. 726 282
Kingsley v. *City of Brooklyn,* 78 N.Y. 200 8
Kinney v. *Mass. B. & Ins. Co.,* 175 N.Y.S. 398 126
Knapp v. *Swaney,* 23 N.W. 162 192, 193
Knickenberg v. *State of New York,* Claim No. 25, 452, New York Court of Claims, not reported in case book. 50
Knight & J. Co. v. *Castle,* 87 N.E. 976 203
Knowles v. *City of New York,* 176 N.Y. 430 92
Koich v. *Cvar,* 110 P. 2d 264 80
Korshoj Constr. Co. Inc. v. *Mills County,* 156 F. Supp. 138 220
Kothe v. *Taylor Trust,* 280 U.S. 224 219, 228
Kramer v. *Gardner,* 104 Minn. 370 166
Krasin v. *Almond,* 290 N.W. 152 76
Kretsch v. *Helm,* 45 Ind. 438 24
Krotts v. *Clark Const. Co.,* 249 F. 181 240
C. J. Kubach Co. v. *McGuire,* 248 P. 676 179
Kutsche v. *Ford,* 192 N.W. 714 76

Labbe v. *Bernard,* 82 N.E. 688 201
LaFrance Electrical Construction & Supply Co. v. *International Brotherhood of Electrical Workers,* 108 Ohio St. 61, 140 N.E. 899. 20
Matter of Ward LaFrance Truck Corp. v. *City of New York,* 7 Misc. 2d 739 84
Laird Norton Yards v. *Rochester,* 134 N.W. 644 9
Lakota Oil & Gas Co., 116 P. 2d 761 140
LaMourea v. *Rhude et al.,* 295 N.W. 304 165
Landsborough v. *Kelly,* 37 P. 2d 293 58
Lane v. *State,* 43 N.E. 244 203
Lawrence v. *Fox,* 20 N.Y. 268 165
Leach et al. v. *Burr,* 188 U.S. 510 23

Leary v. *City of Watervliet,* 222 N.Y. 337 78
Lenart Constructors, Inc. v. *State of New York,* 6 Misc. 2d 473 157
Lent v. *Hodgman,* 15 Barb. 274 154
Lentilhon v. *New York,* 102 App. Div. 548, affirmed 185 N.Y. 549 114
119
Leslie Lumber & Supply Co. v. *Lawrence et al.,* 11 S.W. 2d 458 203
Lewis v. *Board of Education,* 102 N.W. 756 18
Libby v. *L. J. Corporation,* 247 F. 2d 78 61
Lichter v. *United States,* 334 U.S. 742 294
Application of Limitone, 189 N.Y.S. 2d 738 81
Lingler v. *Andrews,* 10 N.E. 2d 1021 193
Little et al. v. *Banks,* 85 N.Y. 258 34, 165
Littler v. *Jayne,* 16 N.E. 374 83
Lonergan v. *San Antonio Loan & T. Co.,* 104 S.W. 1061 115
Long v. *United States,* 127 F. Supp. 623 271
Lord et al. v. *Thomas,* 64 N.Y. 107 141, 143
Lord Const. Co. v. *United States,* 28 F. 2d 340 130, 131
Los Angeles Dredging Co. v. *Long Beach,* 291 P. 839 6
Los Angeles R.R. Co. v. *New Liverpool Salt Co.,* 150 Cal. 21 73
Los Angeles Stone Co. v. *National Surety Co.,* 173 P. 79 203
Louisiana v. *Shaffner,* 78 S.W. 287 83
Lucas County v. *Roberts,* 49 Iowa 159 217
Ludlum v. *Vail,* 166 N.Y. 611 227
Lund v. *Bruflat,* 159 Wash. 89 93
Lyman v. *Lincoln,* 57 N.W. 531 203
Lyman-Richey Sand & Gravel Co. v. *State,* 243 N.W. 891 152
Lynch v. *United States,* 292 U.S. 571 243

MacEvoy v. *United States,* 322 U.S. 102 211, 212
Mach v. *State of New York,* 122 Misc. 86, affirmed 211 App. Div. 825 238
MacKnight & Flintic Stone Co. v. *Mayor of New York,* 160 N.Y. 72, 54 N.E. 661 109, 117, 122
Magoba Construction Co., Inc. v. *United States,* 99 Ct. Cl. U.S. 662 269
Mallon et al. v. *Bd. of Water Commissioners et al.,* 128 S.W. 764 5
Manerad v. *City of Eugene,* 124 P. 662 235
Maney v. *Oklahoma City, 150 Okla.* 77 and 300 P. 642 109, 114
Mansfield v. *N.Y.C. & H.R.R. Co.,* 102 N.Y. 205 230, 231
Marsch v. *Southern New England R. Corp.,* 120 N.E. 120 160
Marsh v. *Kauff,* 74 Ill. 189 227
Marshall & Bruce Co. v. *Nashville,* 71 S.W. 815 17

Martens & Co. v. *City of Syracuse,* 183 App. Div. 622 35, 73
Martindale v. *Shaba,* 51 Okla. 670 93
Maryland Casualty Co. v. *Eagle River Union Free High School District,* 205 N.W. 926 216
Mass. Bonding Co. v. *John R. Thompson Co.,* 88 F. 2d 825 215
Massachusetts Bonding & Ins. Co. v. *Lentz,* 9 P. 2d 408 130
Massachusetts Bonding & Ins. Co. v. *Realty Trust Co.,* 73 S.E. 1053 217
Massachusetts Bonding & Ins. Co. and ano. v. *State of New York,* 259 F. 2d 33 201
Massie v. *Dudley,* 173 Va. 42 93
Massman Construction Co. v. *United States,* 102 Ct. Cl. U.S. 699 247
Matsumato v. *Arizona Sand and R. Co.,* 295 P. 2d 850 169
Mayor etc. of City of Baltimore v. *Flack,* 64 At. 702 58
Mayor etc. of Savannah v. *Collins,* 84 S.E. 2d 454 177
Mazet v. *City of Pittsburgh,* 137 Pa. 548 3
McCall v. *Superior Court,* 1 Cal. 2d 527 73
McCann v. *Albany,* 158 N.Y. 634 78
McCarthy v. *Bloomington,* 127 Ill. App. 215 9
McClung Constr. Co. v. *Muncy,* 65 S.W. 2d 786 111
McConnell v. *Corona City Water Co.,* 85 P. 929 114
McCord v. *United States,* 9 Ct. Cl. U.S. 155 269
McCormick v. *Los Angeles City Water Co.,* 40 Cal. 185 204
McCree & Company v. *State of Minnesota,* 91 N.W. 2d 713 118
McCrum v. *Love,* 58 Pa. Sup Ct. 404 217
McDaniel v. *City of Beaumont,* 92 S.W. 2d 552 116
F. H. McGraw & Co. v. *United States,* 130 F. Supp. 394 267
McHugh v. *City of Boston,* 53 N.E. 905 146
McIntosh v. *Miner,* 53 App. Div. 240 154, 155
McIntosh v. *Pendleton,* 75 App. Div. 621 157
McKegney v. *Illinois Surety Co.,* 180 App. Div. 507 226
McLeod v. *Genius,* 47 N.W. 473 127
McMahon v. *State of New York,* 178 Misc. 865 137
McMaster v. *State of New York,* 108 N.Y. 542 143
McRae v. *Concord,* 6 N.E. 2d 366 140
McRoberts v. *Ammons,* 88 P. 2d 1958 32
Meads & Co. v. *City of New York,* 191 App. Div. 365 233
Meahl v. *City of Henderson,* 209 S.W. 2d 593 23
Memphis v. *Memphis Water Co.,* 5 Heisk, 495 140
Matter of Merriam, 84 N.Y. 596 55
Merwin v. *Chicago,* 45 Ill. 133 197
Meuser v. *Risdon,* 36 Cal. 239 24

Michel v. *Taylor,* 127 S.W. 949 24
P. Michelotti & Son, Inc. v. *Fairlawn,* 152 At. 2d 369 48
Mid-Continental Petroleum Corp. v. *Southern Surety Co.,* 9 S.W. 2d 229 206
Middleton v. *City of Emporia,* 186 P. 981 36
Miller v. *American Bonding Co.,* 158 N.W. 432 207
Miller v. *Arkansas,* 352 U.S. 187 251
Miller v. *Des Moines,* 122 N.W. 226 17
Miller v. *McKinnon,* 124 P. 2d 34 8
Mills v. *Schulba,* 95 Cal. App. 2d 559 73
Milquet v. *Van Straten,* 202 N.W. 670 9
Mining Co. v. *Collins,* 104 U.S. 179 204
Moffett, Hodgkins & Clarke Co. v. *Rochester,* 178 U.S. 373 35, 65, 67, 70, 71, 73, 75
C. N. Monroe Mfg. Co. v. *United States,* 143 F. Supp. 449 70
Montgomery v. *Rief,* 50 P. 623 203
Montrose Const. Co. v. *County of Westchester,* 80 F. 2d 841 109
Moran v. *McLarty,* 75 N.Y. 25 70
Moran v. *Schmidt,* 67 N.W. 323 129
Moran Bros. Co. v. *United States,* 61 Ct. Cl. U.S. 73 235, 269
Morgan-Gardner Elect. Co. v. *Beelick Knob Coal Co.,* 112 S.E. 587 141
Muff v. *Cameron,* 114 S.W. 1125 57
Mumm Contr. Co. v. *Village of Kenmore,* 104 Misc. 268 157
Museum of Fine Arts v. *American Bonding Co.,* 97 N.E. 633 216
Myers et al. v. *Lillard et al.,* 220 S.W. 2d 608 61

National Surety Corp. v. *Fisher,* 317 S.W. 2d 334 202
National Surety Co. v. *Hall-Miller Decorating Co.,* 61 So. 700 203
Nehrbas v. *Incorporated Village of Lloyd Harbor,* 140 N.E. 2d 241 179
New England Electric Co. v. *Shook,* 145 P. 1002 44
New England Ins. Co. v. *Railroad,* 91 N.Y. 153 36
New York v. *Dowd Lumber Co.,* 140 App. Div. 358 35
Newhall v. *Appleton,* 114 N.Y. 140 154, 157
Norcross v. *Wills,* 198 N.Y. 336 234
Norcross v. *Wyman,* 72 N.E. 347 160
Normile, Fastaburd & McGregor v. *United States,* 239 U.S. 344 229
North Pacific Bank v. *Pierce County,* 167 P. 2d 454 201
North-Eastern Const. Co. v. *Town of North Hempstead,* 121 App. Div. 187 52, 84
Northern P.R. Co. v. *Minnesota,* 208 U.S. 583 145

Northrop v. *Cooper,* 23 Kan. 432 25
Northwest Steel Co. v. *School Dist.,* 148 P. 1134 197
Nye-Schneider-Fowler Co. v. *Bridges, H. & Co.,* 151 N.W. 942 207

Oberlies v. *Bullinger,* 132 N.Y. 598 185
O'Brien v. *Fowler,* 11 At. 174 131
O'Brien v. *Town of Greenburgh,* 239 App. Div. 555 178
O'Connell Inc. v. *County of Broome,* 198 Misc. 402 35
O'Donohue v. *Leggett,* 134 N.Y. 40 156
Ohio County v. *Clemens,* 100 S.E. 680 214
Oklahoma City et al. v. *Sanders,* 94 F. 2d 323 252
O'Leary v. *Board of Port Comrs. etc.,* 91 So. 139 126
O'Neill Construction Co. Inc. v. *Philadelphia,* 6 At. 2d 525 120
Orlando v. *Murphy,* 84 F. 2d 531 235
Orpheum Theater v. *Kansas City Co.,* 239 S.W. 841 126
Matter of Ottaviano, Inc. v. *Tallamy, et al.,* 277 App. Div. 929 27
Owen v. *Hill,* 34 N.W. 649 197
Ozark Dam Constructors v. *United States,* 120 Ct. Cl. U.S. 354 279

Pacific Coast Engr. Co. v. *State,* 244 P. 2d 21 152
Pacific Coast Steel Co. v. *Old National Bank of Spokane,* 235 P. 947 202
Packard v. *Hayes,* 51 At. 32 83
Pallas v. *Johnson,* 68 P. 2d 559 18
Palmberg v. *City of Astoria (Oregon),* 199 P. 630 78
Palmer v. *Stockwell,* 9 Gray 237 227
Palsgraf v. *Long Island R. Co.,* 248 N.Y. 339 170
A. J. Paretta Contracting Co. v. *United States,* 109 Ct. Cl. U.S. 324 289
Park v. *Great Lakes Dredge & Dock Co. et al.,* 192 N.W. 1012 128
Pascoe v. *Barlum,* 225 N.W. 506 61
Passaic Valley Sewerage Comrs. v. *Holbrook, Cabot & Rollins Corpn.,* 6 F. 2d 721 120
Passaic Valley Street Comrs. v. *Tierney,* 1 F. 2d 304 114
Patterson v. *Nieverhofer,* 204 N.Y. 96 231
Pearlman v. *City of Pittsburgh,* 155 At. 118 56
Pearlman v. *State of New York,* 18 Misc. 2d 494 51
Pence v. *Dennie,* 182 P. 980 186
Penn Bridge Co. v. *City of New Orleans,* 222 F. 737 109
Penn Iron Co. v. *William R. Trigg Co.,* 56 S.E. 329 146

Matter of Pennie, 15 N.E. 611 — 24
Pennsylvania Cement Co. v. *Bradley Contracting Co.,* 7 F. 2d 822 — 163
Pennsylvania Steel Co. v. *New York City Ry. Co.,* 198 F. 721 — 164
Pennsylvania Turnpike Commission v. *Smith,* 39 At. 2d 139 — 119
People v. *Buffalo County Comrs.,* 4 Neb. 150 — 83
People v. *Buffalo Fish Co.,* 164 N.Y. 93 — 15
People v. *Caldwell-Garvan and Bertini, Inc. et ano.,* 161 Misc. 864 — 51
People v. *Hawkins,* 157 N.Y. 1 — 15
People v. *Santa Clara Lumber Co.,* 213 N.Y. 61 — 137
People v. *Stephens et al.,* 71 N.Y. 527 — 139, 140
People ex rel. Assyrian Asphalt Co. v. *Kent, Commissioner,* 43 N.E. 760 — 82
People ex rel. Coughlin v. *Gleason,* 121 N.Y. 631 — 87
People ex rel. Graves v. *Sohmer,* 207 N.Y. 460 — 140, 150
People ex rel. Martin v. *Dorsheimer et al. Commissioners of the New State Capitol in Albany,* 55 How. Pr. 118 — 86
People ex rel. Ottman v. *Comr. of Highways,* 27 Barb. 94, affirmed 30 N.Y. 470 — 138
People ex rel. Rodgers v. *Coler,* 35 App. Div. 401 — 33
People ex rel. Treat v. *Coler,* 166 N.Y. 144 — 15
Perrault v. *Shaw,* 69 N.H. 180 — 204
Perry v. *Levenson,* 82 App. Div. 94 — 227
Perry v. *United States,* 294 U.S. 330 — 243
Philadelphia v. *Stewart,* 45 At. 1056 — 193
Phoenix Bridge Co. v. *United States,* 211 U.S. 188 — 117, 122
Phoenix Bldg. and Homestead Assn. v. *Meraux,* 189 La. 819 — 48
Picone v. *City of New York,* 176 Misc. 967 — 85
Pidgeon Thomas Iron Co. v. *LeFlore County,* 99 So. 677 — 197
Pitt Const. Co. v. *City of Alliance, Ohio,* 12 F. 2d. 28 — 119, 120
Plumley v. *United States,* 226 U.S. 545 — 126, 271
Plummer v. *Kelly,* 7 N.D. 88 — 36
Pneucrete Corp. v. *U.S. F. & G. Co.,* 46 P. 2d 1000 — 193
Poirier & McLane Corp. v. *United States,* 120 F. Supp. 209 — 288
Polk v. *McCartney,* 73 N.W. 1067 — 24
Porterfield v. *City of Oakland,* 159 P. 202 — 81
Prairie State Natl. Bank v. *United States,* 164 U.S. 227 — 198, 201
Prendergast v. *City of St. Louis et al.,* 167 S.W. 970 — 43
Price v. *Eaton,* 4 B. and Ad. 433 — 164
Priebe & Sons v. *United States,* 332 U.S. 407 — 219, 245
Prudence Co. v. *Fidelity and Deposit Co.,* 2 F. Supp. 454 — 216
Psaty & Fuhrman v. *Housing Authority,* 68 At. 2d 32 — 238
Public Housing Administration v. *Bristol Township,* 146 F. Supp. 859 — 249

Puget Sound Nat. Bank of Tacoma v. *C.B. Lauch Const. Co.*, 245 P. 2d 800 115
Puget Sound Painters Inc. v. *State of Washington*, 278 P. 2d 302 76
Purington Paving Brick Co. v. *Metropolitan Paving Co.*, 4 F. 2d 676 135

Railroad Co. v. *Van Dusen*, 29 Mich. 431 117
Rara Avis Gold & Silver Mining Co. v. *Bouscher*, 9 Colo. 385 205
Rathbun v. *State*, 97 P. 335 191
Red Wing Sewer Pipe Co. v. *Donnelly*, 113 S.W. 1 206
Rice v. *Dwight Manuf. Co.*, 2 Cush. 80 37
Richmond v. *I. J. Smith & Co.*, 89 S.E. 123 114
Rickerson v. *Hartford Fire Ins. Co.*, 149 N.Y. 307 155
Ricketson v. *Milwaukee*, 81 N.W. 864 10
Rigney v. *N.Y.C. Ry. Co.*, 217 N.Y. 31 166
Ripley v. *United States*, 223 U.S. 695 284
Robinson v. *United States*, 80 U.S. 363 153
Robinson v. *United States*, 261 U.S. 486 220
Rollins v. *Salem*, 146 N.E. 795 5
Cf. John E. Rosasco Creameries v. *Cohen*, 276 N.Y. 274 93
Rosemead Co. v. *Shepley Co.*, 207 Cal. 414 73
J. A. Ross & Co. v. *United States*, 126 Ct. Cl. U.S. 323 268
Ross Engineering Co. Inc., v. *United States*, 92 Ct. Cl. U.S. 253 259
Rushlight Automatic Sprinkler Co. v. *Portland*, 219 P. 2d 732 65
Russell v. *DiBandeira*, 13 C.B. N.S. 149 227
Ryan v. *New York*, 159 App. Div. 105 241

Sacramento County v. *Southern P. Co.*, 59 P. 568 9
St. Louis Hay & Grain Co. v. *United States*, 191 U.S. 159 271
St. Louis Second National Bank v. *Grand Lodge*, 98 U.S. 123 164
Saligman et al. v. *United States*, 56 F. Supp. 505 69, 71
Salt Lake City v. *Smith*, 104 F. 457 123, 148
Sampson v. *Commonwealth*, 202 Mass. 326 164, 206
Sanborn v. *City of Boulder*, 221 P. 1077 12
San Christina Invest. Co. v. *San Francisco*, 141 P. 384 7
Schagticoke Powder Co. v. *Greenwich & J. Ry. Co.*, 183 N.Y. 306 206
Schliess v. *City of Grand Rapids*, 90 N.W. 700 123
Schnaier v. *Bradley Constr. Co.*, 181 App. Div. 538 166
Schneck v. *City of Jeffersonville*, 52 N.E. 212 146
School Directors v. *Robinson*, 65 Ill. App. 298 186
School District of Scottsbluff v. *Olson Const. Co.*, 153 Neb. 451 73, 74

Schrey et al. v. *Allison Steel Mfg. Co.,* 255 P. 2d 604 16
Schultz v. *C. H. Quereau Co.,* 210 N.Y. 257 206
Schunnemunk Construction Co. v. *State of New York,* 116 Misc. 770 139, 230
Schwartz v. *Saunders,* 46 Ill. 18 117
Schwasnick v. *Blandin,* 65 F. 2d 354 240
Secrest v. *Delaware County,* 100 Ind. 59 191
Seeger v. *Odell,* 18 Cal. 2d 409 73
Selden Breck Const. Co. v. *Regents of Univ. of Michigan,* 274 F. 982 235, 237
Semper v. *Duffey,* 227 N.Y. 151 114
Severin v. *United States,* 102 Ct. Cl. U.S. 74 268
Seymour v. *Long Dock Co.,* 20 N.J. Eq. 396 117
Sheehan v. *New York,* 37 Misc. 432 7
Shell v. *Schmidt,* 330 P. 2d 817 187
Shelton v. *American Surety Co.,* 127 F. 736 217
Sherwood v. *Wise,* 132 Wash. 295 93
Shields v. *City of New York,* 84 App. Div. 502 121
Shoemaker v. *Buffalo Steam Roller Co.,* 165 App. Div. 836 9
Shore Bridge Corpn. v. *State of New York,* 186 Misc. 1005, affirmed 271 App. Div. 811 230
Silberstein v. *Kittrick,* 169 P. 250 217
Silsby Mfg. Co. v. *City of Allentown,* 153 Pa. 319 15
Silverblatt & Lasker, Inc., v. *United States,* 101 Ct. Cl. U.S. 54 268
Simpson v. *United States,* 172 U.S. 372 117, 122
Singer v. *Grand Rapids Match Co.,* 117 Ga. 94 75
Sinnott v. *Mullin,* 82 Pa. 333 117
Small v. *Lee Brothers,* 61 S.E. 831 186
Smith v. *American Packing & Provision Co.,* 130 P. 2d 951 93
Smith v. *Mackin,* 4 Lans. 41 75
Smith v. *Oosting,* 203 N.W. 131 207
Smith v. *Railroad Co.,* 36 N.H. 459 117
Smyth v. *City of New York,* 203 N.Y. 106 166
Sollitt & Sons Constr. Co. v. *Com.,* 172 S.E. 290 250
South Texas Public Service Co. v. *Jahn,* 7 S.W. 2d 942 140
Southern Lumber Co. v. *Supply Co.,* 89 Mo. App. 141 36
Southern R. Co. v. *Bretz,* 104 N.E. 19 198
Spangler v. *Leitheiser,* 37 At. 832 5
Spence v. *Ham,* 163 N.Y. 220 185
Sperry & Hutchinson Co. v. *ONeill-Adams Co.,* 185 F. 231 135
Standard Gas & Power Corp. v. *N.E. Casualty Co.,* 101 At. 281 203
Standard Supply Co. v. *Vance Plumbing, etc., Co.,* 143 S.E. 248 196

Stanton v. *State of New York,* 103 Misc. 221, affirmed 187 App. Div. 963 132, 182
State v. *Allen,* 107 N.E. 2d 345 177
State v. *Barlow,* 48 Mo. 17 83
State v. *Butler,* 77 S.W. 560 146
State v. *Cherry County,* 79 N.W. 825 25
State v. *Feigel,* 178 N.E. 435 229, 233
State v. *Senatobia Blank Book & Stationery Co.,* 76 So. 258 16
State ex rel. Carpenter v. *Ralston,* 182 Ind. 150 139
State ex rel. City of Beckley v. *Roberts,* 40 S.E. 2d 841 214
State ex rel. Eberhardt v. *Cincinnati,* 1 Ohio Nisi Prius Reports 337 53
State ex rel. Helsel v. *Board of County Commissioners,* 79 N.E. 2d 688 177
State ex rel. Hunt v. *Fronizer,* 82 N.E. 518 9
State ex rel. McMahon v. *McKenzie,* 29 Ohio C.C. 115 5
State ex rel. Shay v. *McCormack,* 167 App. Div. 854 33
State Highway Dept. v. *MacDougald Constr. Co.,* 6 S.E. 2d 570 160
State of Connecticut v. *F. H. McGraw & Co.,* 41 F. Supp. 369 71
State of Ohio ex rel. United District Heating Inc. v. *State Office* Bldg. Commission, 179 N.E. 138 18, 19, 29
State of Utah v. *Union Construction Company and ano.,* 339 P. 2d 241 75
Steffen v. *United States,* 213 F. 2d 266 228
Stehlin-Miller-Henes Co. v. *City of Bridgeport,* 117 At. 811 232
Steinmeyer v. *Schoppel,* 226 Ill. 9 73
Stern v. *Spokane,* 111 P. 231 7, 32
Stewart v. *Barber,* 182 Misc. 91 151
James Stewart & Co. v. *Sadrakula,* 309 U.S. 94 252
James Stewart & Co. v. *United States,* 105 Ct. Cl. U.S. 284 268
Stocking v. *Warren Bros. Co.,* 114 N.W. 789 58
Stratton v. *Allegheny County,* 91 At. 894 5
Straw v. *Williamsport,* 132 At. 804 32
Struck Construction Co. v. *United States,* 96 Ct. Cl. U.S. 186 247
Stuart v. *Cambridge,* 125 Mass. 102 131
W. H. Stubbings Co. v. *World's Columbian Exposition Co.,* 110 Ill. App. 210 237
Stulsaft v. *Mercer Tube & Mfg. Co.,* 288 N.Y. 255 155
Styles v. *F. R. Long,* 57 At. 488 162
Sun Printing and Publishing Assn. v. *Moore,* 183 U.S. 642 219
Sundstrom v. *State of New York,* 213 N.Y. 68, 106 N.E. 924 117, 122
Sutton v. *United States,* 256 U.S. 575 246

Syracuse Intercepting Sewer Board v. *Deposit Co. of Maryland*, 255 N.Y. 288 85

Taber v. *Benton Harbor*, 274 N.W. 324 177
Tacoma Dredging Co. v. *United States*, 52 Ct. Cl. U.S. 447 275
Taylor v. *Board of Education*, 253 App. Div. 653, affirmed 278 N.Y. 641 18, 86
Taylor v. *Buttrick*, 165 Mass. 547 37
Taylor-Fichter Steel Constr. Co. v. *Niagara Frontier Bridge* Commission, 261 App. Div. 288, affirmed 287 N.Y. 669 238
Texas Pacific Coal & Oil Co. v. *Stuard*, 7 S.W. 2d 878 187
Thompson v. *United States*, 130 Ct. Cl. U.S. 1 283
Thomson v. *Union Pacific Railroad*, 9 Wall. 579 250
Thorn v. *London, L.R.* 1 App. Cas. D.C. 120 117
Tifft v. *City of Buffalo*, 25 App. Div. 376 24
Todd Dry Dock Eng. & Repair Corp. v. *City of New York*, 54 F. 2d 490 134
Toner v. *Long*, 111 At. 311 203
Town of Bloomfield v. *New Jersey Highway Authority*, 113 At. 2d 658 178
Trans-World Airlines Inc. v. *Travelers Indemnity Co.*, 262 F. 2d 321 225
Transfer Realty Co. v. *City of Superior*, 147 N.W. 1051 188
Trapp v. *City of Newport*, 74 S.W. 1109 33, 57, 59
The Travelers Insurance Company v. *Village of Illion and others*, 126 Misc. 275 204
Trowbridge v. *Hudson*, 24 Ohio C. C. 76 59
Tucker v. *Collar*, 79 Ariz. 141 170

Underground Constr. Co. v. *Sanitary Dist.*, 11 N.E. 2d 361 240
Underwood v. *Fairbanks, Morse & Co.*, 185 N.E. 118 140
Union & Peoples Nat. Bank v. *Anderson-Campbell*, 256 Mich. 674 73
United States v. *American Surety Co.*, 322 U.S. 96 245
United States v. *Andrews*, 207 U.S. 229 271
United States v. *Atlantic Dredging Co.*, 253 U.S. 1 109, 120
United States v. *Bank of the Metropolis*, 15 Peters 392 243
United States v. *Behan*, 110 U.S. 338 239
United States v. *A. Bentley & Sons Co.*, 293 F. 229 244
United States v. *Bethlehem Steel Co.*, 205 U.S. 105 219

United States v. *Bethlehem Steel Corp.,* 315 U.S. 289 245, 293
United States v. *Beuttas,* 324 U.S. 768 289
United States v. *Binghamton Construction Co. Inc.,* 347 U.S. 171 287
United States v. *Blair,* 321 U.S. 730 231
United States v. *Bostwick,* 94 U.S. 65 138
United States v. *Butler,* 297 U.S. 1 243
United States v. *Callahan Walker Construction Co.,* 317 U.S. 56 268
United States v. *Howard P. Foley Co.,* 329 U.S. 64 278, 282
United States v. *Edward J. Freel, etc.,* 186 U.S. 309 217
United States v. *Gibbons,* 109 U.S. 200 109
United States v. *Jones,* 176 F. 2d 278 71
United States v. *Kanter,* 137 F. 2d 828 220
United States v. *John Kerns Const. Co.,* 140 F. 2d 792 225
United States, to use of Thomas Laughlin Co. v. *Morgan,* 111 F. 474 205
United States v. *Moorman,* 338 U.S. 457 284
United States v. *North American Transp. & Trading Co.,* 253 U.S. 330 246
United States v. *Rice,* 317 U.S. 61 273, 278, 282
United States, to use of Sabine & E.T. Ry. Co. v. *Hyatt et al.,* 92 F. 442 205
United States v. *Smith,* 94 U.S. 214 281
United States v. *L. P. and J. A. Smith,* 256 U.S. 12 109, 114, 119
United States v. *Spearin,* 248 U.S. 132 109, 114, 117, 122
United States v. *Standard Rice Co.,* 323 U.S. 106 245
United States v. *United Eng. & C. Co.,* 234 U.S. 236 224
United States v. *Utah, N. & C. Stage Co.,* 199 U.S. 414 110, 118, 122
United States v. *Walkof,* 144 F. 2d 75 219
United States v. *Wunderlich,* 342 U.S. 98 284
Upington v. *Oviatt,* 24 Ohio St. 232 24
U. S. F. & G. Co. v. *Board of Comrs.,* 145 F. 144 150
U. S. F. & G. Co. for use of Reedy v. *American Surety Co. of* N.Y., 25 F. Supp. 28 203
U. S. F. & G. Co. v. *Tafel Electric Co.,* 91 S.W. 2d 42 193
U. S. F. & G. Co. v. *United States,* 189 F. 339 205
United States Rubber Co. v. *American Bonding Co.,* 149 P. 706 207
Utah Power & Light Co. v. *United States,* 243 U.S. 389 245
Uvalde Contracting Co. v. *City of New York,* 160 App. Div. 284 160

Van Brocklin v. *Anderson,* 117 U.S. 151 138
Van Eaton v. *Town of Sidney,* 231 N.W. 475 140

Vaughan Const. Co. v. *Virginian Ry. Co.*, 103 S.E. 293 126
Vicksburg Waterworks Co. v. *Vicksburg*, 185 U.S. 65 145
Vincennes Bridge Co. v. *Atoka County*, 248 F. 93 9
Vrooman v. *Turner*, 69 N.Y. 280 165

Wakefield Construction Co. v. *City of New York*, 157 App. Div. 35, affirmed 213 N.Y. 633 189
Walla Walla v. *Walla Walla Water Co.*, 172 U.S. 1 145
Wallis et al. v. *Inhabitants of Wenham*, 90 N.E. 396 226
Walls v. *Bailey*, 49 N.Y. 464 155, 156
Walsh v. *United States*, 121 Ct. Cl. U.S. 546 283
Walsh Constr. Co. v. *Cleveland*, 271 F. 701 186
Walter v. *Huggins*, 148 S.W. 148 186
Ward v. *Kilpatrick*, 85 N.Y. 413 182
Warner v. *Hallyburton*, 121 S.E. 756 196
Warren Brother Co. v. *City of New York*, 190 N.Y. 297 14
Water Works Board of Birmingham v. *Stephens*, 78 So. 2d 267 177
Watterson v. *Mayor, etc.*, 61 S.W. 782 126
Weber v. *Collins*, 139 Mo. 501 227
Webster v. *Belote*, 138 So. 721 3
Wedgewood v. *Jorgens*, 190 Mich. 620 93
Weinbaum v. *Algonquin Gas Transmission Co. et al.*, 20 Misc. 2d 276, affirmed 285 App. Div. 818 166
Matter of Weinstein Building Corp. v. *Scoville*, 141 Misc. 902 91
Wells v. *Burnham*, 20 Wis. 112 3, 83
Wells Bros. v. *United States*, 254 U.S. 83 235, 239, 274, 276, 281
West v. *Cruz*, 251 P. 2d 311 170
Weston v. *State of New York*, 262 N.Y. 46 112
Wharton & Co. v. *Winch*, 140 N.Y. 287 241
Wheaton Bldg. & Lumber Co. v. *City of Boston*, 90 N.E. 598 36
White v. *Fresno Bank*, 98 Cal. 166 227
White v. *Little*, 131 Okla. 132 93
White v. *McLaren*, 24 N.E. 91 186
Whiting v. *Mayor*, 172 N.E. 338 140
Whitmore et al. v. *Edgerton*, etc., 87 Misc. 216 15
Wholesale Serv. Sup. Corp. v. *State of New York*, 201 Misc. 56 77
Wicks v. *Salt Lake City*, 208 P. 538 140
Wiley v. *Hart*, 132 P. 1015 130
Williams v. *City of Topeka*, 118 P. 864 88
Williams v. *Gibbes*, 20 How. Pr. 535 246
Williams v. *Markland*, 44 N.E. 562 203

Williams v. *Pacific Coast Casualty Co.,* 140 P. 74 214
Williams et al. v. *Bergin,* 62 P. 59 41
Otis Williams Co. v. *United States,* 120 Ct. Cl. U.S. 249 283
Williams Oil Co. v. *State of New York,* 198 Misc. 907 137
Willis v. *Webster,* 1 App. Div. 301 227
Wilson & English Const. Co. v. *N.Y.C.R.R. Co.,* 240 App. Div. 479 152
Wilson v. *Northwestern Mut. Life Ins. Co.,* 65 F. 38 25
Wilson v. *Oliver Costich Co. Ind.,* 231 App. Div. 346 166
Wilson v. *Salt Lake City,* 174 P. 847 126
Wilson v. *Thompson,* 3 N.W. 699 25
Winters v. *Duluth,* 84 N.W. 788 146
Wise v. *United States,* 249 U.S. 361 219, 221
Wise & Co. Inc., v. *Wecoline Products, Inc.,* 286 N.Y. 365 156
Witherell v. *Lasky,* 286 App. Div. 533 186
Wolff v. *McGavock,* 29 Wis. 290 125
Wood v. *Ft. Wayne,* 119 U.S. 312 125, 128
Wood et al. v. *United States,* 258 U.S. 180 281
Woodward v. *Fuller,* 80 N.Y. 312 182
Woolsey v. *Funke,* 121 N.Y. 87 151
Wright v. *Hoctor,* 145 N.W. 704 17
Wynkoop v. *People,* 1 App. Div. 2d 620, affirmed 4 N.Y. 2d 892 208
Wyoming-Indiana Oil and Gas Co. v. *Weston et al.,* 7 P. 2d 206 61

Yaryan v. *Toledo,* 81 N.E. 1199 84
Matter of Yonkers Contr. Co. v. *Thruway Authority,* 283 App. Div. 749 32
Young Fehlhaber Pile Co. Inc. v. *State of New York,* 177 Misc. 204, aff. 265 App. Div. 61 111

In re Zaephel v. *Russell,* 49 F. Supp. 709 204
Zang v. *Hubbard Bldg. Realty Co.,* 125 S.W. 85 216
Zarthar v. *Saliba,* 185 N.E. 367 129
Zimmerman v. *Judah,* 13 Ind. 286 217

Index

A

Abbreviations, 367
Additional work, 121-135
Advertising for bids, 22-26
Alternate, 57
Anti-Kickback Act, 292
Armed forces procurement
regulations, 254-255
Awards, 27-29, 76
and "best interests," 47
bidder's right to, 29, 87
at caprice of public officer, 29
correct computation of, 27, 87
correction of; statutory, 28, 76
court interference in, 29, 87
determination of, 28
and efficient result, 11
private and public, 88
quality over price in, 11
rejection of all bids in, 29
on single bid, 29
tabulation of bids for, 27
see also Bid; Tie bids

B

Bid; acceptance of as contract, 36
alternate, 57, 60
amount, 27-28
"best interest" served in, 47
bonds; *see* Bid, deposit with
and contract formulation, 37
damages for noncompliance of,
36
deposit with 33-35, 49, 65, *illus. 347-348*
discrepancy in, 45, 52
duty of official toward, 46
firm, 44, 45
form, 37, *illus. 343-344*
informal or irregular, 45-47, 49,
52-53, 87
"no charge," 52
offer and acceptance of, 45
as proposal, 37
qualified, 45
rejection of, 29, 32-33
signature on, 37-44

Bid (*cont.*)
standards for, 46
ties, 30-32
unbalanced, 53-55
undersigned in, 38
withdrawal of, 37, 45, 65
see also Awards; Competitive bidding; Joint venture
Bidder; effect of proposal on, 35
instructions to, *illus. 345-346*
knowledge of law of, 7
neglect of; statutory, 9
prequalification for, 89
presumption of law of, 9
right to award, 29, 87
see also Competitive bidding; Lowest responsible bidder
Bidding; restrictions on, 15-19, 86, 89
see also Bid; Bidder; Competitive bidding
Bonds; labor and material, *illus. 191-192*
performance; *see* Performance bond
see also Surety bonds

C

Calendar day, 158
California contractors law, 94-106
Clauses; special and general, 152
Coercion of contractor, 247
Collusion and extra work, 132
Common law bond, 195
Competitive bidding, 1, 2, 8, 9
control of, 2
exceptions in, 4-6
notice to bidders on, 20, 21
Competitive bidding (*cont.*)
specifications furnished for, 10
see also Awards; Bid; Bidding
Construction contract form, *illus. 260-261*
Context of documents, 147
Continuation sheet (corporate co-surety) forms, *illus. 361-362*
Contractor; federal construction, 255
notice of complaint to, 190
obligation to perform, 181
payments to be returned to, 9
see also Bidder
Contracts, 181
breach of; and extra work, 132, 134, 135, 219
construction, *illus. 260-261*
federal forms, 256-257
implied, 8
invalid but accepted, 8
obligation to complete, 181
offer and acceptance of, 45
parts of, 147
procedural omission in, 7
professional services for, 4
remedy for when invalid, 8
separate classifications of, 11, 12
termination of, 141-145
see also Formation of contracts
Copeland Act, 292
Court of claims, 248-249
Custom and usage, 153-157

D

Damages; liquidated; *see* Liquidated damages
Davis-Bacon Act, 286

Delay, 229-239
  see also Liquidated damages

E

Eight Hour laws, 290
Engineer in charge; decisions by, 159-160
Errors; *see* Mistakes
Experience and financial condition statement form, *illus. 94-103*
Extra work, 121-135

F

Federal construction contracts, 255-257
Federal procurement, 254-256
Fees; *see* Licenses, taxes, and fees
Financial condition and experience statement form, *illus. 94-103*
Fire insurance policies, 174
Firm bid, 44, 45
Formation of contracts, 136-141
  certainty in, 154
  context of, 147, 149
  custom and usage in, 153-157
  decisions by engineer in, 159, 160
  delegating authority in, 139
  fire insurance coverage in, 174
  guide for, 148
  immunity of states from suits in, 169
  indemnification by contractor in, 169
  intention in, 149, 150, 154
  interpretation of, 139, 147, 148
  liability insurance coverages in, 170, 171, 175
Formation of contracts (*cont.*)
  liability for negligence in, 169, 170
  and municipalities, 140
  offer and acceptance in, 136
  and oral agreements, 150
  presumed valid, 136
  printed forms; use of; in, 152
  procedure of, 136
  repudiation of contract in, 144
  rules of construction in, 150, 151
  special clauses and general clauses in, 152
  standards in, 137
  and states, 139
  third party beneficiary, 161-168
  and United States, 138
  words; choice of; in, 149
  "working days" in, 157, 158
  zoning, 176-178
  see also Contracts; Work stoppage
Forms; bid, *illus. 343-344*
  bid bond, *illus. 347-348*
  construction contract, *illus. 260-261*
  continuation sheet (corporate co-surety), *illus. 361-362*
  experience and financial condition statement, *illus. 94-103*
  general provisions, *illus. 262-265*
  individual surety; affidavit of, *illus. 363-364*
  instructions to bidders, *illus. 345-346*
  invitation, bid, and award, *illus. 339-340*
  invitation for bids, *illus. 342*

Forms (*cont.*)
labor and material bond, *illus. 191-192*
labor standards, *illus. 341*
liability insurance certificate, *illus. 175*
liquidated damage clause, 222-224
notice to bidders, 21
payment bond, *illus. 351-352, 357-360*
performance bond, *illus. 183-184, 349-350, 353-356*
proposal, 37-45

I

Immunity of states from suit, 169
Indemnification by contractor, 169
Individual surety form, *illus. 363-364*
Instructions to bidders form, *illus. 345-346*
Insurance; fire, 174
liability, 170-176
Intention of contracts, 149, 150, 154
Interference and delay, 227, 230, 231, 237
Interpretation of contracts, 139, 147, 148
Invitation, bid, and award form, *illus. 339-340*
Invitation for bids form, *illus. 342*

J

Joint venture, 57, 61-63

L

Labor, 203
laws regulating hours of, 158
Labor and material bond form, *illus. 191-192*
Labor standards form, *illus. 341*
Liability, 169-176
insurance certificate form, *illus. 175*
License requirements; local, 250
Licenses, taxes, and fees, 93
summary of state requirements for, 106, 299-337
see also California contractors law
Liens on public property, 190, 192, 203
Liquidated damages, 218-224
and actual damages, 225
and delay, 226, 229, 235, 279
and interference, 227
limitation, 226
and reality, 228
and relationship to damages, 220
rule on, 226
and waiver, 224
Lowest responsible bidder, 80-89

M

Materialmen, 205
"Men and means" clause, 16, 86
Miller Act, 194, 211, 212
see also Surety bonds
Misrepresentation and delay, 237
Mistakes, 64-66
criterion for, 65, 68, 71, 74, 76
due to misinterpretation, 76
not apparent, 70
and notice of no release, 73
obvious, 70-71
by public agency, 77-79
reformation for, 70
rejection of all bids due to, 65-66

Mistakes (*cont.*)
relief due to, 65, 69, 71-76
stating claim of, 66-68
statutory correction of, 76
Municipalities; powers of to engage in contracts, 140

N

Negligence and liability, 169-176
Noncompliance of bid; damages for, 36
Notice to bidders, 20, 21, 33

O

Oral agreements, 131, 150

P

Patented material, 13-15
Payment; for extra work; partial, 129
for work, 181
Payment bond form, *illus. 351-352*
corporate co-surety, *illus. 357-360*
Penalty, 218
Performance bond form, *illus. 183-184*
corporate co-surety; *illus. 353-356*
federal, *illus. 349-350*
Performance of work, 181, 182
equivalent of, 128, 187
remedy of defects in, 182, 186
substantial, 182-189
Permit requirements; local, 251
Plans; warranty of; *see* Warranty of plans
Postqualification, 89
state requirements for, 299-337
Prequalification, 89-92
state requirements for, 299-337
statement form of, 94-103
Procurement regulations; federal, 254-256
Professional services, 4
Progress payments, 239
and contractor's rights, 240, 242
Proposal, 37-45
see also Bid; Bidder; Bidding; Bidding restrictions
Provisions; general; form, *illus. 262-265*
Public works, 146

Q

Quantities; approximate, 118-120

R

Rejection of bid, 29, 32
Renegotiation; in federal contracts, 293, 295-298
Renegotiation Act, 295
Repudiation of contract, 144
Rescission for error, 69-70
see also Mistakes

S

Signatures on bid, 37-44
Specifications, 10-13
see also Warranty of plans
State requirements for licenses, taxes, and fees, 106, 299-337
Statutory bond, 195
Subcontractor, 207
rights of in delay, 234
Subrogation of surety, 197-201

Substantial performance, 182-189
Surety bonds, 190, 193-194, 203-208
distinctions in, 194, 195
function of, 88
influence on awards of, 88
justification of, 195
and labor and materialmen, 194, 203-208
and liens on public property, 190, 192, 203
and material change in contract, 216
and Miller Act, 194, 211, 212
in nature of guaranty, 88, 190, 196, 203, 214
as notice of lien, 217
and omission by public officer, 196
power to require, 193
and public officer liability, 196
release from, 213, 214
and subcontractors, 207, 208, 212
and subrogation of surety, 197-201
Surety obligation, 197, 198, 213, 214

T

Taxes; *see* Licenses, taxes, and fees
Termination of contracts, 141-145
Third party beneficiary, 161-168
Tie bids, 30-32
Time extension, 235, 236
see also Delay

U

Unbalanced bid, 53-55
see also Bid
Undersigned in bid, 38

V

Verbal promise and extra work, 131

W

Walsh-Healey Act, 293
Warranty of plans, 108-118
and breach of contract, 109, 113
contractor's obligation and, 108-113, 181
and contractor's procedure, 114
and contractor's risk, 118, 121
Withdrawal of bid, 37, 45, 65
Words; choice of, 149
Words and figures in proposal, 52
Work; separate classifications of, 11, 12
Work stoppage, 137, 141, 189
and breach of contract, 142, 144
procedure of, 145
Working day, 157, 158
Written order as waiver for extra work, 122, 127, 128, 130, 131

Z

Zoning, 176-178
local regulation on, 249